INTERNATIONAL CENTRE FOR MECHANICAL SCIENCES

COURSES AND LECTURES - No. 294

INFORMATION COMPLEXITY AND CONTROL IN QUANTUM PHYSICS

PROCEEDINGS OF THE 4TH INTERNATIONAL SEMINAR ON MATHEMATICAL THEORY OF DYNAMICAL SYSTEMS AND MICROPHYSICS
UDINE, SEPTEMBER 4-13, 1985

EDITED BY
A. BLAQUIERE, S. DINER, G. LOCHAK

Fondation Louis de Broglie

Springer-Verlag Wien GmbH

Le spese di stampa di questo volume sono in parte coperte da contributi
del Consiglio Nazionale delle Ricerche.

This volume contains 20 illustrations.

Originally published by CISM, Udine in 1987.

ISBN 978-3-211-81992-0 ISBN 978-3-7091-2971-5 (eBook)
DOI 10.1007/978-3-7091-2971-5

PREFACE

From September 4 to 13, 1985, a meeting took place at the International Centre for Mechanical Sciences (CISM), Udine, Italy.

This was the fourth seminar in a series held since 1979, and devoted to dynamical systems, mechanics and microphysics. But with the first one, this was the second seminar almost exclusively devoted to problems of quantum physics and quantum theory. The other seminars, held in 1981 and 1983, were more specifically devoted to problems of mechanics under the subtitles : Geometry and mechanics, Control Theory and Mechanics.

The general philosophy shared by the different organizers of these seminars was that microphysics could have some profit from interacting with the modern theory of dynamical systems and mathematical system theory.

Such an idea was quite natural for people of the CISM, well aware of the huge modern developments of classical mechanics and dynamical system theory, often connected with the needs of technological progress in the field of automatic control and design of complex systems, the field of cybernetics. Such an idea was not and is not yet natural to quantum physicists.

This is why the CISM asked the Fondation Louis de Broglie to organize an international Seminar on concepts and methods of cybernetics as they appear in quantum physics. This is the motivation for the title of this Seminar : Information, complexity and control in quantum physics.

This seminar was not intended to be a conference on the problems of interpretation of quantum mechanics. Nevertheless a number of lectures and papers have a strong bearing on the basic questions in the foundations of quantum theory.

Among the topics which could appear in such a conference were : Quantum probability, Quantum stochastic differential equations, Quantum chaos, Quantum stochastic processes, Open quantum systems and dissipation, Statistical theories for quantum physics, Quantum detection and estimation theory, Theory of measurements, Quantum information theory, Stochastic formulations of quantum mechanics, Quantum mechanics and control theory (Variational problems. Inverse problems), Quantum mechanical computers.

Many of these topics have been discussed from the mathematical physics point of view in different conferences last years.

These topics appear again here in a more physical atmosphere, the originality of this seminar being a very complete panorama of the use of control theory in quantum physics.

The proceedings have been divided into four parts.

Part I deals with Quantum information theory, Quantum probability theory and Quantum symmetry. The papers appearing there, cover some fundamental general problematics in the interpretation and use of quantum theory. The formulation of a quantum information theory is needed for a proper use of quantum optical devices for information transmission. But there are only a few presentation of such a theory in the litterature. A nice book by V.V. Mitiugov -Physical foundations of information theory, was published in russian in 1976 but never translated. It relies heavily on the work of L.B. Levitin, who has given here an extended summary of the theory. This paper provides at the same time a kind of general conceptual framework for the whole seminar.

F. Fer discusses with precise criticisms the interrelations between the basic concepts of statistical physics and information theory, with the conclusion that there is no logical proof which permits to assert that thermodynamical entropy is to be assimilated to a lack of information. I.D. Ivanovic shows how one can get the information allowing to determine the state of a quantum system through a sequence of measurements.

D. Aerts has shown previously that if one considers that the probabilistic character of classical statistical mechanics is due to a "lack of knowledge" about the state of the system, the probabilities of quantum mechanics can be explained as due to a lack of knowledge about the measurement. In his paper he shows why a lack of knowledge about the state of a system gives rise to a classical (Kolmogorovian) probability calculus, and why a lack of knowledge about the measurements give rise to a non classical (non Kolmogorovian) probability model.

L. Accardi studies from a mathematical point of view the fine structure of the states of a composite systems, showing in a sense the universality of the Einstein-Podolsky-Rosen phenomenon.

The paper of G. Lochak illustrates the kind of information one can extract from the quantum mechanical formalism using geometrical -more precisely symmetry- considerations. He shows that Dirac's equation admits not only one local gauge invariance, but two, and only two. The first one leads to the theory of the electron, the second one can be shown to lead to a magnetic monopole. The neutrino can be considered as a special case of this monopole.

All the papers of part I stress the fact that quantum mechanics may be considered as a mathematical system theory which is specific by its original concept of state and where information, probability and symmetry play a central role. In fact quantum mechanics contributes to the elaboration of a general theory of physical systems and one can only regret that the general system theory has not yet systematically included the quantum point of view in its general framework.

One can only remark here that the definition of the concept of state leads to a reflexion on the concept of autonomy and as such on the concept of system. One is faced with the difficult problem of the definitior of a closed system (and the opening of this system). The privileged role played by hamiltonian systems comes from the fact that they provide an ideal model for the evolution of closed systems. The same for self-adjoir operators which play a dominant role for closed systems and reversible evolution. As a general mathematical model, quantum mechanics allows for different kind of mathematical representations. Among them are the representations using stochastic processes.

Part II deals with a class of such representations connected with the theory of optimal stochastic control. Such an approach was advocated by A. Blaquière from 1966 on, with the introduction of the concept of closec loop controlled random process. In the same year Nelson introduced a thec ry based on the consideration of two markovian processes inverse in time and "adjoint" relatively to a common invariant measure, expressed as a scalar product. That Nelson theory could be deduced from a stochastic variational principle and is connected with stochastic control theory has been recognized by K. Yasue and J.C. Zambrini in 1981. Guerra and collaborators have also worked on this variational formulation. In his last book, "Quantum fluctuations", Princeton University Press, 1985, Nelson deduces his theory from a stochastic variational principle. A completely different construction, initiated by E. Schrödinger in 1931, is based on the theory of Bernstein stochastic processes. In his paper J.C. Zambrini develops this approach and shows how Nelson's stochastic mechanics can be reinterpreted in this frame. After hearing J.C. Zambrini's lecture, A. Blaquière found that A. Marzollo and himself had rediscovered this forgotten approach of Schrödinger in 1982 starting from the theory of optimal control (See reference 2 of their paper). His paper in collaboration with A. Marzollo further develops this formulation.

A careful reading of all these papers shows how time reversal, hermiticity, probability amplitudes and closed systems are deeply interconnectec concepts which lie at the foundations of quantum mechanics.

Part III under the heading of "Quantum stochastic processes" is devote to problems of measurement on quantum systems without or with dissipation. The study of successive or continuous measurements requires a specific mathematical formalism which has been developped in last years, leading to the concept of quantum stochastic processes.

S. Albeverio provides an extended introduction to non standard analysi which has become a powerful tool for the systematic study of problems of discretization, scaling, renormalization. All these problems are present in the applications of a stochastic analysis to quantum theory. The theory of Feynman path integral or the study of the classical limit of quantum mechanics have also largely profited from the use of non-standard tools.

G.M. Prosperi reviews the general formalism developed in the Milan group for treating any kind of continuous observation. For that purpose he recalls the more general formulation of Quantum Mechanics based on the idea of effect and operation which has been developed in particular by Haag and Kastler, Ludwig, Davies and Holevo. This allows to associate to the continuous observation of a set of quantities a mathematical structure called Operation valued Stochastic process, which is the fundamental concept of the Quantum Stochastic Calculus. The paper by W. Stulpe presents the similar work made by the group of Hellwig in Berlin. Quantum stochastic processes are defined as one parameter families of instruments.

A. Peres paper is on the "Quantum Zeno paradox" : a quantum system if continuously observed does not evolve. He shows that the Zeno effect is real and can be easily displayed experimentally, but one has to distinguish carefully between continuous monitoring and continuous measurement of quantum systems.

M. Courbage's, and T. Dittrich and R. Graham's papers consider the way dissipation can be handled in quantum mechanics.

Courbage paper is representative of the spirit of the Brussels group work on classical and quantum irreversible and dissipative processes, more specifically B. Misra and I. Prigogine. He shows how to construct in quantum theory a time operator and an entropy functional, giving rise to a formalism well adapted to the description of decay phenomena, measurement processes, and irreversibility.

T. Dittrich and R. Graham include dissipation in the quantum model through the introduction of a specific reservoir. They study the classical limit of a such a model for the kicked rotator.

Part IV is very original, being a small seminar inside the bigger one. It is devoted to "Quantum mechanical control systems", that is the application of the theory of control to quantum dynamical systems. Most of the world specialists of this new area have contributed to the proceedings, which makes this part of the book very complete and self-consistent.

Quantum mechanical control theory is an essential step in the way from quantum physics to quantum technology, as it is characterized in the introduction by S. Diner.

A.J. Van der Schaft studies classical hamiltonian control systems and their properties as controllability and observability. It is shown how these systems can be quantized to quantum mechanical control systems, using ordinary quantization procedures. This is a quantization of the full control system, much in the spirit of Tarn, Huang and Clark who follow the geometric quantization procedures (See reference 3 of Van der Schaft paper).

A quantum mechanical control system is a quantum mechanical system with a time varying part considered as a perturbation. Different kind of problems can be studied on it.

A. Alaoui, following the initial work done by G. Lochak, shows how much one can keep the features of the conservative unperturbed problem for a system interacting with a coherent electromagnetic field. This is done through the definition of the so called "quasi-energies" and "permanent states".

One can be interested in the time varying part as a signal to be extracted from the measurements on the system. This is the quantum filtering problematic, usually associated to the concept of non demolition measurements, which is developped in the papers by V.P. Belavkin and J.W. Clark and T.J. Tarn. At last the time varying part can be considered as a purposive control on the system, a control problem stricto sensu. This is the topic of the paper by A.G. Butkovskiy and Ye.I. Pustil'nykova.

To summarize briefly the content of these proceedings from a cybernetical point of view, one can say that the four parts develop four point of view on quantum systems : System theory and information, Modelling, Measurement, Control. Four corner-stones of quantum cybernetics.

We take this opportunity to express our gratitude to CISM which played a major role in the practical organization of the seminar that inspired this volume. We also wish to acknowledge the financial assistance of UNESCO. Our warmest thanks to Professor P. Serafini and the secretaries of the CISM who helped us so much to enjoy our stay in Udine and to make this meeting successfull.

CONTENTS

PART III — QUANTUM STOCHASTIC PROCESSES

PART IV — QUANTUM MECHANICAL CONTROL SYSTEMS

INTRODUCTION
FROM QUANTUM PHYSICS TO QUANTUM TECHNOLOGY

Simon Diner
Institut de Biologie Physico-Chimique, Paris, France

In the overview of the report on physics published under the direction of W.F. Brinkman : "Physics through the 1990's (1), one can read about quantum mechanics that it illustrates the unpredictable path by which new knowledge in physics can shape society.

> "Based on studies of the properties of matter, the spectra of atoms and the motions of charged particles, quantum mechanics provided an extraordinary new framework for portraying physical reality ...
> It is now recognized that quantum mechanics is basic not only to physics but to chemistry, biology and many of the other sciences.
> Beyond this, quantum mechanics has led to the creation of new industries, such as semiconductors and optical communications, and has opened new paths of technology through the creation of exotic materials and devices like the laser.
> Another more recent example is the 1947 discovery of the transistor, which contributed to the omnipresent computer. Nobody can know how society will ultimately be transformed by this revolution."

The contemporary technology deliberately moves towards a fine mastership of complex phenomena. It tries to assume a precise control of the use of energy, of the memorization and transmission of information, of the creation of new materials, of the biological and medical practice. To attain such objectives one needs more and more miniaturization and rapidity. A systematic recourse to quantum processes seems unavoidable. Systems described by quantum mechanics are looked upon with an active attitude. This contrasts with the passive attitude of the first phase of development of quantum physics when the main goal was to describe the structure of matter. One tries now to use the quantum properties of matter and light to achieve technological goals. As a consequence new requirements appear in quantum physics, leading to new developments in quantum theory in such areas which were not so much explored before. A challenge for the teaching of quantum theory is created. A menace for quantum formalism and it's interpretation could be produced. The technological pressure, as ever, acts as a source of development and renewal. It stimulates the theoretical study of quantum dynamical systems from the point of view of *general cybernetics*. Hence the scientific motivations for this meeting. To bring together people speaking about quantum systems in the language of the modern theory of dynamical systems, information theory, mathematical system theory, control theory. To act purposely on quantum systems one needs a language which could be as "physical" as possible. Many developments are necessary to achieve such goals.

Historically quantum theory has evolved as an independant system of concepts and formalisms. The maximum effort from theoreticians in the last forty years has been devoted to strengthened the internal logic of the theory and to develop new formalisms in the theory itself, relying only on the general principles of quantum methodology. This work can be described as a deepening and a broadening of the mathematical structures of quantum theory. It gives rise to a quantum system theory with the definition and use of fundamental concepts as *events, observables, states* and *operations*. But it is a System Theory of its own which has little to do with the System Theory of Cybernetics.

In the following we recall briefly the main different mathematical structures and axiomatics used for quantum theory. There are very few comparative presentations of all these approaches (2).

CLASSICAL AXIOMATIC	(3)
Dirac. Von Neumann	
Hamiltonian approach. Observables : self-adjoint operators in a Hilbert space Spectral theory. The Dirac approach (use of formal eigenvectors) is given a rigourous basis in the "Gelfand triplet" approach	

ALGEBRAIC AXIOMATIC	(4)
Jordan, Von Neumann, Wigner, Segal	
Observables : elements of a C* algebra State : mean value functional on the C* algebra	

LOGICAL AXIOMATIC	(5)
Birkhoff, Von Neumann, Mackey, Jauch, Piron	
Building stones of the axiomatic : system of propositions about the physical system. Quantum logic : proportional lattices Generalized probabilities	

ORDERED VECTORIAL SPACE AXIOMATIC (6)
Ludwig, Davies and Lewis
The states of the system build a convex set.

OPERATIONAL APPROACH (7)
Randall, Foolis

FEYNMAN QUANTIZATION (8)
Feynman
Lagrangian approach.
Functional integral point of view.

PHASE SPACE QUANTIZATION (9)
Grossman, Berezin
Operators are represented by their symbols : functions on phase space.
Theory of pseudo-differential operators.
Wigner function.

GEOMETRIC QUANTIZATION (10)
Souriau, Kostant
The relationship between canonical transformations and operators on Hilbert space is expressed in the framework of the geometry of phase space.

QUANTIZATION by DEFORMATIONS OF ALGEBRA (11)
Flato, Lichnerowicz
Quantization manifests itself as a deformation of the structure of the algebra of classical observables, rather than a change in the nature of the observables.

STOCHASTIC QUANTIZATION (12)
Fenyes, Nelson
Stochastic processes are used to simulate quantum mechanical properties.

QUANTIZATION of DISSIPATIVE SYSTEMS (13)
Dekker, Gisin

QUANTIZATION of OPEN SYSTEMS Davies	(14)
QUANTIZATION of CHAOTIC SYSTEMS QUANTUM CHAOS Zaslavsky	(15)
QUANTUM STATISTICAL DECISION THEORY Helström, Holevo	(16)

All these developments constitute an impressive amount of mathematical work and will continue to provide mathematicians with active working fields. But physics has not gained so much from all that !

The physicists who meet regularly at the bi-monthly Seminar of The Fondation de Broglie in Paris have in common the feeling that Quantum Theory, with all these different approaches, is a strong building, but that we still need some "more physical explanation" for the quantum formalism. This formalism appears to them a as series of "lucky guessing" in the choice of mathematical objects reflecting incomprehensible peculiarities of microphysics. They dream of classical alternatives to quantum mechanics, be it a theory of hidden variables or something else. They follow the road open by Louis de Broglie in the fifties when he tried to give some strength to his theory of the double solution and the pilot wave. He was convinced, with his pupils, as Francis Fer and Georges Lochak, who are here, that there is still enough extraordinary properties to be discovered in classical dynamical systems, to explain the quantum behaviour. The wealth of phenomena in non-linear mechanical systems seemed to them full of promise for microphysics. They were right for non-linear mechanics, and one must recall that G. Lochak has discovered a soliton wave some years before Kruskal and others (17). But so far there has not been decisive reward for microphysics from the explosion of non-linear dynamics.

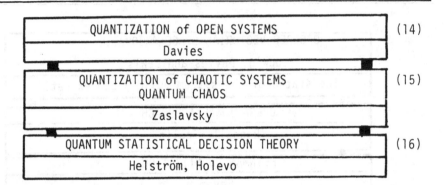

One must recall that this attitude of the de Broglie School continued a tradition of thinking which can be traced to the founders of the modern theory of dynamical systems. Poincaré had much interest for quantum theory in its infancy. Birkhoff (Georges David) wrote some papers on the foundations of quantum mechanics and tried even to elaborate an alternative theory (18). A.A. Andronov, the russian founder of the modern theory of non-linear oscillations, had in the beginning of his career a double interest for quantum mechanics and non-linear vibrations, following his teacher L.I. Mandelshtam.Mandelshtam was a deep propagandist for the theory of vibrations. He spoke of the international language of the theory of vibrations and said that fundamental physical discoveries are in essence "vibrational". One can only say that quantum mechanics brings also some evidence for that. There seems to be some model of universal harmonic oscillator at the basis of Q.M., not to speak of the de Broglie wave. Andronov thought of an explicit application of the theory of non-linear vibrations to quantum physics. In his wonderful lecture of 1933 on the Mathematical problems of the theory of auto-oscillations, he explicitely set the hypotheses that atoms could be non conservative auto-oscillating systems (19).

All that explains the enthusiasm shown by different friends of the Fondation Louis de Broglie to participate to the Dynamical Systems and Microphysics meetings in Udine.

In 1979, at the first meeting, it was a time when some of us thought that one can develop hidden variable theories or alternative theories taking advantage of all the advances in the modern theory of dynamical systems, more precisely in the understanding of the chaotic behaviour of deterministic systems.

In fact this classical concept of chaotic systems has led to the study of what is called "quantum stochasticity or chaos" that is the study of what happens in the quantum theory when the classical system "corresponding" to the quantum one has chaotic properties (is a K-system for example). "Corresponding", meaning here either the classical system used as a basis for quantization or the classical limit. This is a very interesting field which has strong bearing on the foundations of quantum statistical mechanics and on problems of molecular mechanics (dissociation of a molecule for example)(15).

More interesting for the foundations of quantum theory seems the work of Benettin, Galgani and Giorgilli on the coexistence of ordered and chaotic motion in dynamical systems as it appears in the KAM (Kolmogorov-Arnold-Moser) theory, and more precisely in the Nekhoroschev theorem. This theorem provides a general framework for understanding Boltzmann's conjectures about distinction between mechanical and thermodynamical energy and the existence of oscillators either frozen or freely exchanging their energy (20).

S. Diner has stressed the fact that the wave-particle dualism involves the coexistence of order and chaos (21) The recent philosophy of non-linear phenomena stresses that pure order and pure chaos are limit situations and that the coexistence of both is a general phenomenon (22). According to Hasegawa (23) one can find physical systems for which there is an ordered structure for one physical quantity letting other physical quantities take care of the increase of entropy. A selective auto-organization. This has to be compared to the existence of coherent structures in turbulence (24). It appears that chaotic phenomena in deterministic classical systems provide an incredible spectra of complex behaviour. One can only regret that active research of classical models for microphysics has already stopped after the failure of elementary trials in Stochastic Electrodynamics (25). One knows that a major reason for this discouragment is due to the negative influence of the success of Aspect's experiment. But on the other hand more and more "reality" is attributed to the Vacuum (26). This reality of the zero-modes of the electromagnetic field has been demonstrated in a beautiful experiment by Hulet, Hilfer and Kleppner on the inhibition of spontaneous emission by a Rydberg atom in a cavity (27). *The vacuum exists if one can prevent it to act on matter !* The atoms exist if one can count them said Poincaré after Perrin's experiment.

In fact this "action on vacuum" is representative of an evolution which is taking place in the quantum domain.

Powerful advances in technology open new fields of experimentation with a new requiring for interpretative formalisms. A challenge to quantum theory, more powerful than discussion of paradoxes or design of crucial experiments. Rauch experiments on neutron interferometry (28) ; the manipulation of single atoms and the basic experiments it allows : the inhibition of spontaneous emission (27), the possibility of observation of quantum jumps (29); the trapping of atoms in laser beams (30) are just some examples of these new breathtaking developments.

As Daniel Kleppner says : "This is a new ball park, and the whole history of physics shows that when you move into areas that are different by order of magnitudes, there are always surprises". *Quantum physics is entering a new area of fundamental experiments and development of a quantum technology based on fundamental effects*. This requires from quantum theory the development of extended considerations about measurements, statistics of the results of measurements, control of the experiments. This is the source of an overall development of quantum theory at the three semiotical levels characteristic of any language : *syntactics, semantics* and *pragmatics*.

SEMIOTICAL LEVELS OF QUANTUM THEORY

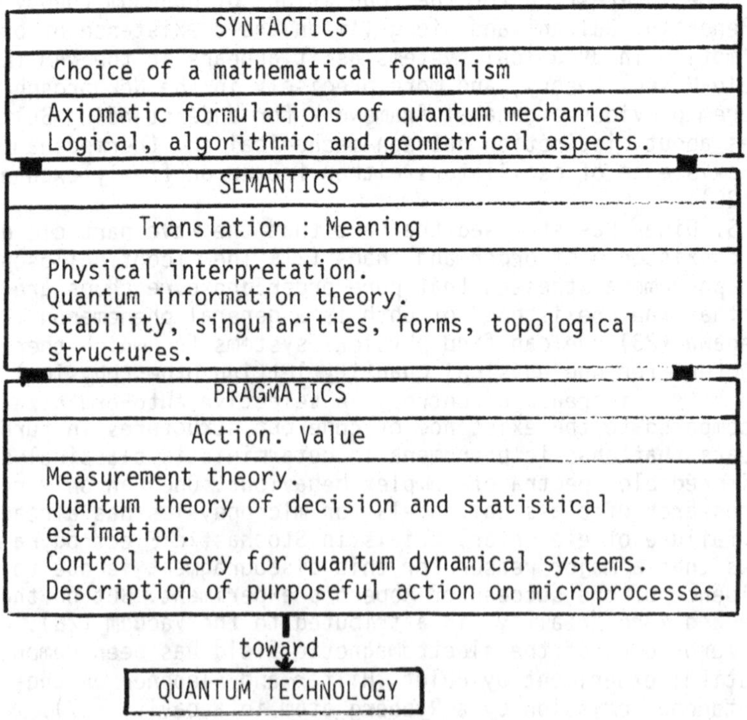
SYNTACTICS
Choice of a mathematical formalism
Axiomatic formulations of quantum mechanics
Logical, algorithmic and geometrical aspects.
SEMANTICS
Translation : Meaning
Physical interpretation.
Quantum information theory.
Stability, singularities, forms, topological structures.
PRAGMATICS
Action. Value
Measurement theory.
Quantum theory of decision and statistical estimation.
Control theory for quantum dynamical systems.
Description of purposeful action on microprocesses.
toward
QUANTUM TECHNOLOGY

As it occurs generally, technology develops more rapidly than science and fundamental understanding. This is due to the fact that some simplified models are sufficient for an extended development of a technology. But in the quantum domain it could not be the case, because pictures and simplified models are lacking at a deep level of quantum physics.

This conference was precisely devoted to those new development of quantum theory which could be of great help for a mastership of quantum technology.

Without any claim for completness we shall conclude this introduction with a panorama of quantum technology. We shall not give a systematic list of references but shall only quote some representative papers and sources of information.

PANORAMA OF QUANTUM TECHNOLOGY

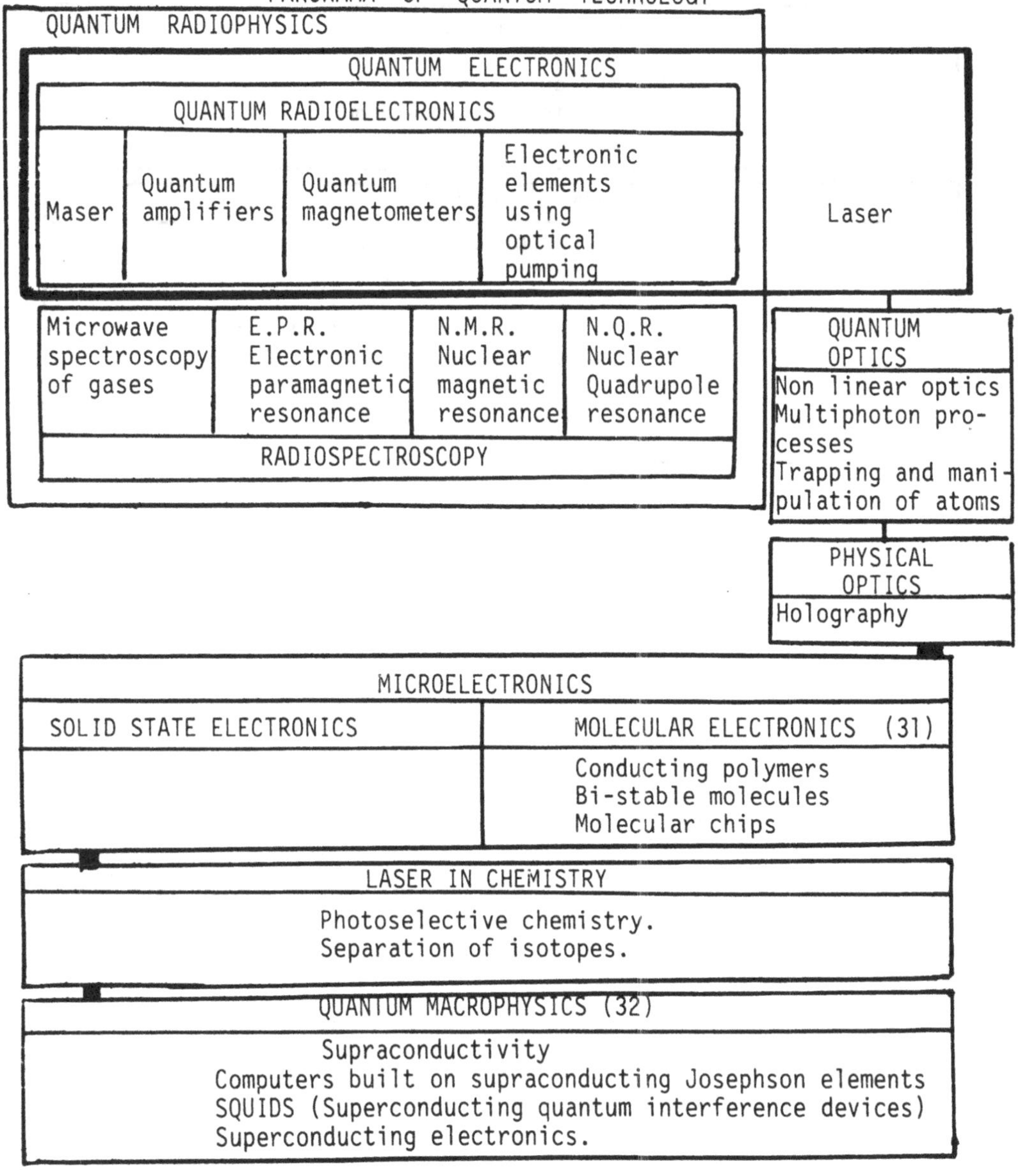

CORPUSCULAR OPTICS
Electronic microscopy. Tunneling microscopy.(33) Electron and neutron diffraction

QUANTUM METROLOGY (34)
Laser used to measure length, position, speed. Quantum gyroscopy (laser, nuclear and electronic paramagnetism). Quantum clocks.

QUANTUM CYBERNETICS
Laser in transformation, conservation or transmission of information. Laser control. Quantum noise. Squeezed states of light. Non demolition measurements. Optical bistability. Quantum automata. Quantum computers (35).

REFERENCES

(1) Physics Through the 1990s. National Research Council, 1986. A review of this survey is given in Physics ToDay. April 1986.

(2) Whigtman, A.S. : Hilbert's sixth problem. Mathematical treatment of the axioms of physics, in, Amer. Math. Soc. Symposia on pure mathematics vol. 28, 147-240, 1976.
Gudder, S.P. : A survey of axiomatic quantum mechanics in Hooker, C.A., ed. The logico-algebraic approach to quantum mechanics. Vol. II. Contemporary consolidation. Reidel. 1979.
Ludwig, G. : Connections between different approaches to the foundations of quantum mechanics, in, Neumann, H., ed, Interpretations and foundations of quantum theory, B.I. Manheim. 1981.

(3) Dirac, P.A.M. : The principles of quantum mechanics, Clarendon, Oxford, 1930.
Von Neumann, J. : Mathematical foundations of quantum mechanics, Princeton U.P. Princeton, 1955.
Prugovecki, E. : Quantum mechanics in Hilbert space, Academic Press N.Y. 1971.

(4) Segal, I.E. : Ann. Math. 48, 930, 1947.
Thirring, W. : Quantum mechanics of atoms and molecules (A course in mathematical physics, T.3), Springer 1981.
Emch, G.G. : Mathematical and conceptualfoundations of 20th century physics, North-Holland, Amsterdam 1984.

(5) Piron, C. : Foundations of quantum physics, Benjamin 1976.
Piron, C. : Cours de mécanique quantique, Université de Genève, 1985.
Beltrametti, E.G. and Cassinelli, G. : The logic of quantum mechanics, Addison Wesley-Reading 1981.

(6) Davies, E.B. and Lewis, J.T. : Comm. Math. Phys. 17, 239, 1970.
Ludwig, G. : Foundations of quantum mechanics, I. Springer 1983.
Ludwig, G. : An axiomatic basis for quantum mechanics, I. Derivation of Hilbert space structure, Springer 1985.

(7) Randall, C.H. and Foulis, D.G. : The operational approach to quantum mechanics, in, Hooker, C.A., ed, Physical theory as logico-operational structure, Reidel, 1979.

(8) Feynman, R.P. and Hibbs, A. : Quantum mechanics and path integrals, MacGraw Hill, N.Y. 1965.
Schulman, L.S. : Techniques and applications of path integration, Wiley, 1981.

Glimm, J. and Jaffe, A. : Quantum physics. A functional integral point of view. Springer 1981.

(9) Grossman, A., Loupias, G. and Stein, E.M. : Ann. Inst. Fourier Grenoble, 18, 343, 1968.
Berezin, F. and Shubin, M. : The Schrödinger equation, Reidel, 1986.

(10) Woodhouse, N.M. : Geometric quantization, Oxford U.P., 1980.

Sniatycki, J. : Geometric quantization and quantum mechanics, Springer, 1980
Kirillov, A.A. : Geometric quantization, in, Arnold, B.I., and Novikov, S.P., eds, Contemporary problems of mathematics, Fundamental directions, Vol. 4, Vinniti an SSSR, Moscow 1985. (in russian, announced to be published in english by Springer in 1987 in a series under the title "Encyclopedia of Mathematical Sciences").

(11) Bayen, F., Flato, M., Fronsdal, C., Lichnerowicz, A., and Sternheimer, D. : Ann. of Phys. 111, p. 61-110 and 111-151, 1978.
Lichnerowicz, A. : Deformations and quantization, in, Avez, A., Blaquière, A., and Marzollo, A. : Dynamical systems and microphysics, Geometry and mechanics, Academic Press, 1982.

(12) Nelson, E. : Dynamical theories of brownian motion, Princeton U.P., 1967.
Nelson, E. : Quantum fluctuations, Princeton U.P., 1985.
Guerra, F. : Structural aspects of stochastic mechanics and stochastic field theory, in, De Witt-Morette, C., and Elworthy, K.D., eds. New stochastic methods in physics, Physics Reports 77, 3, 1981.
Zambrini, J.C. : this volume.

(13) Dekker, H. : Physics Reports 80, 1-112, 1981.
Gisin, N. : Un modèle de dynamique quantique dissipative. Thèse, Genève, 1982.
Gisin, N. : Found. of Physics, 13, 643, 1983.
Courbage, M. : this volume.

(14) Davies, E.B. : Quantum theory of open systems, Academic Press, 1976.
Exner, P. : Open quantum systems and Feynmann integrals, Reidel, 1984.

(15) Zaslavsky, G.M. : Chaos in dynamical systems, Harwood, N.Y., 1985.
Ackerhalt, J.R., Milonni, P.W. and Shih, M.L. : Chaos in quantum optics, Physics Reports, 128, 205-300, 1985.

(16) Helstrom, C.W. : Quantum detection and estimation theory, Academic Press 1976.
Holevo, A.S. : Probabilistic and statistical aspects of quantum theory, North Holland 1982.

(17) Lochak, G. : Comptes Rendus Acad. Sci. Paris, 250, 1985 and 2146, 1960.

(18) Birkhoff, G.D. : Collected mathematical works, AMS Providence 1950.

(19) Andronov, A.A. : Mathematical problems of the theory of self-oscillations (in russian) in 1st national conference on vibrations. Gostechteorizdat 1933, reprinted in Selected works of A.A. Andronov, Editions of the Academy of Science of SSR 1956, p. 85-124.

(20) Benettin, G., Galgani, L. and Giorgilli, A.: On the persistance of ordered motions in hamiltonian systems and the problem of energy partition, in Diner, S., Fargue, D. and Lochak, G., eds., Dynamical Systems. A Renewal of Mechanism, World Scientific Publisher, Singapore, 1986.

(21) Diner, S. in Diner, S., Fargue, D., Lochak, G. and Selleri, F. eds.: The wave-particle dualism, Reidel 1984.

(22) Gaponov-Grekhov, A.V. and Rabinovitch, M.I. : Nonlinear physics, Stochasticity and Structures, in Twentieth Century Physics, Development and perspectives, Naouka, Moscow, 1984 (in russian).

(23) Hasegawa, A. : Advances in Physics 34, 1-42, 1985.

(24) Lesieur, M. in Tatsumi, T. ed. : Turbulence and chaotic phenomena in fluids, North-Holland, 1984 and some other papers in that book.

(25) For a review of Stochastic Electrodynamics and an analysis of it's failure see (21) and de la Pena, L. in Gomez, B., Moore, S.M., Rodriguez-Vargas, Rueda, A. : Stochastic processes applied to physics and other related fields, World Scientific Publ. Singapore 1983.

(26) Boyer, T.H. : The classical vacuum, Scientific American, August 1984.
Maddox, J. : How empty is the vacuum ? Nature 305, 273, 1983.
Podolnyi, R. : Something called nothing, Znanie-Moscow 1983 (in russian).

(27) Hulet, R.G., Hilfer, E.S. and Kleppner, D. : Phys. Rev. Lett. 55, 2137, 1985.

(28) Rauch, H. : Contemporary Physics 27, 345, 1986.

(29) Nagourney, W., Sandberg, J. and Dehmelt, H. : Phys. Rev. Lett. 56, 2797, 1986.
Maddox, J. : Nature 323, 577, 1986.

(30) Chu, S., Bjorkholm, J., Ashkin, A., Cable, A. : Phys. Rev. Lett. 57, 314, 1986.
Maddox, J. : Nature, 322, 403, 1986.

(31) Since 1986 a new journal is published by Wiley : Journal of molecular electronics.
See also : Carter, F.L., ed. : Molecular electronic devices, Dekker N.Y. 1982.
Haddon, R.C. and Lamola, A.A. : Proc. Nat. Acad. Sci. U.S.A. 82, 1874, 1985.

(32) The March 1986 issue of Physics ToDay is entirely devoted to surperconducting technology.

(33) Binnig, G. and Rohrer, H. : Scientific American, August 1985 p. 50.
Wolf, E.L. : Principles of electron tunneling spectroscopy, Oxford U.P. 1985.

(34) Cutler, P.H., ed. : Quantum metrology and fundamental physical constants, Plenum, 1983.

(35) Feynman, R.P. : Quantum mechanical computers, Optics News, February 1985.
Peres, A. : Phys. Rev. A. 1985.

PART I

QUANTUM INFORMATION, PROBABILITY AND SYMMETRY

INFORMATION THEORY FOR QUANTUM SYSTEMS

Lev B. Levitin
Boston University, Boston, U.S.A.

ABSTRACT

Basic concepts and results of physical information theory are presented. The entropy defect and Shannon's measure of information are introduced and the entropy defect principle is formulated for both quasiclassical and consistently quantum description of a physical system. Results related to ideal physical information channels are discussed. The entropy defect and the amount of information coincide in the quasiclassical case, but the latter quantity is, in general, smaller than the former in quantum case due to the quantum-mechanical irreversibility of measurement. The physical meaning of both quantities is analyzed in connection with Gibbs paradox and the maximum work obtainable from a non-equilibrium system. Indirect (generalized) vs. direct (von Neumann's) quantum measurements are considered. It is shown that in any separable infinite-dimensional Hilbert space direct and indirect quantum measurements yield equal maximum information.

1. INTRODUCTION

The origin of the physical information theory ascends to L. Boltzmann and L. Szilard [1], who attributed an information meaning to the thermodynamic notion of entropy long before the quantitative measure of information was rigorously introduced by C.E. Shannon [2]. However, the information theory leading off with fundamental Shannon's work developed at first as a pure mathematical branch of science. An impression arose that the laws of transmission and processing of information were not physical and the concepts of the information theory could not be defined on the base of physical concepts. The erroncousness of such views was noted as far back as 1950 by D. Gabor [3], who pointed out that "the communication theory should be considered as a branch of physics." Later in the remarkable book of L. Brillouin [4] a profound relationship between physical entropy and information was suggested in the form of the "negentropy principle of information". Though the term "negentropy" was not rigorously defined, the principle implied that any information is represented by a certain state of a physical system and associated with its deviation from the thermodynamic equilibrium. The "entropy defect principle" [5,6] implements this basic idea in a mathematically consistent and rigorous form. According to this principle, any information is represented by an *ensemble* of states of a physical system and associated with the *average* deviation from the thermodynamic equilibrium caused by the *choice* of one of those states. Thus the information properties of real systems can be described in purely physical terms and a way is open to develop a consistent physical information theory.

In the first applications of information theory the communication systems were described on macroscopic level and their physical properties, in so far as they were taken into account, were considered from the point of view of classical physics.

It should, however, be borne in mind that the physical foundations of information theory must be essentially based on quantum theory. Indeed, in classical statistical physics the entropy dimension is that of logarithm of action, and entropy includes, as was noted by Planck [7], an infinite additive constant, so that only the difference of two values of entropy has physical meaning, but not the absolute value of entropy. Such a definition of entropy corresponds to the so called "differential entropy" (the term coined by Kolmogorov [8]) in information theory. But differential entropy depends on the choice of variables and does not allow calculating (without additional assumptions) the amount of information in the continuous message (formally the amount of information becomes infinite, which is a nonsense from a physical point of view). In the conventional theory the properties of continuous message ensembles are usually described in terms of ε-entropy. Such a description, however, is purely phenomenological and is valid only in the range where quantum effects make no appearance. According to quantum theory, every real physical system has a well-determined finite ("absolute" by Planck) entropy which leads to a natural absolute measure of information associated with statistical ensembles of quantum macrostates.

On the other hand, the vigorous development of quantum electronics, which brought into use the infrared and optical range of electromagnetic waves and allowed us to build systems with extremely low thermal noise, calls for consideration of the quantum nature of information transmission processes.

The physical information theory developed historically along the following three lines:

1. Investigation of the interrelations between the basic concepts of statistical physics and information theory.
2. Investigation of the physical nature of information transmission and processing.
3. Application of information-theoretical concepts and approaches to the problems of statistical physics.

Though numerous papers have been devoted to the first two areas (e.g., [9-19]), the ony book on this subject is [20] (in Russian). The use of quantum carriers of information brings into consideration the fundamental properties of quantum measurements (e.g. [21-27]). Results on quantum detection and estimation theory have been systemized in the comprehensive monograph by C. Helstrom [28]. It was found that a consistent quantum-mechanical treatment of information transmission calls for generalization of the basic information-theoretical concepts [6,29-32]. The effect of the quantum-mechanical irreversibility of the measurement on the information transfer has been revealed in the quantum-mechanical formulation of the entropy defect principle [6,29]. The controversial problem of information efficiency of direct and indirect quantum measurements has been solved in [33].

Investigations in the third of above-mentioned areas (e.g. [4,34-39]) promise to shed a new light on the foundations of statistical physics, in particular, regarding the origin and the meaning of the Second Law of Thermodynamics. It seems that there exist fundamental relationships between information, energy and work which manifest itself in such important and still controversial physical situations as Gibbs' paradox, Maxwell's demon, etc. Results obtained in [37] lead to a conjecture that in the case of irreversible processes the change of information rather than that of entropy determines the amount of work produced by a non-equilibrium system.

Information-theoretical concepts seem to be fruitfully applicable to some classical problems of statistical physics, such as the Ising model [40,41].

In general, one can conclude that physical information theory is, today, a well-established area of science with a number of rigorous general results and important applications in communication engineering.

2. ENTROPY DEFECT PRINCIPLE (QUASICLASSICAL FORMULATION)

The entropy defect principle was first formulated for the quasi-classical case in [5], with the view to obtain a general expression for information (in Shannon's sense) in terms of physical entropy. The quasi-classical description of a physical system implies, as usual, that all the operators corresponding to physical observables commute, which leads to two important premises:

1. all different microstates ("complexions", by Planck) of the system are inequivocally distinguishable by the same (complete) measurement;
2. any macrostate of the system is uniquely represented by a statistical ensemble of the microstates.

Consider an ensemble $X = \{x_i, p_i\}$ of signals x_i with corresponding probabilities p_i (i is the subscript of a signal; for the sake of simplicity the set of signals is presumed to be denumerable) and let this ensemble act on a physical system. The signal x_i brings the system to the state s_i.

As a result, we obtain an ensemble of macrostates $S = \{s_i, p_i\}$ which is in one-to-one correspondence with the ensemble of signals X. Each s_i is, generally, a certain macrostate of the system, i.e., a certain ensemble of various microstates defined by probabilities of microstates w_{ik} (k is the subscript of a microstate; the set of microstates is presumed to be denumerable). Macrostate s_i has some physical entropy H_i defined by the following expression:

$$H_i = -\sum_k w_{ik} \ln w_{ik}$$

(Amount of information and entropy are henceforth expressed in natural units-nats.)

We introduce now the *average* entropy of macrostates s_i

$$\overline{H} = \sum_i p_i H_i$$

Note that $\overline{H}$ does not have, in general, the meaning of the entropy of a particular macrostate but it is the conditional expected value of the entropy of a macrostate arising under the action of a determined (prescribed) signal.

On the other hand, it makes sense to speak about the macrostate s of the physical system, arising when the signal is chosen from the ensemble at random. The corresponding probabilities of microstates are

$$w_k = \sum_i p_i w_{ik} ,$$

and the entropy is

$$H = -\sum_k w_k \ln w_k .$$

We call the quantity

$$I_0 = H - \bar{H} \qquad (1.1)$$

"entropy defect".

The entropy defect shows how far (on the average) the state of the system, arising under the action of a determined signal, is from thermodynamic equilibrium (when entropy is a maximum and probabilities of microstates are given by Gibbs' distribution) in comparison with the state arising under the action of signal chosen completely at random. In short, the entropy defect shows how far the *determinate choice* of the signal deflects the system from its thermal equilibrium state.

The entropy defect principle claims the following: The amount of information about the signal *A* obtained by a physical system is equal to its entropy defect

$$I = I_0 . \qquad (1.2)$$

Indeed, since distinctions between the microstates are the finest possible distinctions between the states of a physical system, the maximum information about the signal is obtainable by measuring the microstate, i.e. by performing a complete measurement that distinguishes between all the microstates. Let $M = \{m_k\}$ be the random variable representing the microstate of the system. The joint probability distribution of two random variable - the signal X and the microstate M - is given by

$$\Pr\{X = x_i, M = m_k\} = \Pr\{\mathcal{S} = s_i, M = m_k\} = p_i w_{ik} . \tag{1.3}$$

Therefore,

$$I_0 = \mathcal{H}(M) - \mathcal{H}(M|X) = I(M;X), \tag{1.4}$$

where $I(M;X)$ is Shannon's information about the signal X in the microstate M.

Though the equality (1.2) seems to be almost evident, it is far from being trivial. Indeed, its validity is based on two crucial assumptions 1. and 2. about the nature of the physical world. These assumptions are not valid for a consistent quantum-mechanical description, where the equality (1.2) turns into inequality, as will be shown in section 4. The importance of the entropy defect principle is that it allows us to interpret information as a measure of the deviation of a physical system from the thermodynamic equilibrium state and to establish not only mathematical analogy but also identity of some fundamental concepts of information theory and statistical physics.

3. IDEAL PHYSICAL INFORMATION CHANNELS

The entropy defect principle has been used to analyze properties of certain systems which are, physically speaking, the simplest information transmission systems - ideal physical channels. We define an ideal physical information channel to be a channel in which the transmitter uniquely specifies the microstate of the transmitted signal - the physical agent that carries the information, and the receiver uniquely determines the microstate of the received signal. Thus, the signal ensemble in an ideal physical channel is the ensemble of microstates of a certain physical system.

The following results have been obtained in [42].

Theorem 3.1. The capacity of a nonideal channel does not exceed the capacity of an ideal channel with the same transition probabilities, obeying the same constraints on the ensemble of microstates of the transmitted signal.

Consider in more detail an ideal physical channel with statistically independent additive noise. In a channel of this type, the received signal is formed by addition of the transmitted signal and noise (the latter being of the same physical nature as the signal). The received signal and the noise may, therefore, be regarded as two different states of the same physical system. We shall assume that the microstate of this system is uniquely determined by the energies assigned to the various degrees of freedom (for example, to the various quantum states of the particles constituting the system). Additivity means here that the energy assigned to any degree of freedom of the received signal is the sum of the respective energies for the transmitted signal and the noise. (In particular, if one is using a description in terms of occupation numbers of quantum states of the particles constituting the system, the occupation numbers of the received signal are the sums of the respective occupation numbers of the transmitted signal and the noise.) Statistical independence means that the distribution of noise microstates is independent of the transmitted signal.

An ideal physical channel with statistically independent additive noise is an idealized model of systems in which the information carriers are particles obeying Bose-Einstein statistics (such as photons or elementary acoustical excitations - phonons). If the information carriers are particles obeying Fermi-Dirac statistics, this description is apt only in the limiting case of Boltzmann statistics (i.e., small average occupation numbers).

Let the ensemble average of the energy of the transmitted signal be E_0, that of the noise E_1. Then, by virtue of additivity, the average energy of the received signal is $E_2 = E_0 + E_1$. Assume, moreover, that the noise is thermal, i.e., the distribution of noise microstates corresponds to thermodynamic equilibrium of the physical system in question (i.e., of the agent transmitting the information) at a temperature $T_1 = T(E_1)$, where $T(E)$ is the temperature of the system as a function of its average energy. Under these conditions the entropy H_1 of the noise has the maximum value possible given the average energy E_1. We then have the following.

Theorem 3.2. The maximum amount of information that can be transmitted over an ideal physical channel with statistically independent additive noise is

$$I_{max} = \int_{E_1}^{E_2} \frac{dE}{kT} = \int_{T_1}^{T_2} \frac{1}{kT} \frac{dE(T)}{dT} \, dT \tag{3.1}$$

and this maximum is achieved when the distribution of microstates of the received signal corresponds to thermodynamic equilibrium at the temperature T_2 determined by

$$\int_{T_1}^{T_2} \frac{dE(T)}{dT} \, dT = E_0 \, . \tag{3.2}$$

Here E(T) is the average energy of the system as a function of temperature; $E(T_2) = E_2$.

Corollary. The minimum amount of energy necessary to transmit one natural unit of information over a channel with statistically independent additive noise satisfies the inequality

$$E_{min} = - \frac{E_0}{I} \geq kT_1 \tag{3.3}$$

where T_1 is the temperature of the noise. Equality holds asymptotically in case of a weak signal, i.e. when

$$\frac{T_2 - T_1}{T_2} \ll 1. \tag{3.4}$$

(For the usual physical systems, in which E(T) is a monotonic and smooth function, condition (3.4) is equivalent to $E_0/E_1 \ll 1$).

This corollary is a rigorous specialization of the well-known conjecture of Brillouin [4] that the minimum energy necessary to obtain one bit of information at temperature T_1 is $kT_1 \ln 2$.

In particular, the minimum energy per nat has been calculated in [5,10] for a broadband one-dimensional photon channel:

$$E_{min} = kT + \frac{3}{2\pi^2} hR. \qquad (3.5)$$

Here h is Planck's constant, R is the rate of information transmission (nat/s). This elegant formula clearly shows both the effect of thermal noise and the limitation imposed by the quantum structure of the electromagnetic field: the faster one transmits information, the more energy required per each unit of information. Other types of ideal physical channels have been investigated in [5,9-11,43,44]. For corpuscular channels (with particles of non-zero rest mass as information carriers) the specific effects of degeneracy (Bose-Einstein condensation and Fermi saturation) which affect channel capacity have been demonstrated [44].

4. ENTROPY DEFECT PRINCIPLE (QUANTUM FORMULATION)

In this section we generalize the concept of information (by Shannon) for quantum statistical ensembles, obtain the consistently quantum-theoretical formulation of the entropy defect principle and analyze the effect of the irreversibility of a quantum measurement on information transmission.

In quantum mechanics a macrostate s of a physical system is described by a density operator (density matrix) $\hat{\rho}$ which is a self-adjoint nonnegative definite operator with $\mathrm{Tr}\hat{\rho} = 1$ in a separable Hilbert space $\mathbb{H}$. A microstate (pure state) of a physical system is described by a state vector (wave function) $\psi \in \mathbb{H}$ of a length equal to 1, or, alternatively, by a density operator $\hat{\rho}$ of a special form, namely such that any matrix element $\rho_{nn'} = \psi_n \bar{\psi}_{n'}$ (the bar denotes complex conjugation). A complete set of observables is such that all the observables can take on simultaneously definite values, and a set of those values, determines uniquely the microstate of the physical system. A measurement whose outcome is a set of values of a complete set of observables is called a complete measurement. We consider here observables which take on a countable set of values. Then any complete direct quantum-mechanical measurement is associated with a certain countable orthonormal basis L in the space $\mathbb{H}$, the probabilities of the m-th outcome being given by the diagonal element $\rho_{kk}(L)$ of the density matrix $\hat{\rho}$ in the basis L:

$$\rho_{kk}(L) = \mathrm{Tr}(\hat{\rho}\hat{P}_k) = \sum_n \sum_{n'} \rho_{nn'}(L)\sigma_{kn'}\sigma_{n'n} \; . \tag{4.1}$$

Here $\hat{P}_k$ is a projection operator, having in the basis L a form:

$$\|P_{k,nn'}(L)\| = \|\sigma_{kn}\sigma_{nn'}\| . \tag{4.2}$$

Obviously, $\hat{P}_k$ is a nonnegative definite operator, and the set of all $\hat{P}_k$ ($k = 1,2,\ldots$) forms an orthogonal resolution of the identity $\hat{1}$ in the space $\mathbb{H}$:

$$\hat{P}_k \geq \hat{0}; \quad \sum_k \hat{P}_k = \hat{1}; \quad \mathrm{Tr}\hat{P}_k\hat{P}_{k'} = 0 \quad (k \neq k') \; . \tag{4.3}$$

Note that each $\hat{P}_k$ is a density operator corresponding to a microstate m_k described by a state vector ψ_k which is one of the basis vectors: $\psi_k \in L$. Thus, the physical meaning of the m-th outcome of the measurement is that the system is found in the microstate m_k. A complete set of observables can include physical quantities which take on a continuum of values (e.g., coordinates or momenta of particles). In a separable Hilbert space, however, any measurement of continuous variables can be arbitrarily closely approximated by measurements with a countable set of outcomes. Therefore henceforth we restrict ourselves to the measurements of type (4.3).

The entropy of a quantum system in a macrostate s described by a density matrix $\hat{\rho}$ is defined as follows [45,46]

$$H = -\,\mathrm{Tr}\,\hat{\rho}\ln\hat{\rho}. \tag{4.4}$$

A quantum-mechanical measurement changes the state of a system in an irreversible way: the entropy of the ensemble of microstates obtained as a result of a measurement is, generally speaking, larger than the entropy of the initial state. Namely, by Klein's lemma [45,46]

$$-\,\mathrm{Tr}\,\hat{\rho}\ln\hat{\rho} \leq -\sum \rho_{kk}(L)\ln\rho_{kk}(L) \; , \tag{4.5}$$

where the equality holds if $\hat{\rho}$ is diagonal in basis L.

Consider now an ensemble $S = \{s_i, p_i\}$ of macrostates of a physical system, each macrostate s_i occuring with probability p_i and being described by a density matrix $\hat{\rho}^{(i)}$. (The preparation of a certain macrostate s_i can be interpreted as "physical encoding" of a signal x_i taken from an ensemble of signals $X = \{x_i, p_i\}$). The a priori state of the system s chosen at random from the ensemble S is described by a density matrix $\hat{\rho}$:

$$\hat{\rho} = \sum_i p_i \hat{\rho}^{(i)} \tag{4.6}$$

Definition 4.1. The entropy defect of a system described by the ensemble of macrostates S is the quantity

$$I_0 = - \mathrm{Tr}\, \hat{\rho} \ln \hat{\rho} + \sum_i p_i \; \mathrm{Tr}\, \hat{\rho}^{(i)} \ln \hat{\rho}^{(i)} . \tag{4.7}$$

Entropy defect characterizes the average decrease of the entropy of the system, when it becomes known, which macrostate (out of the ensemble S) has been chosen.

Theorem 4.1.

$$0 \leq I_0 \leq - \sum_i p_i \ln p_i \; . \tag{4.8}$$

The lefthand equality holds if all $\hat{\rho}^{(i)}$ are identical, the right-hand equality holds if all $\hat{\rho}^{(i)}$ are orthogonal (i.e., $\mathrm{Tr}\, \hat{\rho}^{(i)} \hat{\rho}^{(i')} = 0$, $i \neq i'$). Thus, entropy defect does not exceed the entropy of the signal

$$H(X) = - \sum_i p_i \ln p_i \; .$$

We have seen in section 2 that, in the quasiclassical case, the entropy defect is equal to the information about the macrostate of the system obtained by measuring its microstate. The situation is different and more complex in quantum theory. Information about the macrostate (or about the signal, since the random macrostate S and signal X are in one-to-one correspondence) depends on the choice of the complete set of observables to be measured. Using Shannon's concept of information, we come to the following definition:

Definition 4.2. Information about the macrostate S of a physical system obtained in a complete measurement associated with a countable orthonormal basis L in the Hilbert space $\mathbb{H}$ is the quantity

$$I_L = -\sum_k \rho_{kk}(L) \ln \rho_{kk}(L) + \sum_i \sum_k p_i \rho_{kk}^{(i)}(L) \ln \hat{\rho}_{kk}^{(i)}(L) . \quad (4.9)$$

Here $\rho_{kk}(L)$ and $\rho_{kk}^{(i)}(L)$ are diagonal elements of $\hat{\rho}$ and $\hat{\rho}^{(i)}$, respectively, in basis L.

Since the quantity I_L depends on L, it determines not an absolute, but a conditional maximum of amount of data transmittable over a quantum channel under a fixed choice of L.

Definition 4.3. Information about the macrostate S of a physical system obtainable by direct measurements is the quantity

$$I = \sup_L I_L , \quad (4.10)$$

where the least upper bound is taken over all possible choices of L (or, in other words, over all possible thogonal resolutions of identity in $\mathbb{H}$).

The quantity I plays the same role in information transmission as information by Shannon in the classical theory. (Note that here we restrict ourselves to direct measurements only, i.e. to the usual quantum-mechanical measurements in the sense of von Neumann [46] performed in the Hilbert space $\mathbb{H}$ of the system. The problem of so called indirect (or generalized) measurements will be considered in Section 7). Indeed, the following theorem is valid:

Theorem 4.2. Let $X(\tau) = \{x_i(\tau), p_i(\tau)\}$ be a set of signals of duration τ which can be transmitted over a channel, the signal $x_i(\tau)$ being used with probability $p_i(\tau)$. Let $S(\tau) = \{s_i(\tau), p_i(\tau)\}$ be a set of macrostates of a physical system carrying the information, each macrostate $s_i(\tau)$ corresponding to the signal $x_i(\tau)$ and being described by a density matrix $\rho^{(i)}(\tau)$. Let $I(\tau)$ be defined according to (4.10) with $p_i(\tau)$ and $\rho^{(i)}(\tau)$ substituted for p_i and $\rho^{(i)}$, respectively. Then the capacity C of the channel is equal to

$$C = \lim_{\tau\to\infty} \frac{I(\tau)}{\tau} \tag{4.11}$$

(It implies, of course, that the limit exists).

A crucial question is, what is the relation between information I and entropy defect I_0.

Theorem 4.3.

$$I \leq I_0 , \tag{4.12}$$

and the equality holds if all the matrices $\hat{\rho}^{(i)}$ commute. In this case

$$I = I_{L_0} = I_0 \tag{4.13}$$

for a basis L_0 in which all $\hat{\rho}^{(i)}$ are diagonal.

Theorem 4.3 is the quantum-mechanical counterpart of the entropy defect principle (cf. Section 2). It shows that the irreversibility of a quantum measurement results in inevitable loss of information (except for the case when all the density matrices are diagonal, and the quasi-classical approximation is applicable).

Note that I_0 can be written in a form:

$$I_0 = - \sum_n \lambda_n \ln \lambda_n + \sum_i \sum_n p_i \lambda_n^{(i)} \ln \lambda_n^{(i)} , \tag{4.14}$$

where λ_n and $\lambda_n^{(i)}$ are eigenvalues of $\hat{\rho}$ and $\hat{\rho}^{(i)}$, respectively, i.e. the probabilities of orthogonal (i.e. perfectly distinguishable) microstates whose mixtures are the macrostates described by $\hat{\rho}$ and $\hat{\rho}^{(i)}$. (But, of course, when $\hat{\rho}^{(i)}$ do not commute,

$$\lambda_n \neq \sum_i p_i \lambda_n^{(i)} .)$$

Thus, I_0 can be interpreted as information in the macrostate about the microstate. On the other hand, I is the information in the microstate (specified by the measurement) about the macrostate. Since, in general, $I \neq I_0$, it means that the quantum measurement breaks the symmetry between input and output of a channel (in the classical theory $I(X;Y) =$

I(Y;X), where X and Y are input and output variables, respectively).

Suppose now that the information carrier is a closed system with a Hamiltonian H. Then the time evolution of the system is described by a unitary transformation

$$\hat{\rho}_t^{(i)} = \exp\left(-\frac{iHt}{\hbar}\right)\hat{\rho}_0^{(i)}\exp\left(\frac{iHt}{\hbar}\right), \tag{4.15}$$

where $\hat{\rho}_0^{(i)}$ and $\hat{\rho}^{(i)}$ are density matrices corresponding to the signal x_i at the initial moment of time and at the moment t, respectively. Since both I and I_0 are unitary invariant, the following theorem is valid:

Theorem 4.4.

$$\frac{dI}{dt} = \frac{dI_0}{dt} = 0 . \tag{4.16}$$

Thus, information and entropy defect are integrals of motion for a closed system. One should not forget, however, that the operators associated with the optimal measurement changes in time in the same way as the density matrices. Therefore, generally speaking, the optimal measurement becomes more and more complicated with the increase of time.

5. ENSEMBLE OF TWO PURE QUANTUM STATES

In the general case it seems not possible to derive an explicit expression for I or to determine explicitly a basis L for which $I_L = I$ (when such a basis exists). This is a rather difficult problem, even in simple special cases. (Some related results have been obtained in [32]). In this section explicit expressions will be found for the amount of information I and for an optimal basis L in the simple case of two pure states. We shall also present some results for the case of two mixed states described by second-order matrices (such as spin polarization matrices). Consider a quantum system which has a probability p of being in a state $\hat{\rho}^{(1)}$ and a probability 1 - p of being in a state $\hat{\rho}^{(2)}$. Suppose that these are pure states, i.e., their density matrices $\hat{\rho}^{(1)}$, $\hat{\rho}^{(2)}$ are of the form

$$\rho^{(1)}_{nn'} = \psi^{(1)}_n \bar{\psi}^{(1)}_{n'} \; ; \quad \rho^{(2)}_{nn'} = \psi^{(2)}_n \bar{\psi}^{(2)}_{n'} \tag{5.1}$$

where $\psi^{(1)}$, $\psi^{(2)}$ are wave functions (state vectors). The following theorem holds.

Theorem 5.1. The amount of information I_L for an ensemble of two pure states, occurring with probabilities $p_1 = p$, $p_2 = 1 - p$, achieves its maximum $I_L = I$ in a basis $L = \{\phi_i\}$ $(i = 1,2,\ldots)$ such that two basis vectors (say ϕ_1 and ϕ_2) lie in the plane spanned by the state vectors $\psi^{(1)}$, $\psi^{(2)}$, and the following condition holds: $\hat{\sigma} = p\hat{\rho}^{(1)} - (1 - p)\hat{\rho}^{(2)}$ is a diagonal matrix.

This result means that in the optimal basis L only two components of the vectors $\psi^{(1)}$ and $\psi^{(2)}$ are different from zero (say $\psi^{(i)}_1$, $\psi^{(1)}_2$), and moreover

$$p\psi^{(1)}_1 \bar{\psi}^{(1)}_2 = (1 - p)\psi^{(2)}_1 \bar{\psi}^{(2)}_2 \; . \tag{5.2}$$

In terms of density matrices, this means that in the optimal basis

$$\begin{aligned} &\rho^{(1)}_{ll'} = \rho^{(2)}_{ll'} = 0 \quad \text{for } l > 2,\ l' > 2 \\ &p\rho^{(1)}_{ll'} = (1-p)\rho_{ll'} \quad \text{for } l \neq l' \; . \end{aligned} \tag{5.3}$$

Since the phase of a wave function (a constant factor of the form $e^{i\alpha}$) may be chosen arbitrarily (it does not affect the form of the density matrix), condition (5.2) implies that the phases of states $\psi^{(1)}$, $\psi^{(2)}$ and the basis vectors ϕ_1, ϕ_2 can be chosen in such a way that all the components of $\psi^{(1)}$, $\psi^{(2)}$ in this basis are real. Condition (5.2) may then be given a simple geometric meaning. Let (x_1, x_2) be the Cartesian coordinate system formed by the basis vectors ϕ_1, ϕ_2. Then the endpoints of the vectors $\sqrt{p}\psi^{(1)}$ and $\sqrt{1-p}\psi^{(2)}$ must lie on a hyperbola $x_1 x_2 = \text{const}$ (see Figure 1).

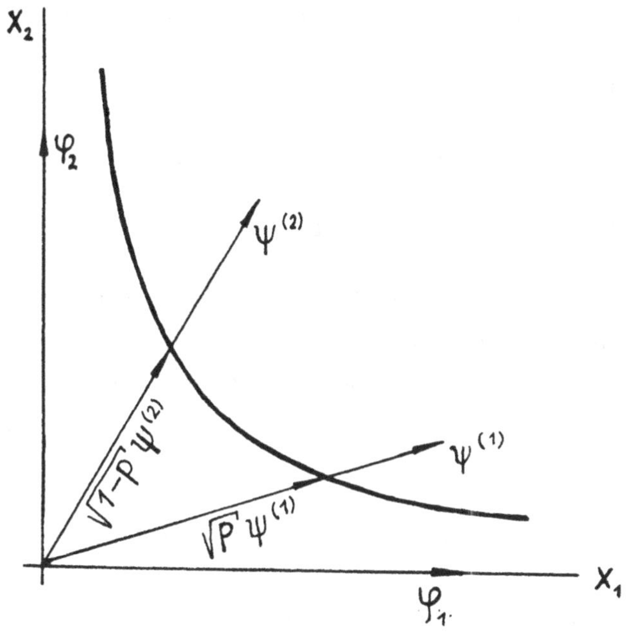

Fig. 1. Two nonorthogonal pure quantum states and optimal basis.

For two pure states, the values of I_0 and I depend only on the probability p and on the only joint invariant of $\hat{\rho}^{(1)}$ and $\hat{\rho}^{(2)}$ - the trace r of their product:

$$\mathrm{Tr}\,\hat{\rho}^{(1)}\,\hat{\rho}^{(2)} = r = |s|^2$$

where $s = \langle\psi^{(1)},\psi^{(2)}\rangle$ is the scalar product of the wave vectors $\psi^{(1)}$ and $\psi^{(2)}$. The calculations give the following results.

$$I_0 = -\mathrm{Tr}\,\hat{\rho}\ln\hat{\rho} = \ln 2 - \frac{1}{2}\,[(1+\sqrt{1-4p(1-p)(1-r)})\ln(1+\sqrt{1-4p(1-p)(1-r)}) + (1-\sqrt{1-4p(1-p)(1-r)})\ln(1-\sqrt{1-4p(1-p)(1-r)})] \quad (5.4)$$

$$I = \frac{1}{2}\,\{p[(\sqrt{1-4p(1-p)r}+1-2(1-p)r)\ln(\sqrt{1-4p(1-p)r}+1-2(1-p)r) + (\sqrt{1-4p(1-p)r}-1+2(1-p)r)\ \ln(\sqrt{1-4p(1-p)r}-1+2(1-p)r)] +$$

$$+(1-p)\ [(\sqrt{1-4p(1-p)r} - 1+2pr)\ \ln(\sqrt{1-4p(1-p)r} - 1+2pr) +$$

$$+(\sqrt{1-4p(1-p)r} + 1 - 2pr)\ \ln(\sqrt{1-4p(1-p)r} + 1 - 2pr)]-$$

$$-(\sqrt{1-4p(1-p)r} + 1 - 2p)\ \ln(\sqrt{1-4p(1-p)r} + 1 - 2p) -$$

$$-(\sqrt{1-4p(1-p)r} - 1 + 2p)\ \ln(\sqrt{1-4p(1-p)r} - 1 + 2p)\}\ (1-4p(1-p)r)^{-\frac{1}{2}}\ . \qquad (5.5)$$

In the general case, the formulae for I_0 and I are very cumbersome. We shall compare them for the case $p = \frac{1}{2}$. It is convenient to introduce the angle α between the vectors $\psi^{(1)}$ and $\psi^{(2)}$; then $r = \cos^2\alpha$. In this case the vectors $\psi^{(1)}$ and $\psi^{(2)}$ are symmetric about the bisectrix of the angle between the basis vectors ϕ_1, ϕ_2:

$$I_0 = \ln 2 - \frac{1}{2}\ (1+\cos\alpha)\ln(1+\cos\alpha) + (1-\cos\alpha)\ln(1-\cos\alpha) \qquad (5.6)$$

$$I = \frac{1}{2}\ (1+\sin\alpha)\ln(1+\sin\alpha) + (1-\sin\alpha)\ln(1-\sin\alpha)\ . \qquad (5.7)$$

The following interesting relationship is noteworthy:

$$I\ (\frac{\pi}{2} - \alpha) = \ln 2 - I_0(\alpha)\ .$$

The difference $I_0 - I$ achieves its maximum at $\alpha = \pi/4$. In that case $I_0 + I = \ln 2$ and $I \approx \frac{2}{3} I_0$. Plots of I_0 and I for the case $p = \frac{1}{2}$ are shown in Figure 2.

In the general case of two mixed states described by second-order density matrices (for example, different spin states) the quantities I_0 and I depend on invariants of the matrices $\hat{\rho}^{(1)}$ and $\hat{\rho}^{(2)}$: their determinants d_1 and d_2 and the trace r of their product. In the special case when $d_1 = d_2 = d$ and $p = \frac{1}{2}$ we obtain:

$$I_0 = \frac{1}{2}\ [(1+\sqrt{1-4d})\ln(1+\sqrt{1-4d}) + (1-\sqrt{1-4d})\ln(1-\sqrt{1-4d}) -$$

$$-(1+\sqrt{r-2d})\ln(1+\sqrt{r-2d}) - (1-\sqrt{r-2d})\ln(1-\sqrt{r-2d})]. \qquad (5.8)$$

$$I = \frac{1}{2}[(1+\sqrt{1-r-2d})\ln(1+\sqrt{1-r-2d}) + (1-\sqrt{1-r-2d})\ln(1-\sqrt{1-r-2d})]\ . \qquad (5.9)$$

For fixed r, the values of I_0 and I decrease with increasing d.

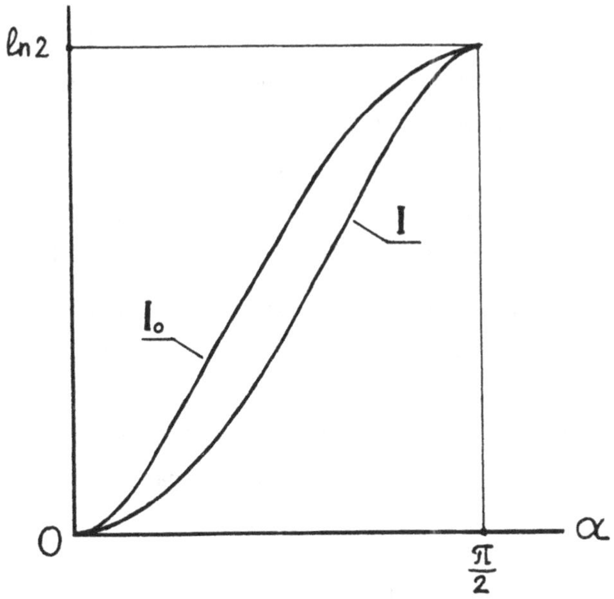

Fig. 2. Entropy defect and amount of information for two pure states, $p = \frac{1}{2}$. The point (ln2/2,π/4) is the center of symmetry of the graph.

It is interesting that in the case of two pure states the condition for maximum amount of information coincides with the condition for optimal detection, guaranteeing a minimum mean error probability [24,28]. However, this is not true in the general case of two mixed states.

6. THE GIBBS PARADOX AND MAXIMUM WORK IN MIXING OF GASES

In their original sense, entropy defect and amount of information characterize the properties of physical systems as carriers of information. The question arises whether these two quantities have a direct physical sense, whether they can be associated, say, with the work that can be produced by a non-equilibrium system in the process of relaxation. In this section we shall obtain a positive answer to this question, based on consideration of phenomena related to the Gibbs paradox. It will be shown that the maximum amount of heat convertible into mechanical work by isothermic mixing of gases which are in different but nonorthogonal quantum states, is determined not by the entropy defect but by the amount of information. This process is accompanied by an increase in total entropy of the gas and the thermostat, i.e., it is in principle irreversible. There exists, however, a more complex, reversible process, in which the amount of work is determined by the entropy defect.

Consider a vessel divided by a partition into two parts of equal volume V_0, filled with different gases 1 and 2 at the same pressure and temperature (it is assumed that these are ideal Boltzmann gases, i.e., sufficiently rarefied). Let the number of molecules of each gas be N. After the partition has been removed and the gases mixed, the entropy (in natural units) increases by an amount

$$\Delta S = 2N \ln 2 \ . \tag{6.1}$$

However, if the gases on either side of the partition are the same, there is no increase of entropy when diffusion occurs. This is the classical Gibbs paradox. The entropy increase occurring when different gases are mixed indicates that, when separated, the gases constitute a non-equilibrium system which can be used to obtain work. Indeed, if the partition is replaced by two movable semipermeable partitions (contiguous to each gas is a partition permeable to that gas but impermeable to the other), the gases are forced to push on the partitions, owing to the difference between the total pressure of the gases in the space between the partitions and the pressure of one of the gases on the other side of the partition. A simple calculation shows that under isothermic conditions the work obtained in this way is equal to

$$R = 2NkT \ln 2 \tag{6.2}$$

where k is the Boltzmann constant and T the absolute temperature. This work is derived from the heat removed from the thermostat, since the internal energy of the gases remains unchanged when they are mixed. This process is reversible, and the entropy increase of the gases is exactly equal to the entropy decrease in the thermostat:

$$\Delta S + \frac{\Delta Q}{kT} = \Delta S - \frac{R}{kT} = 0$$

Of course, when identical gases are mixed one cannot obtain any work. Thus, the possibility of converting heat into mechanical work is intimately bound up with the possibility of distinguishing between the molecules of the gases, i.e., of using the information contained in each molecule about the fact, in which half of the vessel the molecule was located. From the contemporary standpoint, the essence of the Gibbs paradox is that, whereas the entropy increase (or maximum obtainable mechanical work) when two identical gases are mixed is zero, this quantity is abruptly increased when the identical gases are replaced by different gases, however similar their properties may be. Lyuboshits and Podgoretskii [47] considered the case in which the molecules of gases 1 and 2 are the same particles, but in different quantum states $\psi^{(1)}$ and $\psi^{(2)}$ (e.g., with different spin orientations). If these states are not orthogonal, the molecules of the two gases can be distinguished only with a certain probability, and the entropy increase of the system when the gases are mixed is determined not by (6.1), but by the formula

$$\Delta S = 2\,N\,I_0 \tag{6.3}$$

where I_0 is the quantity determined by (5.4). Thus, these authors' measure of non-equilibrium for a system of two gases is precisely the entropy defect of the system for the particular case of two pure states. (Note that in this context, the states of all molecules of the gas are assumed to be statistically independent.) In the general case of mixed states, the measure of non-equilibrium is the entropy defect, defined by (4.7). According to the physical meaning of entropy, the quantity I_0 should define the maximum work (per molecule of the gas) that can be obtained from heat removed from the thermostat in isothermic reversible mixing of gases; in other words, we should have

$$R_0 = 2\,NkT\,I_0\,. \tag{6.4}$$

But the situation is not so simple. It turns out that, if the density matrices describing the states of the molecules of gases 1 and 2 do not commute, in particular, if they correspond to two pure but nonorthogonal states $\psi^{(1)}$ and $\psi^{(2)}$, the work R_0 cannot be derived directly by mixing the gases. The point is that any procedure of discrimination between the molecules of the two gases involves a certain quantum-mechanical measurement applied to the molecules, thus introducing a fundamental irreversibility into the evolution of the entire system.

Let us consider our previous device for obtaining work with the help of two semipermeable partitions (Figure 3).

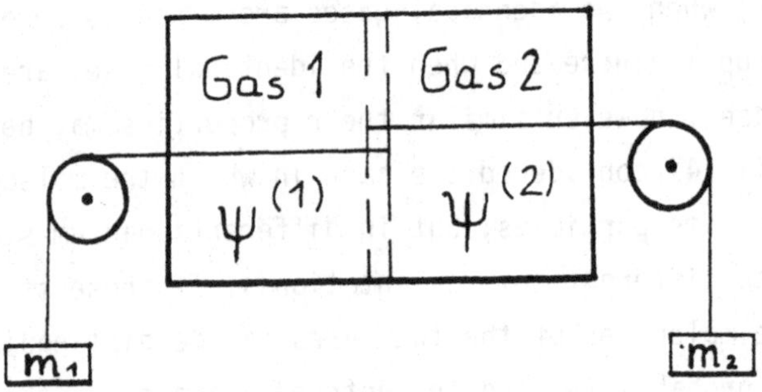

Fig. 3. Scheme for obtaining work by mixing gases. The excess pressure of the gases between the semipermeable partitions pushes them apart and lifts loads m_1 and m_2. Due to interaction with the partitions, the initial states $\psi^{(1)}$ and $\psi^{(2)}$ of the gas molecules change to a mixture of states ϕ_1 and ϕ_2.

Suppose that the partition facing gas 1 passes only molecules which are in a certain pure state ϕ_1, while that facing gas 2 passes only molecules in a state ϕ_2 orthogonal to ϕ_1. Thus, we are actually trying to distinguish between the molecules of the gases by measuring a certain physical quantity (denote it by L), the eigenfunctions of whose operator are our basis wave functions ϕ_1 and ϕ_2. Let $\rho_{11}^{(1)}$, $\rho_{22}^{(1)}$ and $\rho_{11}^{(2)}$, $\rho_{22}^{(2)}$ denote the diagonal elements of the matrices $\hat{\rho}^{(1)}$ and $\hat{\rho}^{(2)}$ in the basis $L = \{\phi_1, \phi_2\}$. Upon collision with the partition, there is a probability $\rho_{11}^{(1)}$ that a molecule of gas 1 will pass into state ϕ_1 and go through the partition, and a probability $\rho_{22}^{(1)}$ that it will pass into state ϕ_2 and remain on the same side, say to the left, of the partition as before. The molecules of gas 2 behave similarly with respect to the right-hand partition, which passes molecules in state ϕ_2. Thus, the density matrices are reduced - their off-diagonal elements vanish - and instead of $\hat{\rho}^{(1)}$ and $\hat{\rho}^{(2)}$ the states of the gases are described now by the matrices

$$\rho^{(1')}_{ll'} = \rho^{(1)}_{ll'}\delta_{ll'} \; ; \qquad \rho^{(2')}_{ll'} = \rho^{(2)}_{ll'}\delta_{ll'} \tag{6.5}$$

As a result of this reduction, the states of the gas to the left, to the right and in between the partitions are different mixtures of state ϕ_1 (call it gas A) and of state ϕ_2 (gas B). Consider the instant when the left partition moves to the left by a volume V and the right partition to the right by a volume V'. Gas A, "produced" from gas 1, will then fill the volume $V_o + V'$ and gas B ("produced" from gas 1) the volume $V_o - V$. Similarly, gas A (from gas 2) will fill the volume $V_o - V'$ and gas B (from gas 2) the volume $V_o + V$). The pressures of the gases to the left, to the right and in between the partitions are, respectively,

$$\begin{aligned} P_l &= NkT\left(\frac{\rho^{(1)}_{11}}{V_o + V'} + \frac{\rho^{(1)}_{22}}{V_o - V}\right) \\ P_d &= NkT\left(\frac{\rho^{(2)}_{22}}{V_o + V} + \frac{\rho^{(2)}_{11}}{V_o - V'}\right) \\ P_m &= NkT\left(\frac{\rho^{(1)}_{11}}{V_o + V'} + \frac{\rho^{(2)}_{22}}{V_o + V}\right) \end{aligned} \tag{6.6}$$

The work produced by the gas in the quasi-static isothermic process associated with the measurement of the quantity L is

$$R = \int_0^{V_1} (P_m - P_l)dV + \int_0^{V_2} (P_m - P_d)dV' \tag{6.7}$$

where V_1 and V_2 are defined by the conditions

$$P_m(V_1) = P_l(V_1) \ ; \quad P_m(V_2) = P_d(V_2) \ . \tag{6.8}$$

From (6.6), (6.7) and (6.8) we obtain

$$R_l = NkT\, [\rho_{11}^{(1)} \ln \rho_{11}^{(1)} + \rho_{22}^{(1)} \ln \rho_{22}^{(1)} + \rho_{11}^{(2)} \ln \rho_{11}^{(2)} + \rho_{22}^{(2)} \ln \rho_{22}^{(2)} -$$

$$- (\rho_{11}^{(1)} + \rho_{11}^{(2)}) \ln \frac{\rho_{11}^{(2)} + \rho_{11}^{(2)}}{2} - (\rho_{22}^{(1)} + \rho_{22}^{(2)}) \ln \frac{\rho_{22}^{(1)} + \rho_{22}^{(2)}}{2}] =$$

$$= 2NkT\, I_l \ . \tag{6.9}$$

Thus the maximum amount of work obtainable in this process is

$$R = 2\, NkT\, I \tag{6.10}$$

where I is determined, in the case of pure initial states, by (5.7). Why does this work fall short of R_0? The reason is that the process is irreversible. It is accompanied by an increase in the total entropy of the gas and the thermostat. Indeed, as the initial states of the gas were pure, the entropy increase of the gas is

$$\Delta S = -\, 2N \sum_l \rho_{ll} \ln \rho_{ll} = 2N \ln 2$$

and

$$\Delta S + \frac{\Delta Q}{kT} = \Delta S - \frac{R}{kT} = 2N\, (\ln 2 - I) \geq 0 \tag{6.11}$$

(equality holding only when the initial states are orthogonal). The reason for this irreversibility is the increase of entropy in the measurement process, due to reduction of the density matrices [46] (when we pass on from the matrices $\hat{\rho}^{(1)}$, $\hat{\rho}^{(2)}$ to $\hat{\rho}'^{(1)}$, $\hat{\rho}'^{(2)}$). Note that the final state of the mixture of the gases is described not by the density matrix $\hat{\rho} = \frac{1}{2}\, (\hat{\rho}^{(1)} + \hat{\rho}^{(2)})$ but by the matrix $\rho'_{ll'} = \rho_{ll'}\, \delta_{ll'}$. Thus the maximum work obtained by direct isothermic mixing of gases is determined not by the entropy defect but by the amount of information I. The question now arises, whether an isothermic process of mixing of the gases exists in which the amount of heat equal to R_0 can be converted to work. If such a process were in principle impossible, this would imply the need for a drastic revision of the physical meaning of entropy. However, this is not the case. Work corresponding to the entropy defect I_0 may indeed be obtained, but indirectly. Let us again consider the case of pure initial states whose vectors $\psi^{(1)}$ and $\psi^{(2)}$ form an angle α. The final density matrix $\rho_{ll'} = \frac{1}{2}(\psi_l^{(1)}\bar{\psi}_{l'}^{(1)} + \psi_l^{(2)}\bar{\psi}_{l'}^{(2)})$ when these states are mixed is a mixture of the two orthogonal pure states ψ_1 and ψ_2 defined by

$$\psi_1 = \frac{\psi^{(1)} + \psi^{(2)}}{2\cos\frac{\alpha}{2}} \; ; \quad \psi_2 = \frac{\psi^{(1)} - \psi^{(2)}}{2\sin\frac{\alpha}{2}} \tag{6.12}$$

The states ψ_1, ψ_2 occur with probabilities equal to the eigenvalues of the matrix $\hat{\rho}$: $\lambda_1 = \cos^2\frac{\alpha}{2}$, $\lambda_2 = \sin^2\frac{\alpha}{2}$. We now proceed as follows: without removing the initial partition, we divide the whole volume of the vessel by an additional partition into two parts, $2V_0\cos^2\frac{\alpha}{2}$ and $2V_0\sin^2\frac{\alpha}{2}$. All molecules located in the first part of the vessel we shall bring into the state ψ_1, and all those in the second part into the state ψ_2 (see Figure 4).

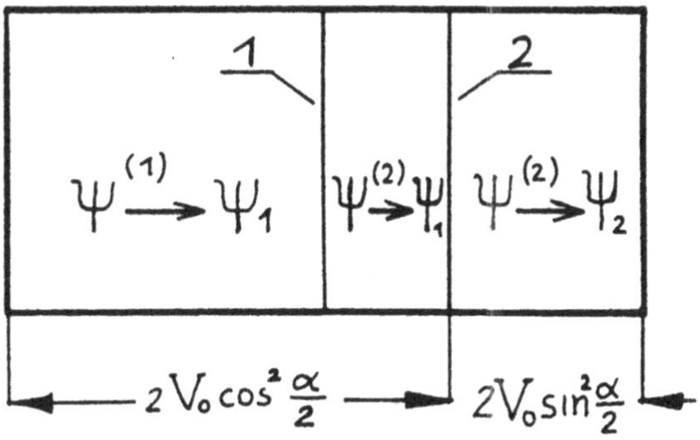

Fig. 4. Preliminary transformation of states of gas molecules to obtain work proportional to the entropy defect. After the transformation, partition 1 is removed and partition 2 is replaced by two membranes which are semipermeable for states ψ_1 and ψ_2, respectively, as in the scheme of Figure 3.

Since both the initial and the final states of the molecules are pure, the entropy is not changed thereby, and the process may, in principle, be reversible. Note that this process is described by a unitary transformation - a rotation of the state vectors. This may be accomplished physically be some appropriate interactions - different interactions for molecules initially in the state $\psi^{(1)}$ and in $\psi^{(2)}$. After this transformation the usual scheme with semipermeable partitions must be used. As a result one obtains a reversible process in which the amount of heat converted to mechanical work is given by formula (6.4).

7. INFORMATION OBTAINABLE BY INDIRECT QUANTUM MEASUREMENTS

The concept of quantum measurements can be extended by introducing indirect measurements. Consider again a quantum system with an ensemble of macrostates $S = \{s_i, p_i\}$, each macrostate s_i being described by a density matrix $\hat{\rho}^{(i)}$ in a Hilbert space $\mathbb{H}_1$. Consider an auxiliary quantum system (called in [28] "ancilla") whose state is independent of the state of the system carrying information and described by a density operator $\hat{\rho}^{(o)}$ in a separable Hilbert space $\mathbb{H}_2$. Then the state of the compound system consisting of the two systems together is described by a tensor-product density operator

$$\hat{\hat{\rho}}^{(i)} = \hat{\rho}^{(i)} \otimes \hat{\rho}^{(o)} \tag{7.1}$$

in the tensor-product Hilbert space

$$\mathbb{H} = \mathbb{H}_1 \otimes \mathbb{H}_2 \ .$$

Denote also

$$\hat{\hat{\rho}} = \hat{\rho} \otimes \hat{\rho}^{(o)} = \sum_i p_i \hat{\rho}^{(i)} \otimes \hat{\rho}^{(o)} \tag{7.2}$$

A complete indirect (generalized) measurement is a usual quantum measurement performed over the compound system and associated with a certain countable orthonormal basis K in the space $\mathbb{H}$.

Definition 7.1. Information about the macrostate S of the quantum system for a given state of the ancilla in the outcome of an indirect measurement associated with a basis K in the space $\mathbb{H}_1 \otimes \mathbb{H}_2$ is the quantity

$$J_K = - \sum_k \sum_n \rho_{kn,kn}(K) \ln \rho_{kn,kn}(K) + \sum_i \sum_k \sum_n \rho^{(i)}_{kn,kn}(K) \ln \rho^{(i)}_{kn,kn}(K), \tag{7.3}$$

where $\rho_{kn,kn}$ and $\rho^{(i)}_{kn,kn}$ are diagonal elements of density matrices $\hat{\rho}$ and $\hat{\rho}^{(i)}$, respectively, in the basis K.

Definition 7.2. Information about the macrostate S of a physical system obtainable by indirect measurements is the quantity

$$J = - \sup_K \sup_{\hat{\rho}^{(o)}} J_K \tag{7.4}$$

It has been shown by Naimark [48] that any direct measurement, i.e. any orthogonal resolution of identity in $\mathbb{H}$ is equivalent to a resolution of identity (generally speaking, non-orthogonal) in $\mathbb{H}_1$, and, conversely, any non-orthogonal resolution of identity in $\mathbb{H}_1$ can be considered as induced by an orthogonal resolution of identity in a certain tensor-product space $\mathbb{H}$. By a non-orthogonal resolution of identity we mean a set of self-adjoint operators $\{B_k\}$ such that each B_k is nonnegative definite

$$\hat{B}_k \geq \hat{0}, \quad \text{and} \quad \sum_k \hat{B}_k = \hat{1}, \tag{7.5}$$

but, in general, $\mathrm{Tr}\, B_k B_{k'} \neq 0$.

Indirect quantum measurements were discussed in a number of works (e.g. [20-23,25-28,30-33]). It was shown that randomized, successive and adaptive quantum measurements are equivalent to some indirect measurements [26], and, thus, indirect measurements are the most general kind of quantum measurements. It was also shown [31] that

$$J \leq I_o, \tag{7.6}$$

where the equality holds if all the density matrices $\hat{\rho}^{(i)}$ commute.

However, it remained a controversial problem for more than ten years, whether indirect measurements can yield more information than the direct ones. In [30] an example was constructed, where, indeed, $\mathcal{J} > I$. This example, however, uses "state vectors" in a two-dimensional Euclidean space and therefore it seems to be not relevant to quantum systems with states described by operators in a Hilbert (infinite-dimensional) space. Intuitively, the phenomenon that $\mathcal{J}$ can exceed I in a finite-dimensional space is due to the fact that in such cases the cardinality $|S|$ of the set of states exceeds the dimensionality d of the space in which they are embedded. (It has been shown by Davies [32] that the cardinality of the set $\{B_k\}$ of self-adjoint operators used in the optimal indirect measurement obeys an inequality

$$d \leq |\{B_k\}| \leq d^2 , \tag{7.7}$$

but, obviously, this result is not applicable in the infinite-dimensional case.)

Recently it was proved [33] that the situation is, indeed, different in an infinite-dimensional Hilbert space.

Lemma 7.1. For any set of density operators $\hat{\rho}^{(i)}$, probabilities p_i, and the density operator of the ancilla $\hat{\rho}^{(o)}$, and for any basis K in the product Hilbert space $\mathbf{H}_1 \otimes \mathbf{H}_2$ there exists a pure state $\rho_1^{(o)} = |\psi_1\rangle\langle\psi_1|$ of the ancilla such that

$$\mathcal{J}_K\{\hat{\rho}^{(i)};\ p_i;\ \hat{\rho}_i^{(o)}\} \geq \mathcal{J}_K\{\hat{\rho}^{(i)};\ p_i;\ \hat{\rho}^{(o)}\}. \tag{7.8}$$

This pure state can be always considered as the first basic vector in the initial basis of the Hilbert space $\mathbf{H}_2$ of the ancilla.

Lemma 7.1 implies that the variation over $\hat{\rho}^{(o)}$ can be omitted in Definition 7.1.

Lemma 7.2. For any set of $\hat{\rho}^{(i)}$ and p_i and for any basis K in $\mathbb{H} = \mathbb{H}_1 \otimes \mathbb{H}_2$ there exists a set of density operators $\hat{\sigma}^{(i)}$ and a basis R in $\mathbb{H}_1$ such, that

$$I_R\{\hat{\sigma}^{(i)};\ p_i\} = \mathcal{J}_K\{\hat{\rho}^{(i)};\ p_i\} . \qquad (7.9)$$

Lemma 7.3. If all the density operators $\hat{\rho}^{(i)}$ can be represented as mixtures of a finite number of pure states, then for any set of p_i and for any basis K in $\mathbb{H}$ there exists a basis R in $\mathbb{H}_1$ such that

$$I_R\{\hat{\rho}^{(i)};\ p_i\} = \mathcal{J}_K\{\hat{\rho}^{(i)};\ p_i\} . \qquad (7.10)$$

Theorem 7.1. For any set of density operators $\hat{\rho}^{(i)}$ in a separable (infinite-dimensional) Hilbert space, for any set of probabilities p_i such that $\mathbb{H} = -\sum_i p_i \ln p_i < \infty$ and for any ancilla with a separable Hilbert or a Euclidean space

$$I\{\hat{\rho}^{(i)};\ p_i\} = \mathcal{J}\{\hat{\rho}^{(i)};\ p_i\} . \qquad (7.11)$$

Theorem 7.1 shows that in the case of a quantum system associated with a separable Hilbert space direct measurements are as effective as more general indirect measurements.

8. CONCLUSION

Information theory for quantum systems is based on two important concepts: the entropy defect and the information obtainable in quantum measurements. The consistent quantum-mechanical formulation of the entropy defect principle (Section 4) shows that these two quantities, being identical in the classical case, split apart from each other in the realm of quantum theory, which reflects the effect of the irreversibility caused by a quantum-mechanical measurement. This expresses the fundamental limitations imposed on information transmission and storage by quantum nature of information carriers even in the absence of any external noise. On the other hand, the concepts of entropy defect and quantum information have an important thermodynamical meaning, as established in Section 6. These results point to the close relationship between the use of non-equilibrium physical systems to obtain work, on the one hand, and their use to transmit information, on the other. To obtain work one must carry out a quantum-mechanical measurement on the system, and this implies, as well as in the case of information transmission, essential irreversibility and unavoidable losses. These losses may be avoided only through a preliminary state transformation, i.e., through replacing the original physical system by another system with the same entropy defect. (In the information-transmission case, this means that one uses as signals an ensemble of orthogonal states, described by the same density matrix as the original ensemble of nonorthogonal states.) It is to be expected that the equivalence relation (6.10) between amount of information and work is valid in the general case and not only for mixing of gases.

The results of Section 7 solve the important problem concerning the information efficiency of direct and indirect quantum measurements. Though indirect measurements (such as in homodyne or heterodyne reception of an electromagnetic signal) may be practically convenient, in principle, we cannot gain, compared to direct measurements, when we deal with a quantum system described by an infinite-dimensional Hilbert space.

REFERENCES:

1. Szilard, L.: Über die Entropieverminerung in einem thermodynamischen System bei Eingriff intelligenter Wesen, Z. Physik, B. 53, No. 5 (1929), 840-856 (in German).
2. Shannon, C.E. and Weaver, W.: The Mathematical Theory of Communication, Urbana Univ. Press., Chicago, Ill. 1949.
3. Gabor, D.: Communication theory and physics, Phil. Mag., 41 (1950), 1161-1187.
4. Brillouin, L.: Science and Information Theory, Acad. Press, New York 1956.
5. Lebedev, D.S. and Levitin, L.B.: Information transmission by electromagnetic field. Inform. Contr. 9, No. 1 (1966), 1-22.
6. Levitin, L.B.: On the quantum measure of the amount of information, in: Proc. of the IV National Conf. on Information Theory, Tashkent, 1969,

111-115 (in Russian).
7. Planck, M.: Wärmestrahlung, Berlin 1913 (English translation: Theory of Heat Radiation, Dover, New York 1959).
8. Kolmogorov, A.N.: On the Shannon theory of information in the case of continuous signals, IRE Trans. on Inform. Theory, IT-2 (1956), 102-108.
9. Gordon, J.P.: Quantum effects in communication systems. Proc. IRE 60, No. 9 (1962), 1898-1908.
10. Lebedev, D.S. and Levitin, L.B.: The maximum amount of information transmissible by electromagnetic field, Soviet Physics-Doklady, 8 (1963), 377-379.
11. Takahasi, H.: Information theory of quantum-mechanical channels, in: Advances in Communication Systems, v. 1, Acad. Press, New York 1965, 227-310.
12. Stratonovich, R.L.: Information transmission rate in some quantum communication channels, Probl. Info. Transm., 2 (1966), 45-57.
13. Mityugov, V.V.: On quantum theory of information transmission, Probl. Info. Transm., 2 (1966), 48-58.
14. Ingarden, R.S.: Quantum information theory, Institut of Physics, N. Copernicus Univ., Torun, Poland, 1975.
15. Vainshtein, V.D. and Tvorogov, S.D.: Some problems of measurement of quantum observables and determination of joint entropy in quantum statistics, Comm. Math. Phys., 43 (1975), 273-278.
16. Drikker, A.S.: Homodine reception of a quantum electromagnetic signal, Probl. Info. Transm., 12 (1976), 57-68.
17. Poplavskii, R.P.: Thermodynamic models of information processes, Sov. Phys. Uspekhi, 115 (1975), 222-241.
18. Pierce, J.B., Posner, E.C. and Rodemich, E.C.: The capacity of the photon counting channel, IEEE Trans. on Info. Theory, IT-27 (1981), 61-77.
19. Kosloff, R.: Thermodynamic Aspects of the Quantum-Mechanical Measuring Process, in: Advances in chemical physics, v. 46, eds. I. Prigogine, S.A. Rice, Wiley and Sons, 1981, 153-193.
20. Mitiugov, V.V.: Physical foundations of information theory, Sovietskoe Radio, Moscow, 1976 (in Russian).
21. Arthurs, E. and Kelly, J.L., Jr.: On the simultaneous measurement of a pair of conjugate observables, Bell System Tech. J., 44 (1965), 725-729.
22. Gordon, J.P. and Louisell, W.J.: Simultaneous measurement of non-commuting observables, in: Physics of Quantum Electronics (Eds. P. Kelley, M. Lax, B. Tannenwald), McGraw-Hill, New York 1966, 833-840.
23. Levitin, L.B., Mitiugov, V.V.: Reception of coherent signals by splitting of the beam, 1st Conf. on Problems of Information Transmission by Laser Radiation, Kiev, 1968.
24. Helstrom,C.W.: Detection theory and quantum mechanics, Inform. Contr. 10 (1967), 254-291.
25. Davis, E.B. and Lewis, J.T.: An operational approach to quantum probability, Comm. Math. Phys., 17 (1970), 239-260.

26. Benioff, P.A.: Decision procedures in quantum mechanics, J. Math. Phys., 13 (1972), 908-915.
27. Helstrom, C.W. and Kennedy, R.S.: Noncommuting observables in quantum detection and estimation theory, IEEE Trans. Inform. Theory, IT-20 (1974), 16-24.
28. Helstrom, C.W.: Quantum Detection and Estimation Theory, Academic Press, New York 1976.
29. Levitin, L.B.: Amount of information and the quantum-mechanical irreversibility of measurement, in: Proc. of the II Intern. Symp. on Inform. Theory, Yerevan, 1971, 144-147.
30. Holevo, A.S.: Informational aspects of quantum measurements, Probl. Info. Transm., 9 (1973), 31-42.
31. Holevo, A.S.: Certain estimates of information transmissible over a quantum communication channel, Probl. Info. Transm., 9 (1973), 3-11.
32. Davies, E.B.: Information and quantum measurement, IEEE Trans. on Inform. Theory, IT-24 (1979), 596-599.
33. Levitin, L.B.: Direct and indirect quantum measurements yield equal maximum information, 1981 IEEE Intern. Symp. on Information Theory, Santa Monica, CA., USA, 1981.
34. Jaynes, E.T.: Information theory and statistical mechanics, Phys. Rev., Part I, 106 (1957), 620-630; Part II, 108 (1959), 171-190.
35. Katz, A.: Principles of Statistical Mechanics. The Information Theory Approach, W.H. Freeman, 1967.
36. Ingarden, R.S.: Information Theory and Thermodynamics. Part I, Torun, Poland, 1974. Part II, Torun, Poland, 1976.
37. Levitin, L.B.: Quantum amount of information and maximum work, in: Proc. of the 13th IUPAP Conf. on Statistical Physics (Eds. D. Cabib, D.G. Kuper and I. Riess), A. Hilger, Bristol, England, 1978.
38. Poplavskii, R.P.: Maxwell demon and correlations between information and entropy, Sov. Phys. Uspekhi, 128 (1979), 165-176.
39. Mityugov, V.V.: Entropy, information and work in quantum statistics, Probl. Control and Inform. Theory, 2 (1973), 243-256.
40. Berger, T.: Communication theory via random fields, IEEE Intern. Symp. on Inform. Theory, St. Jovite, Quebec, Canada, 1983.
41. Levitin, L.B.: Information-theoretical approach to Ising problem, IEEE Intern. Symp. on Inform. Theory, Brighton, England, 1985.
42. Levitin, L.B.: A thermodynamic characterization of ideal physicsl information channels, Journal of Information and Optimization Sciences, 2 (1981), 259-266.
43. Levitin, L.B.: Information transmission in an ideal photon channel, Probl. Inform. Transm., 1 (1965), 55-62.
44. Levitin, L.B.: Ideal corpuscular information channels, IEEE Intern. Symp. on Inform. Theory, St. Jovite, Quebec, Canada, 1983.
45. Klein, O.: Zur quantenmechanischen Begründung des zweiten Hauptsatzes der Wärmelehre, Z. Phys., 72 (1931), 767-775.
46. Neumann, J.: Mathematische Grundlagen der Quantenmechanik, Springer-Verlag, Berlin, 1932 (English translation: Mathematical Foundations of Quantum Mechanics, Princeton Univ. Press, Princeton, NJ, USA, 1955).

47. Lyuboshitz, V.L., and Podgoretzkii, M.I.: Entropy of a polarized gas and Gibbs paradox, Soviet Physics-Doklady, 194 (1970), 547-550.

THERMODYNAMIC AND INFORMATIONAL ENTROPIES IN QUANTUM MECHANICS

Francis Fer
Ecole des Mines, Paris, France

It is to-day a widespread opinion that entropy (in its thermodynamical sense) and information (in Shannon's one) are closely related notions (or quantities). As is known, the idea was first initiated by L. Szilard[1] as early as 1929, resumed in 1931 by J. Von Neumann[2] in his treatise about Quantum Mechanics, and it was resumed again and more broadly developed by L. Brillouin[3] in 1956 after the publication of the work of C.E. Shannon[4].

To express briefly the afore-mentioned connection, the best is to quote the last two authors. From Brillouin ([3],p. 155) :"...entropy measures the lack of information about the true structure of a system" ; and, from Von Neumann ([2], p. 400) : "The time variations of the entropy are then based on the fact that the observer does not know everything, that he cannot find out (measure) everything which is measurable in principle".

One can be tempted to repel these statements by a classical argument: how can a physical objective property such as the value of the entropy of a system be influenced by the more or less deep knowledge -a notion somewhat subjective- we have about the system ? As attractive as it may seem, such an argument cannot lead to anything else than an endless discussion on the "reality of the things", the fact that reality "does not exist" unless we observe it, and so on.

My purpose is different. The two statements quoted above are the concise expression of a line of thought which is, obviously, of *physical* importance. Consequently they must be discussed, not from a philosophical point of view (or, worse, from personal preferences), but on the basis of *physical and mathematical coherence* as well as *physical inferences*.

1. SOME PRELIMINARY ENLIGHTENMENTS

We must begin by clarifying the vocabulary, namely the words : entropy, information, and their interconnection, since they could be pregnant with confusions.

a) First we must emphasize the leading role played in this whole affair by macroscopic Thermodynamics. The reason for that is not its historical anteriority, but the fact that it is a field of Physics which is fairly verified by experiment -within certain limits of validity, of course- and that any theory which claims to deepen its results must begin by agreeing with them on the macroscopic level.

In particular it must be emphasized that entropy is a *macroscopic* quantity attached to every state of a *macroscopic* system, and that this system must be itself *well-specified and well-delimited*. This precaution is omitted sometimes, as we shall see later on (§ 3).

There is yet a most important point about entropy. The second law of Thermodynamics does not reduce itself, as is often said, to Carnot's principle or Kelvin's one ; these are only particular cases. The general, and powerful, formulation of the second law lies in this statement : for any infinitesimal transformation, the entropy S of a given system undergoes a variation dS of the form[5]

$$dS = d_eS + d_iS \tag{1}$$

where d_eS results from the relation of the boundary of the system with its surroundings (for a closed system it is the familiar $\int \frac{dQ}{T}$), and d_iS is the entropy production due to internal phenomena. d_eS can have any sign according to the external circumstances, but d_iS is strictly positive as soon as the system is the seat of irreversibilities, which happens always in all the known real processes. It is this property which is considered to-day as the fundamental feature of the second law[5].

Finally it must be pointed out that the value of an entropy calculated from the rules of statistical Thermodynamics must coincide with the value that macroscopic Thermodynamics derives from measurements.

b) The word "information" needs too precisions. It should not be taken in the vague sense of "acquisition of some knowledge" or of experimental data. I shall use it uniquely in the shannonian sense, that is : given a set of possible outcomes E_i ($i = 1$ to q) of an event, each of which has a definite probability p_i, known previously to any measurement, the "shannonian entropy" of the set $\{E_i\}$ is defined as $-K \sum_{i=1}^{q} p_i \ln p_i$, K being a constant which can be chosen at will. If we perform an experiment which enables us to see what subset of outcomes has happened, the information brought by the experiment is the diminution of the shannonian entropy.

When all the probabilities p_i are equal, and then equal to $1/q$, the entropy is $K \ln q$, and if the experiment shows that only $q' < q$ outcomes have appeared, the information is $K(\ln q - \ln q')$.

c) After these precisions, we come to the central question, namely the connection between entropy and information. In fact this question divides into two distinct ones.

First, does the thermodynamical entropy of any system, calculated from the rules of statistical Thermodynamics, have the same value as the shannonian entropy calculated in an adequate manner ? It must be noticed that a numerical equality between these two quantities does not signify compulsorily an identity between the two properties.

Secondly, and this is the important point, do thermodynamical entropy and informational entropy relative to a given system interfere, or, in other words, does a loss of information imply a variation of the entropy and conversely ?

As we shall see later on, these two questions are independent.

2. EQUALITY BETWEEN THE TWO ENTROPIES

A. Macroscopic system in equilibrium

Let us consider a macroscopic system constituted of N particles which, for simplicity, we suppose to be identical. We assume that this system is in equilibrium for given values of external parameters and, in a first step, that it is isolated ; hence its energy U is fixed.

For this value of the energy, Quantum Mechanics determines, from its own rules, a set of independent eigenstates among which we select the "accessible" states, that is which obey the laws of symmetry or antisymmetry according to the kind of particles we are dealing with. Let Ω the number (finite in principle) of these states.

To these properties statistical Thermodynamics adds, since the system is in equilibrium, the ergodic postulate, namely that the independent accessible states have equal probabilities, hence $\frac{1}{\Omega}$ for each of them.

Starting from this basis, the method of Darwin-Fowler[6] and Khinchin[7] shows that, in order to find again the macroscopic equation $dS = dQ/T$ for reversible processes, statistical Thermodynamics must take, for the entropy of the system, the expression

$$S = k(\ln Z + \alpha N + \beta U) \tag{2}$$

where $Z = Z(\alpha,\beta)$ is the grand partition function, k the Boltzmann constant, and α and β are determined as functions of N and U by the two equations

$$\frac{\partial \ln Z}{\partial \alpha} + N = 0, \qquad \frac{\partial \ln Z}{\partial \beta} + U = 0 \tag{3}$$

with $\beta = 1/kT$ (in fact Z depends also on external parameters, but we do not explicit them as long as it is not necessary). This result is valid as well for interacting particles as for non-interacting ones.

Now, when the particles are non-interacting, it is readily shown (see[7], p. 142) that the value (2) is practically (within an error of order 1/N) equal to

$$S = k \ln \Omega \tag{4}$$

One recognizes here the well-known expression of the thermodynamical entropy as related to the number of accessible eigenstates.

Now let us look at the same system from the point of view of the theory of information. We may regard the accessible eigenstates of the system as the possible outcomes of a random event, each of which has the same probability. Hence the shannonian entropy of this set of outcomes is $K \ln \Omega$, and identifies itself to the entropy (4) if we choose $K = k$. We have therefore a practically perfect equality between thermodynamical and informational entropies.

This result can be easily extended to the case when the system, always in equilibrium under given external constraints, is no more isolated but connected with a heat-bath. The formula (2) remains valid, under the only condition of replacing the exact value U of the energy by its average $\bar{U}$; but the equation (4) is unchanged, and so is the equality between the two entropies.

Lastly it must be noticed that, if we used classical Mechanics instead of Quantum Mechanics, the equations (2) and (3) should be replaced by

$$S = k(\ln Z + \beta U), \qquad \frac{\partial \ln Z}{\partial \beta} + U = 0 \tag{5}$$

(the number N of the particles being fixed once for all and the parameter α turning useless) ; but one shows ([8], p. 91) that the formula (4) remains still valid (with Boltzmann's fashion of defining Ω), as well as the equality of the two entropies.

B. Macroscopic system in a non-equilibrium state

Consider now such a system and suppose, for simplicity, that it is isolated, with an energy U. Macroscopic Thermodynamics defines its entropy as follows (in a way which goes back as far as Gibbs[9], and was resumed axiomatically by Duhem[10], vol. II, p. 48).

Let us divide the volume D occupied by the system into cells D_ρ (ρ = 1 to ν), each of which is sufficiently small to assimilate discrete summations to integrals over the volume, and sufficiently large to contain a great number -say 10^8- of particles (under current circumstances cells of 1 cubic-micron are adequate). Macroscopic Thermodynamics assumes that the matter contained in each cell D_ρ can be considered as if it were in equilibrium ; therefore, besides its energy U_ρ which is defined without any question (with $\sum_\rho U_\rho = U$), the matter of D_ρ has an entropy S_ρ, and we adopt, for the entropy of the whole system

$$S = \sum_{\rho=1}^{\nu} S_\rho \tag{6}$$

Now statistical Thermodynamics, according to the so-called hypothesis of "local equilibrium" just quoted, must state $S_\rho = k \ln \Omega_\rho$, Ω_ρ being the number of accessible eigenstates of the matter filling the cell D_ρ with the energy U_ρ. Therefore (6) yields

$$S = \sum_{\rho=1}^{\nu} k \ln \Omega_\rho = k \ln\left(\prod_{\rho=1}^{\nu} \Omega_\rho\right) \tag{7}$$

But $\Pi_\rho\Omega_\rho$ is nothing else than the total number of the eigenstates of the whole system which are compatible with the distribution of the energies U_ρ among the different cells D_ρ in the considered state of non-equilibrium.

Now, if we consider the things from the point of view of information theory, $\Pi_\rho\Omega_\rho$ is also the number of possible outcomes of the random event "occurence of an eigenstate of the whole system". Thus the shannonian entropy of this set of outcomes could be considered as equal to the thermodynamical entropy of the system.

I say only "could be", since this second result is not so far reaching as it appears at first sight. In effect we must not forget that this counting of the number of possible eigenstates has a significance -either for thermodynamical or for informational entropy- only *if all these states have the same probability*, viz. $1/\Pi_\rho\Omega_\rho$. But there is a serious doubt whether we have *the right of assuming such probabilities* -in other words of applying the ergodic postulate- in the case of non-equilibrium, for the following reason.

The set E_i (i : irreversible) of eigenstates which represent a given macroscopic state of non-equilibrium for given values of external parameters and a given energy U is also a proper subset of the total number E_e (e : equilibrium) of eigenstates which represent the equilibrium of the same system under the same conditions. When regarded as belonging to the total set E_e, the states of E_i are always considered as fluctuations about the (large) subset of average eigenstates, and it is known that the probability of E_i is then very low (for a gas for example, it is of the order of $N^{-\frac{1}{4}}$) ; thus the states of E_i should be rarely observed. But experiment shows that nothing is easier than inducing an irreversible process, and consequently we should attribute to its successive states probabilities far from being negligible. Thus we are faced with a contradiction which seems difficult to remove unless we change the law of probability according to what we know : equilibrium or non-equilibrium, about the macroscopic behaviour of the system. This problem is not the subject of this paper ; but in any case we can ask whether the equality that we found between thermodynamical and informational entropy in the course of an irreversible process is anything else than a purely formal relation.

C. Conclusion

From the above considerations we can conclude that, at least in the case of a system in equilibrium, the shannonian entropy is numerically equal to the thermodynamical one ; for systems in states of non-equilibrium the answer is more doubtful.

But, in any case, we must remark at once than calling an entropy informational instead of thermodynamical *is only a change of name*, which is of a very small interest. The true problem is elsewhere, and we are entering upon it now.

3. RELATION BETWEEN INFORMATION AND ENTROPY VARIATION

The real problem is the following : does an information -that is, a *change* of the shannonian entropy- implies a *variation* of the thermodynamical entropy, and conversely ? Only these concomitant variations, if they exist, have a physical interest.

First let us recall briefly the argument of Brillouin, which is the most explicit which has been enunciated at the present time. It is, textually ([3], p. 159 and 161),

neg.entropy → information → neg.entropy

If we translate this formula in terms of entropy, in order to avoid the awkward use of the minus sign of the neg.entropy, it becomes

an increase (>0) of the entropy allows information ; (8)
the possession of an information allows a decrease of entropy.

One or the other formula is not so clear as would be desirable, since *they do not precise to what system are attached the variation of entropy as well as the information*. This is a major point we have to clear up through our own resources.

A. Entropy production needed for an information

For getting an information about a given system Σ, we must make a measurement on it, which implies to bring it in connection with a measurement apparatus M, and let them evolve during an adequate time. This operation is a real physical process and, as such, comprises some irreversibilities. Therefore it implies an entropy production -in the sense precised on equation (1)- which is always strictly positive. This does not result compulsorily in an increase of the total entropy $S_\Sigma+S_M$ of the whole system Σ+M, since the variation of this entropy may depend on the surroundings ; but, if we have taken Σ and M separately isolated before the measurement and preserved the isolation of Σ+M during the measurement, we have $\Delta(S_\Sigma+S_M) > 0$, Δ being the symbol of variation during the process.

Even in this case, the inequality $\Delta(S_\Sigma+S_M) > 0$ does not imply $\Delta S_\Sigma > 0$ (nor $\Delta S_M > 0$), and it is easy to find out measurements for which $\Delta S_\Sigma < 0$. For example if we measure the temperature of a region of Σ with a thermometer colder than this region, an elementary calculation shows that S_Σ diminishes (whatever be the size of the thermometer), although there has been a part of the entropy production in Σ itself.

Then we can conclude that *the "cost" of an information is not an increase of the entropy of such or such system, but an entropy production in all systems which take a part into the acquisition of the infor-*

mation. And indeed it is a cost, because recovering an entropy production needs above all mechanical work, a precious thing for users of Thermodynamics.

B. Influence on an information on the entropy

We come now to the question enclosed in the second line of the statements[8], which is the most difficult question.

Let us comment first the famous example of Szilard and Von Neumann ([2], p. 400) about a molecule which expands irreversibly and at constant energy U from a volume $\frac{V}{2}$ to a volume V. The argument of Von Neumann is the following : if we only know that the value of the volume passes from $\frac{V}{2}$ to V, then the entropy increases by k ln2 ; but, if we knew the initial position and momentum of the particle and thus could calculate its coordinates at every time, the entropy would not vary. Hence the conclusion : a gain of information makes the entropy decrease.

But the second statement of Von Neumann is contrary to the usual rules of statistical Thermodynamics. The easiest to see it is to place ourselves in the frame of classical Mechanics. The hamiltonian of the particle has the form H(q,p;v), the volume v (external parameter) appearing compulsorily in the expression of the hamiltonian. For given values of the volume v and of the energy U, the phase-point q,p lies on the hypersurface of the phase-space H(q,p;v) = U. When, at constant U, the volume varies from $\frac{V}{2}$ to V, this hypersurface changes from a geometrical position Σ_1 to another Σ_2. The fact that we could know (*in a purely mental manner, we must remark*) the trajectory of the phase-point q,p during the process of expansion does not prevent this point from going from Σ_1 to Σ_2.

Now the entropy of the molecule is given by equations (5), the partition function $Z(\beta,v)$ being determined by the well-known equation

$$Z(\beta,v) = \int \exp[-\beta H(q,p;v)]dq\, dp \qquad (9)$$

It is seen immediately on this expression that Z does not depend at all on q,p, but only on β and v, that is, taking into account the second equation (5), on U and v. Hence the entropy depends only on U and v, exactly as in macroscopic Thermodynamics (which is not at all surprising, since statistical Thermodynamics was devised for that).

The same would be obviously true if, instead of considering a single molecule (which is not a thermodynamical object, properly speaking), we should consider a macroscopic gas in the same process of expansion. The partition function depends only on U and the external parameter v, and, *whether we know or not (always mentally) the position of the phase-point on the hypersurface of constant energy for a given volume does not change anything to the value of entropy*.

The reasoning is extended readily to Quantum Mechanics ; the entropy is then given by (2) and (3), where the grand partition function is determined by

$$Z(\alpha,\beta;v) = \sum_k e^{-\alpha k}\ \mathrm{Tr}[\exp(-\beta\ {}_k)] \qquad (10)$$

${}_k(q,v)$ being the hamiltonian operator of a system of k particles enclosed in the volume v. Z, hence S, remains yet a function of N, U and v only, and knowing (purely mentally) the state-vector of the system does not in the least influence the value of the entropy.

But now we must give up this purely theoretical case, in which we suppose -but only by a mere operation of mind- that we know something about the microscopic state of the system. Physically we cannot get any information unless we make a real measurement, and a measurement implies always a perturbation of the observed system (of whatever kind it may be, quantal or not). Then the true problem is : *assuming we have acquired some information about a system* Σ *by such or such means, what can we say about the final entropy of* Σ *if we take into account the information itself*? (it is here no more question of the entropy production during the operation, but of the proper entropy of the observed system).

Consider a system Σ *in equilibrium* under given external constraints (volume for a gas for example) ; we suppose moreover that Σ is initially isolated. Be U_0 its energy, S_0 its entropy, T_0 its temperature under these conditions. The number of its eigenstates compatible with the energy U_0 and the external constraints is Ω_0, related to S_0 by (4), namely

$$S_0 = k \ln \Omega_0 \qquad (11)$$

Let us now make a measurement on the system (which of course interrupts for a while the isolation), the measurement apparatus interacting with a small sub-system σ of Σ ; then let the system, *anew isolated, come back to equilibrium*. It has a well-determined energy $U_1 = U_0 + \Delta U$, and a well-determined entropy $S_1 = S_0 + \Delta S$. *We do not prejudge at all the sign of* ΔU *nor of* ΔS ; we know only (if the perturbation is sufficiently small) that $\Delta S = \frac{\Delta U}{T_0}$ at the first order. The new number of eigenstates of Σ compatible with the energy U_1 and the external constraints (supposed to be unchanged) is Ω_1 related to S_1 by (4), that is $S_1 = S_0 + \Delta S = k \ln \Omega_1$. From this equation and (11) we get

$$\Delta S = k(\ln \Omega_1 - \ln \Omega_0) \qquad (12)$$

Let us come now to the information. The indications given by the measurement apparatus enable us to calculate the number of outcomes (eigenstates of Σ) which remains still possible after the knowledge obtained from the measurement ; let us call it Ω'. It must be noted by the way that Ω' cannot be known without a careful analysis of the measurements data. Now, according to the axiomatics of Shannon, the information I is to be defined with reference to the initial set of eigenstates, whose number is Ω_0. Then we must take

$$I = k(\ln \Omega_0 - \ln \Omega') \qquad (13)$$

Now if we look for a relation between ΔS and I *which be independent of the initial situation* Ω_0 -and it seems difficult to look for anything else- the only thing we can do is to eliminate $\ln \Omega_0$ between (12) and (13), so that we obtain

$$\Delta S + I = k(\ln \Omega_1 - \ln \Omega') \tag{14}$$

Such as it is, this equation does not teach us very much : *it shows only that the interesting quantity which comes out from the comparison of* ΔS *and* I *is their sum*. As for the right-hand member, it does not lead apparently to general conclusions : we see a priori no relation between Ω_1, which represents a physical property of the perturbed system Σ, and Ω', which is the result of a calculation based on the measurement data. All that we can do is to evaluate this right-hand member on examples.

Let us reason on a gas constituted of N identical particles (bosons or fermions) enclosed in a volume V, and consider the case when the information deduced from the observation of the sub-system σ is *maximum, that is when we know exactly the eigenstate of* σ ; thus we shall have an upper bound of I and ΔS + I (note that we do not mind in any case the Heisenberg uncertainties which could arise in such measurement : we merely reason on the maximum available information). As a consequence we know in particular the number ν of particles of σ, its volume v and its energy u. We suppose $\nu \ll N$, which leaves open the possibility for σ to be either a microscopic system or a small macroscopic one.

Since the observation of σ is complete and thus reduces to 1 the number of its possible outcomes, Ω' is the number of the possible eigenstates of the system Σ-σ constituted of N-ν particles enclosed in the volume V-v and having the energy U-u. Owing to the smallness of $\frac{\nu}{N}$, $\ln \Omega' - \ln \Omega_0$ is nearly equal to the differential of $\ln \Omega$ for $dV = -v$, $dN = -\nu$, $dU = -u$. Now from the equations (4) and (2) we derive the general formula (recall that Z is a function of V,α,β)

$$d \ln \Omega = \frac{\partial \ln Z}{\partial V} dV + \frac{\partial \ln Z}{\partial \alpha} d\alpha + \frac{\partial \ln Z}{\partial \beta} d\beta + d(\alpha N) + d(\beta U)$$

hence, by virtue of (3)

$$d \ln \Omega = \frac{\partial \ln Z}{\partial V} dV + \alpha\, dN + \beta\, dU$$

which gives, in the present case

$$\ln \Omega_0 - \ln \Omega' = \frac{\partial \ln Z}{\partial V} v + \alpha\nu + \beta u$$

But it is well-known that for a gas $\ln Z$ is proportional to V and consequently $\frac{\partial \ln Z}{\partial V} v = \ln Z \frac{v}{V}$. Finally we obtain

$$I = k(\ln Z \frac{v}{V} + \alpha\nu + \beta u) \tag{15}$$

Now remark that the energy variation ΔU of the whole system Σ after its return to equilibrium results essentially from the energy communicated to the sub-system σ during the measurement, since the duration of this latter is much shorter than the time of relaxation of the whole system, so that a small fraction of energy goes directly to -or comes directly from- the complementary system $\Sigma-\sigma$. Then, as the energy of σ cannot become negative, we must have $\Delta U + u \geqslant 0$, and finally, as $\Delta S = \frac{\Delta U}{T_0}$, we have

$$\Delta S + \frac{u}{T_0} \geqslant 0 \tag{16}$$

Eliminating u between this inequality and (15), where $\beta = \frac{1}{kT_0}$, we obtain

$$\Delta S + I \geqslant k(\ln Z \frac{v}{V} + \alpha\nu) \tag{17}$$

Thus we possess a fair analytical lower bound for $\Delta S + I$, but the problem would be rather to convert it into numerical value and, from this point of view, the results are erratic, as we shall see.

If the sub-system σ has a microscopic size (and letting aside the the real possibility of measurements in this case), the values of $\frac{v}{V}$ and ν are random, and we can say nothing about the value of the right-hand member of (17).

To try to arrive to a definite conclusion, let us suppose that σ, although small, is of macroscopic size. As Σ was initially in equilibrium, hence homogeneous, we have

$$\frac{\nu}{N} = \frac{v}{V} = \frac{u}{U}$$

and (17) becomes

$$\Delta S + I \geqslant \frac{\nu}{N} k(\ln Z + \alpha N)$$

Now a well-known formula of statistical Thermodynamics states that $-kT_0(\ln Z + \alpha N) = U_0 - T_0S_0$, the free energy of Σ ; therefore

$$\Delta S + I \geqslant - \frac{\nu}{N} \frac{U_0 - T_0S_0}{T_0} \tag{18}$$

Let us show that, even in this favourable case, we cannot draw any general conclusion from this formula, since the right-hand member can have any sign. If the gas is a Bose-Einstein one, the expression of the molar free energy is (see for example[11], p. 84 or [6], p. 67), at the classical approximation and in M.K.S. units, $U_0 - T_0S_0 = RT_0(19{,}2 - \ln p + \frac{3}{2} \ln M + \frac{5}{2} \ln T_0)$ (p : pressure, M : molar mass, R : constant of ideal gases). For hydrogen, at room temperature and ordinary pressure, we find $\frac{U_0 - T_0S_0}{T_0} = -12{,}6$ R J/mole.K, and therefore $\Delta S + I > 0$. On the contrary, for a strongly degenerate Fermi gas, as electrons in a metal, we have ([10], p. 248 or [6], p. 459) for the molar free energy, $U_0 - T_0S_0 =$

$\frac{3}{5} R\theta\left[1 - \frac{5\pi^2}{12}\left(\frac{T_0}{\theta}\right)^2\right]$, where θ is the degeneracy temperature, which is of order 10^5 K ; hence, at ordinary temperatures, the inequality (18) shows that $\Delta S + I$ is larger than a strictly negative quantity.

Summarizing the whole discussion, we can conclude that

a) *in the case when we draw from the measurement the maximum available information, we can find for* $\Delta S+I$ *a well-definite analytical lower bound; but unfortunately the numerical value of this lower bound varies so widely, in magnitude and sign, according to the particular case considered, that be cannot arrive at any definite conclusion ;*

b) *if the measurement made does not afford the maximum information, the value of* I *is smaller than the right-hand member of* (15) *; consequently the lower bound exhibited in* (17) *for* $\Delta S+I$ *is lessened by an unknown quantity, so that we can conclude even less than in the preceding case.*

Finally let us remark that we did not come across any reason which would prescribe a definite sign for the entropy variation of the *observed* system.

4. THE MAXWELL DEMON

Contrary to these conclusions, Brillouin has stated that there is a strong interference between entropy and information, whose general form has been transcribed above ((8),§3), and that this interference is the only means which permits to remove the well-known paradox of the Maxwell demon.

Let us recall briefly the argument of Brillouin concerning the action of the demon ([3], p. 159 ff.).

a) The observation of the position of a particle of a gas by the demon implies an increase ΔS_d of the entropy of the gas (by the means of a torch which illuminates the particle, thus giving heat or energy to the gas).

b) Owing to this entropy increase, the demon can get data which yield an information* I (obviously > 0), which is less than the entropy increase : $\Delta S_d - I > 0$.

c) Owing to this information, the demon knows whether the particle is sufficiently "hot" to be sent from the up-stream versel towards the down-stream one ; if such is the case, he opens the shutter, and accordingly the entropy of the two vessels decreases. Thus an information has turned into a lowering of entropy.

As this argument has appeared as the only possible explanation of the paradox and reinforced the idea of a link between entropy and information, it deserves some deeper analysis.

*The notation of Brillouin is ΔN_i, variation of neg.entropy.

A. Criticism of Brillouin's proof

In fact this proof is not satisfactory.

First it is inexact that a measurement which is intended to provide an information must necessarily increase the entropy of the observed system. We have already seen (§3A) the case of a thermometer which cools a system and makes its entropy decrease, but we can find numerous other examples : thus the radiation emitted by a light source may provide an information (in the strict shannonian sense) about the source, and yet this latter loses energy and its entropy decreases. As we have seen above (§3A), *the only quantity which is surely strictly positive when we get an information is the entropy production*, but we cannot draw any more precise conclusion from this fact for two reasons : the first one is that *we do not know a priori how this entropy production is shared between the observed system and the measurement apparatus*, and in fact this sharing depends strongly on each particular case ; the second reason is that the entropy variation of the observed system contains, according to equation (1), a term due to the external action of the measurement apparatus, and this too depends strongly on the particular device.

Secondly the result of Brillouin, $\Delta S_d - I > 0$, can be considered as valid only when we measure uniquely the position of the particle at *one* time. But, in order to know whether he must open the shutter or not, the demon must know also the direction and the magnitude of the impulse of the particle. Whatever be the manner in which he operates to get all this knowledge, the total resulting information is considerably higher than the increase ΔS_d : according to the very calculations of Brillouin one has, after 2×3 measurements of position, $\Delta S_d = 6kb$, where k is the Boltzmann constant and b a number slightly > 1 ; on the other hand, for an ordinary gas and on the basis of the equality (15), it is easily shown than $I > 13k$. Hence $\Delta S_d - I < 0$.

Thus the first two statements of Brillouin related above do not have any general validity. As for the third one, we shall see now that *it is not at all necessary to call upon an information for explaining why the entropy of the whole gas is decreasing.*

B. A purely thermodynamical solution of the Maxwell paradox

As is known, the paradox of the demon lies on the following statement : assuming that a) the motion of the shutter is free from any friction ; b) the whole system (gas + demon + shutter) is completely isolated ; and, c) the system (demon + shutter)(which we call simply the demon, for brevity) is itself completely isolated from the gas ; then the demon can transfer heat from the cold part of the gas to the hot one without expenditure of mechanical work -or, in other words, makes the entropy of an isolated system decrease- which is contrary to the second law of Thermodynamics.

In fact this alleged infraction to the second law is a mere appearance : it rests on an error of reasoning, *the error which consists in forgetting to take into account the entropy of the demon*. Of course if this latter were a supernatural being escaping the laws of the physical world, it would be not at all surprising that the whole system, and therefore the gas, does likewise. But the demon is supposed to be a natural* system, and consequently must possess an entropy. There, if we denote by S_d this entropy and by S_g the entropy of the gas, *the only relation prescribed by Thermodynamics* during the course of the experiment is (since the whole system is isolated) $\Delta(S_d + S_g) > 0$. We cannot split this inequality into two similar ones concerning S_g and S_d, although the two sub-systems, gas and demon, are thermally isolated. In effect the gas does not undergo a macroscopic process (it would be in any case impossible, even if the demon exerted directly a mechanical action on the gas) ; the action of the demon is more subtle, and we cannot draw the conclusion $\Delta S_g > 0$. On the contrary the demon, when opening and shutting the window, acts as a macroscopic system, which implies $\Delta S_d > 0$. Hence it remains possible that $\Delta S_g < 0$, provided that ΔS_d has a sufficiently positive value.

But this answer to the paradox, however valid it may be with regard to the second law, may seem insufficiently convincing to people who would like to understand *by virtue of what physical process* the entropy of the demon is constrained to increase. This question can be answered too.

First remark that, for each passage of a molecule, the shutter and accordingly the demon *must start from a state of equilibrium and come back to a state of equilibrium*, for the simple reason that the shutter is to be handled *at will*. If there were not a finite stage of rest (hence of equilibrium) between two successive openings of the shutter, the motion of the latter would be periodic, or almost -periodic, or random, and the demon could not fit the opening to the convenient molecules.

Now we can state the following general (macroscopic) theorem. Let us consider an *isolated* system Σ free from any solid friction and from any hysteresis** (mechanical, dielectric or magnetic), but which can be affected by any other factor of irreversibility (fluid viscosity, electric resistance, internal heat transfer, diffusion, chemical affinity). We suppose moreover -but uniquely for simplicity, for the generalization is plain- that there is no thermal barrier inside the system.

I assert that if such a system starts from an equilibrium (evidently by removing or modifying some external constraints) and comes back to an equilibrium (different from the first one, but this point is irrelevant), its entropy has strictly increased.

I only sketch the proof because, to be complete, it would need a very long mathematical study and many precautions. It rests essentially on the three following propositions.

a) In order that a state E of an isolated system Σ would be (under fixed constraints, such as the volume for example) a state of equilibrium, it is necessary and sufficient that E should be a state of maximum entropy over all states of the same energy U(E) as E.

*as natural as can be for such a "Gedanken experiment".

**If these two conditions were not fulfilled, there would be no need of a further discussion : the entropy of Σ would obviously increase strictly.

b) Conversely, in order that E would be a state of equilibrium, it is necessary and sufficient that E should be a state of minimum energy over all states of the same entropy S(E) as E.

c) If E is a state of maximum entropy over all states having the energy $U = U(E)$, E is also a state of minimum energy over all states having the entropy $S = S(E)$; and conversely.

These three propositions were enunciated by Gibbs ([9], p. 56) as early as 1875. The sufficient condition a or b is well known and easy to show from the second law ; the necessary condition is more difficult to prove, but it can be done for a large class of systems by calling upon convexity properties of thermodynamical functions, derived themselves from the second law.

Then let U_0, S_0 be a state of equilibrium of the system Σ under given constraints. Σ cannot leave this equilibrium unless we remove or modify some constraints. Suppose this is done, once for all, and Σ begins to move. Let $U_m(S_0)$ be the minimum energy over all states of entropy S_0, these states being defined according to the new constraints. Note that, owing to proposition c, S_0 is the maximum entropy over all states of energy $U = U_m(S_0)$, in symbols $S_0 = S_M[U_m(S_0)]$.

By definition of a minimum we have $U_m(S_0) \leqslant U_0$; but we cannot have $U_m = U_0$, since in this case the system would remain in equilibrium even with the new constraints (sufficient condition b) ; hence $U_m < U_0$. On the other hand we have during all the motion and since the system is isolated $U + K = U_0$ (K : kinetic energy). Then if Σ comes back to an equilibrium ($K = 0$), it has anew the energy U_0 and its entropy has become, according to proposition a (necessary condition), $S = S_M(U_0)$, maximum entropy over all states of energy U_0.

Consider now the values U of the energy which belong to the interval $[U_m, U_0]$. For each of these values there exists a maximum entropy $S_M(U)$, which determines a state of equilibrium and consequently of uniform temperature T. But, for systems in equilibrium, we have the classical equality $\frac{dS_M}{dU} = \frac{1}{T}$. Thus it follows that $S_M(U_0) - S_M(U_m) > \frac{U_0 - U_m}{T_M}$, T_M being the maximum temperature over all states of equilibrium for energies $U \in [U_m, U_0]$; as $S_M(U_0)$ is the final entropy in the new equilibrium and as we have seen that $S_M(U_m) = S_0$, initial entropy, we obtain finally

$$S_f - S_0 = \Delta S \geqslant \frac{U_0 - U_m}{T_M} > 0 \qquad \text{Q.E.D.} \tag{19}$$

This strict increase of the entropy of an isolated system during its transition from an initial equilibrium to a final one is perfectly understandable, from a physical point of view, if we think that a system in motion must, in order to end in a state of rest, perform a self-damping. This increase explains as well why the descent of a mountain is nearly as tiring as climbing it.

Lastly let us remark that the inequality (19), together with the equation $U + K = U_0$, yields a convenient estimate of the lower bound of ΔS. It is easily shown that

$$\Delta S \geqslant \frac{K_{max}}{T_M} \tag{20}$$

where K_{max} is the maximum kinetic energy that the system would reach if the motion were reversible, and may be replaced by the maximum kinetic energy that it reaches really.

Let us apply the inequality (20) to one opening of the shutter. For an ordinary gas, a very small shutter (1 micron in diameter, mass of 1 μg), a full turn of the shutter alone during the mean time of flight of the molecule requires an entropy increase of the demon of $\frac{10^{-21}}{T_0}$ J/K (T_0 : temperature of the demon) ; simultaneously a change of temperature of one molecule of 20 K implies an entropy decrease of the gas of $\frac{4.10^{-22}}{T_0}$ J/K (T_0 is supposed to be the same temperature for the gas as for the demon). If we take into account the kinetic energy of the demon, who is considerably larger and heavier than our micro-shutter, it is clear that the demon is wasting his neg-entropy for a negligible result.

5. CONCLUSION

The essential conclusions may be summarized as follows.

a) It is perfectly right that, for a *physical system in equilibrium*, the thermodynamical and the informational entropies are equal. But this equality is of no importance at all : it rests uniquely on the fact that (in the field of physics, we do not deal with communication here) we give two different names to the same thing, namely the $k \ln \Omega$ of formula (4).

b) The physical cost of an information, or more generally of gathering experimental data, is a *strictly positive entropy production*, taking place in both observed system and measurement apparatus. This does not imply *any determined sign of variation for the separate entropies* of the two systems.

c) As the entropy of a system depends only on its energy (or temperature) and its external parameters, *any knowledge* (it is a misuse of words that to speak of information) *about the microscopic behaviour of the system* -a purely mental knowledge, besides- *does not change anything to the value of the entropy.*

d) There is *no noteworthy relation between the information* we get about a system by the means of measurement and *the entropy variation* of this system which results from the measurement.

e) The well-known argument in favour of the equivalence of entropy and information, namely the explanation of the paradox of the Maxwell demon, is worthless : *the paradox is easily removed with the only help of macroscopic Thermodynamics.*

In conclusion there is no logical proof which permits to assert that thermodynamical entropy is to be assimilated to a lack of information.

REFERENCES

[1] Szilard, L., : Ueber die Entropieverminderung in einem thermodynamischen System bei Eingriffen intelligenter Wesen, Zeit. für Phys., 53 (1929), 840-856.

[2] Von Neumann J., : Mathematical foundations of Quantum Mechanics, Princeton Univ. Press, Princeton 1955.

[3] Brillouin L., : La science et la théorie de l'information, Masson, Paris 1959. Translation of the original american edition : Science and information theory, Acad. Press, N.Y. 1956
My quotations refer to the french edition.

[4] Shannon, C.E. and Weaver, W., : The mathematical theory of communication, Univ. of Illinois Press, Urbana 1949.

[5] De Groot, S.R. and Mazur, P., : Non-equilibrium Thermodynamics, North-Holland, Amsterdam 1962.

[6] Fowler, R.H. and Guggenheim, M.A., : Statistical Thermodynamics, Cambridge Univ. Press, London 1939.

[7] Khinchin, A.Y., : Mathematical foundations of Quantum Statistics, Graylock Press, Albany, N.Y. 1960.

[8] Khinchin, A.Y.,: Mathematical foundations of Statistical Mechanics, Dover, N.Y. 1949.

[9] Gibbs, J.W., : On the equilibrium of heterogeneous substances, in : The scientific papers of J.W. Gibbs, Vol. I, Dover, N.Y. 1961.

[10] Duhem, P., : Traité d'Energétique, Gauthier-Villars, Paris 1911.

[11] Pacault, A., : Eléments de Thermodynamique Statistique, Masson, Paris, 1963.

OPTIMAL STATE DETERMINATION: A CONJECTURE

Igor D. Ivanović
Faculty of Sciences, Belgrade, Yugoslavia

We consider state determination of simple quantum systems in non-relativistic quantum mechanics. Assuming that the ensemble in some unknown state is available in a sufficient number of replicas, it is possible to perform a state determination from the resultes of different, mutually noncommutative, measurements each one performed on a replica of the ensemble. In particulare,state determination is possible from the resultes of measurements of spin, position and energy. In the case of a finite collection of quantum systems any state determination is a finit sequence of measurements and their results and we conjecture that an optimal procedure may exist

1. Introduction.

The problem of quantal state determination (SD) is a very old one [1] and its aim is to determine a state W, $W \geqslant 0$ $tr(W) = 1$ of an ensemble of quantum systems from the results of different measurements,each one performed on a replica of the inspected ensemble. The very sence of SD depends on the interpretation of quantum mechanics. It may range from the point of view in which SD determines a set of parameters of the preparational procedure for the inspected ensemble to the point of view in which SD determines a property of the inspected ensemble itself. For our purposes this difference is irrelevant and we will not discuss it.

We restrict our attention to finit dimensional Hilbert spaces and stationary states over it. Such SD-s has been inspected especially by Park and Band [2], [3], [4] following the approach proposed by Fano [5]. The approach, adopted in this paper differs because of the explicite use of projection postulate [6] as a description of quantum measurements.

The paper is organized as follows. In section II. a description of SD is given assuming that all operators are observables.We show that a quorum (set of measurements which allows SD) can be choosen to be minimal and orthogonal in the set of zero trace operators. Section III. contains the proof that SD is possible for any simple quantum system for which spin, position and energy measurements are possible. In section IV. we discuss the choice of a representative state in an incomplete SD, comparing the standard, maximum entropy state and an expected state, obtained from some probability density over a set of admissible states. Finally in section V. we propose arguments which should justify a conjecture that a SD for a collection of quantum systems allows an optimal procedure.

II. Formal state determination.

In this section we consider the case of SD in a finite dimensional Hilbert space assuming that all operators are observables. Unfortunately, the only such example is C^2 space for spin s= 1/2 (or photon polarization) so that the content of this section is only a formal one. Still, it can serve to introduce some well known but slightly modified facts about the set of states for finite dimensional spaces and measurements over it.

For an n-dimensional Hilbert space H, the set of all Hermitian operators over it $V_h = \{A | A^+ = A\}$ is an n^2-dimensional real vector space. With scalar product $(A,B) = tr(AB)$ and norm $||A|| = (tr(A^2))^{1/2}$, V_h becomes real Euclidian space. The set of states in V_h, $V_W = \{W | W \geq 0, tr(W) = 1\}$ is convex set in V_h.

If $\{P_k\}$ is an orthogonal, ray-resolution of the identity in H i.e. if

$$\Sigma P_k = I \ , \quad P_k P_j = \delta_{kj} P_k \quad \text{and} \quad tr(P_k) = 1$$

it deffines an n-dimensional sub-space of V_h, $V_h(\{P_k\})$ which is also a "commutative" subset. The set of states in $V_h(\{P_k\})$ $V_W(\{P_k\}) = V_W \cap V_h(\{P_k\})$ is maximal set of mutually commutative states. $V_W(\{P_k\})$ is in fact an n-dimensional simplex, having $P_k \in \{P_k\}$ as its extremal points and $W_o = (1/n)\ I$ as its baricenter. Applying all possible unitary transformations on a single $V_W(\{P_k\})$ one obtains V_W which is, in some sence, a rotational simplex.

Another way to characterize a "commutative" subset in V_h is the following one. Let A be a non-degenerate observable. Then $A^o, A, A^2, \ldots, A^{(n-1)}$ defines the same "commutative" subset as $\{P_k\}$ (where P_k are eigen-projectors of A) i.e. $V_h(\{P_k\})$.

The importance of $V_W(\{P_k\})$ follows from the description of quantum measurement. As already said.a measurement of an observable e.g. $A = \Sigma_k a_k P_k$ performed on an ansemble in a pre measurement state W is described as

$$M_{\{P_k\}} W' = \Sigma_k P_k W P_k = \Sigma_k w_k P_k \ , \quad w_k = tr(WP_k)$$

Therefore, the result of the measurement of $A = \Sigma_k a_k P_k$ on W is the orthogonal projection of W into $V_h(\{P_k\})$ i.e. into $V_W(\{P_k\})$.

In this description SD becomes a determination of a point from n^2dimensional space V_h from its orthogonal projections into n-dimensional "commutative" subspaces $V_h(\{P_k\})$. Obviously, if a set of operators e.g. $\{A^{(m)}\}$ is a quorum of observa bles it means that the set of its eigen projectors $\{P_k^{(m)}\}$ is a basis in V_h. Irrelevantly of the fact that projectors corresponding to different observables will not commute, every measurement will give n-1 data about an unknown state due to the fact that $\Sigma_k P_k^{(m)} = I$. and $tr(W) = 1$. As a result a minimal quorum contains(n + 1) observables.

Assuming that all operators are observables the existence of a minimal quorum is a trivial fact. Still, the only example for this assumption i.e. in C^2 for spin s = 1/2 posseses the following property: if W_x,W_y and W_z are the results of spin measurements along the x,y and z axis, the unknown state is

$$W_u = W_x + W_y + W_z - I$$

while eigen projector; P_{xk}, P_{yk}, P_{zk} ; k= +1/2,-1/2 satisfies $tr(P_{xk}P_{yk'}) = 1/2$. In the hyper plane $h = \{A|tr(A) = 1\}$ these projectors becomes mutually orthogonal and a question may arise : "Is it possible to find similar, minimal quorums which are orthogonal in the hyperplane of zero trace operators, for n>2?". The answer is affirmative [7] and here we give an example for n = 4.

$|1^{(1)}> = (1,0,0,0); |2^{(1)}> = (0,1,0,0); \dots 14^{(1)}> = (0,0,0,1).$

$|1^{(2)}> = (1/2)(1,1,1,1); \quad |2^{(2)}> = (1/2)(1,1,-1,-1);$
$|3^{(2)}> = (1/2)(1,-1,1,-1); \quad |4^{(2)}> = (1/2)(1,-1,-1,1).$

$|1^{(3)}> = (1/2)(1,i,1,-i); \quad |2^{(3)}> = (1/2)1,i,-1,i);$
$|3^{(3)}> = (1/2)(1,-i,1,i); \quad |4^{(3)}> = (1/2)1,-i,-1,-i).$

$|1^{(4)}> = (1/2)(1,1,i,-i); \quad |2^{(4)}> = (1/2)(1,1,-i,i);$
$|3^{(4)}> = (1/2)(1,-1,i,i); \quad |4^{(4)}> = (1/2)(1,-1,-i,-i).$

$|1^{(5)}> = (1/2)(1,i,i,-1); \quad |2^{(5)}> = (1/2)(1,i,-i,1);$
$|3^{(5)}> = (1/2)(1,-i,-i,-1); |4^{(5)}> = (1/2)(1,-i,i,1).$

The set of projectors, projecting on vectors listed above is the mentioned basis in $V_h^{(4)}$ which satisfies

$$tr(P_i^{(k)}P_j^{(r)}) = (1/4) \quad \text{and} \quad P_i^{(k)}P_j^{(k)} = \delta_{ij}P_i^{(k)}.$$

A simple consequence of this minimal, "orthogonal" quorums is the fact that if $W^{(k)}$ is the result of the measurement in k-th subspace $V_h(\{P_r^{(k)}\})$ the unknown state is

$$W_u = \Sigma_k W^{(k)} - I$$

It is interesting to notice that minimal "orthogonal" quorums can be easyly constructed when n is a prime number [7] while a general procedure for non-prime numbers is not known to this author.

III. State determination: some operators are observables.

In this section we consider a more realistic problem of state determination when only some of the operators are observables. We show that SD is possible from the measurements of position and energy.

In this case an important fact is our inabbility to prepare all states we may wish to, which may result in an impresion that Hilbert space formulation is, in a way a redundant one. One of the results in this section should remove such an impresion.

We start with the spin state determination. For a given value s the corresponding space is (2s+1) dimensional while the corresponding V_h is $(2s+1)^2$ dimensional. As already said we assume that only proper spin operators are observables. Which means that, after establishing e.g. s_z operator and its eigen-projectors P_{zm} ,$-s \leq m \leq s$ every other spin operator can be writen as

$$P_m(\alpha,\beta) = U(\alpha,\beta,0)P_{zm}U^+(\alpha,\beta,0)$$

where $U(\alpha,\beta,0)$ is Wigner matrix while α, β, and $\gamma = 0$ are Euler angles of rotation.

A way to establish a minimal quorum is the following one From the operators s_z,s_+ and s_- one may construct the following basis [8]

$$A_{ko} = s_z^k \qquad 0 \leq k \leq 2s,$$

$$A_{kr} = s_-^r s_z^{(k-r)} + s_z^{(k-r)} s_+^r \quad ; \quad B_{kr} = i(s_-^r s_z^{(k-r)} - s_z^{(k-r)} s_+^r)$$

$$1 \leq r \leq k \leq 2s.$$

For a fixed k operators A_{kr},B_{kr} will deffine (2k+1) dimensional subspace V_k decomposing $V_h^{(2s+1)} = \Sigma \oplus V_k$. On the other hand it is easy to chek that only $s^k(\alpha,\beta)$ will have a non zero component in V_k. V_k, having the largest dimension is V_{2s} and $\dim(V_{2s}) = 4s+1$ and this is exactly the minimal number of measurements which allows SD in spin case. A proof that it can be done for any s one can find in [8].

Another example we need is SD for one-dimensional system. Again an operator basis in terms of x and p is useful

$$A_{mn} = x^m p^n + p^n x^m \quad \text{and} \quad i(x^m p^n - p^n x^m) = B_{mn}$$

For a given state W, expected values $\langle A_{mn} \rangle$ and $\langle B_{mn} \rangle$ one may calculate from $d^n x^m/dt^n$ [4]. On the other hand we assume that energy of the system is finit and that a cut-off procedure can be applied i.e. that a projector P exists satisfying

$$\mathrm{tr}(P)<\infty, \quad PW = W \text{ and } [P,H] = 0.$$

In this way proper operators x and p are replaced by

$$x' = PxP \quad \text{and} \quad p' = PpP$$

The measurement of position distribution $\rho(x') \sim \rho(x)$ at several different moments $t_1,\dots,t_k$ allows one to, approximately of course, calculate $PA_{mn}P$ and $PB_{mn}P$ and then to determine the unknown state. The most importante assumption is, however, that evolution of the system i.e. Hamiltonian of the system is known.

As a result, for any system prepared in some state

$$W = W_o \otimes W_s$$

one can perform a sucsesful SD. If $W \neq W_o \otimes W_s$ (where "o" stands for orbital part and "s" stands for spin part) it means that system itself is a subsystem of a larger system. Even in that case SD is formally possible if coincidence measurements on the other subsystems are possible. Therefore any state allowed by the Hilbert space formalism can be determined by appropriately choosen measurements of spin, energy and position.

IV. Representative state in SD.

In this section we consider some preliminaries concerning the problem of an incomplete state determination [9,10]

The problem is the following one. The set of measurements performed in order to obtain inaf data for SD is inadequat and state determination resultes not in a point but in a proper set of admissible states. Perhaps the simpliest example is if one assumes that single measurement e.g. that of $A = \Sigma_k a_k P_k$ is an incomplete SD. Therefore if the result is

$$W' = \Sigma_k P_k W_u P_k = \Sigma_k w_k P_k$$

the set of admissible states is $V_w^{ad} = \{W \mid \Sigma_k P_k W P_k = W'\}$. It is easy to show that $V_W(W')$ is convex subset in V_W and that W' is its baricenter. Without some further assumptions any state from $V_W^{ad}(W')$ may be a representative state. This problem is usually resolved through the maximum entropy principle [9,11]. Nevertheless, this principle assumes that all pure states are equally probable and maximum entropy stete coincides with the most probable state under this assumption. The aim of this section is to show that an alternative representative state may be helpfull in some cases.

Formally, it is easy to assume that a probability density $\rho(W)$ is deffined over V_W^{ad} satisfying

$$\rho(W) > 0 \qquad \int_{V_W^{ad}} \rho(W)\, dV = 1$$

The most probable state (and maximum entropy as a special case) satisfies W_{MP}: $\rho(W_{MP}) > \rho(W) \quad \forall\, W \in V_W^{ad}$ where V_W^{ad} denotes the set of admissible states. The alternative choice is an expected state

$$W_E = \int_{V_W^{ad}} \rho(W)\, dv$$

A simple example in which an expected state can be more appropriate than the maximum entropy state is the following one. Assume, that for spin $s = 1/2$ case one obtains that the set of admissible states is deffined by $\langle s_z \rangle > 0$.

The maximum entropy state is simply, $W=(1/2)I$ which lies on the boundary of the set of admissible states. An expected state e.g. assuming that all states W with $tr(Ws_z)>0$ are equally probable will lie in the interior of the set of admissible states. An example for composite systems in which maximum entropy state and expected state differs, one can find in [10].

This is not an attempt to question allmost universal validity of the maximum entropy principle but to offer an alternative which in some cases may be more appropriate.

V. Optimal state determination: a conjecture.

In this section we will apply the above mentioned results to propose a conjecture that in the case of SD for a collection of quantal systems (not an ensemble) an optimal procedure may exist. The reason to consider collections of systems is obvious, but its consequences are quite important ones. The first one is that the result of measurement e.g. $W = \Sigma(n_k/N)P_k$ is only one of the possible realisations. If some collection is described by $W = \Sigma(n_{ko}/N)P_{ko}$ the only result which can be predicted with certainity is the one for an observable which commutes with W. For any other set of measurements one can speak only of probabilities of obtainig some after-measurement state. Therefore, after one obtains certain result W' the only conclusion is that the proper state of the collection, befor the measurements lies with some probability p_o in the set of states which can be obtained in the following way. The result W' together with the chosen p_o and with the number of systems in the collection will define a set of probable results e.g. a ball B of radius $r = r(p_o, N, W')$ centered at W'. The set of admissible states is inter section of the cilinder in V_h, of which B is base, and V_W. The important point is dimensionality of the set of admissible states which is n^2 dimensional convex set, irrelevantly of the number of measurements used to obtain it.

In an ideal case, two SD can be compared only on the basis of number of linearly independent operators used i.e. if $\{P_r^k\}$ and $\{Q_s^m\}$ are eigen projectors of measurements used in SD-s then e.g. $\{P_r^k\}$ is better than $\{Q_s^m\}$ if $\dim(V(\{P_r^k\})) > \dim(V(\{Q_s^m\}))$. E.g. in an ideal case, any two quorums are equivalent etc. As mentioned above, the assumption of finit collections of quantum systems allows one to introduce the volume of the set of admissible states as a simple criterium for comparation between different SD-s due to the fact that all such sets will be equally dimensional i.e, n^2dimensional.

The following scheme may be helpfull. Let N be the number of systems in a collection; let $\{P_r^k\}$ be the set of measurement eigen-projectors where $1<r<n$ and $1<k<s$, i.e. n is dimensionality of the underlying Hilbert space and s is number of different measurements. Any SD procedure can be represented as a sequence of events

$$f = (k_1r_1, k_2r_2, \dots, k_ir_i, \dots, k_Nr_N)$$

where k_ir_i denotes that i-th system from the collection is subjected to the k_i-th measurement inwhich result r_i has been obtained. The set of sequences we will denote by Ω. For given probability p_o one can define the volume $v_m(p_o,f)$ where f is a sequence from Ω and m denotes the first m events in the sequence. E.g. $v_N(p_o,f)$ denotes the volume of admissible states for a completed SD given by f. The unknown state for sequence f is the baricenter of $v_N(p_o,f)$.

Now, the set of final volumes $v_N(p_o,f)$ is a discrete set of positive numbers, hence, completely ordered set, and one may say that SD described by f is better than a SD described by f' if $v_N(p_o,f) > v_N(p_o,f')$.

Which sequence will occur depends on the state of the inspected collection but also on the choice of measurements which is experimenters responsibility. E.g. if the result, after j-th step is $(k_{i1}r_{i1},...,k_{ij}r_{ij})$ the choice of the measurement to which j+l-st system will be subjected i.e. of $k_{i,j+1}$ will exclude certain set of sequences to occur as a final one. If one excludes the set of sequences in which lies the maximal value of $v_N(p_o,f)$ an optimalization is acchieved. Due to the fact that all volumes are well defined, that all sets are in fact a discrete ones, the existence of an optimal procedure is established. At this moment, this author is not aware of some exact properties of an optimal procedure , except in the case of C^2 and $V_h^{(2)}$. Only the part of difficulties stems from the "strange" shape of V_W for n>2(cf.e.g. [12]) which affects dependence of optimal procedure on p_o. One must expect that optimal procedure is also dependent on N (number of systems in colection).

A minimal conclusions, however, that for fixed p_o and N an optimal procedure exists. One must also notice that irrelevantly of the initial state this optimal procedure will prefer the states near boundaries of V_W,where the volume is smaler, then in the interior of V_W,e.g. if the initial state is $W_o = (1/n)I$ there is no way to obtain a result i.e. a volume of states for which W_o is the baricenter.

VI. Summary.

As mentioned in introduction, SD can be made a proper part of the standard quantum formalizm, which can describe either a set of preparational parameters or a set of properties of quantum systems. Another importante point is that SD is a proper quantal counterpart of the classical term measurement.However, one should not mix a SD procedure with any kind of hidden variable models. A SD assumes a sufficient number of replicas of the inspected ensemble, in ideal case or a large collection of quantal systems in a more realistic case. Finally the SD is a rough description of our way to gain some new data about the world we live in.

References:

1. Pauli,W.,Die Allgemeinen Prinzipien der Wellenmechanik, Handbuch der Physik V. Springer-Verlag,Berlin,1958,p.17
2. Park,J.L. and W.Band.,Found.Phys.,1,211(1971)
3. Band,W and J.L.Park, Found.Phys.,1,339(1971)
4. Band,W and J.L.Park, Am.J.Phys.,47,188(1979)
5. Fano,U.,Rev.Mod.Phys.,29,74(1957)
6. von Neumann,J. Mathematical Foundations of Quantum Mechanics, Princeton Univ.Press, Princeton 1955,p.351
7. Ivanović,I.D. J.Phys.A 14,3241(1981)
8. Ivanović,I.D. J.Math.Phys.,24,1199(1983)
9. Wichmann,E.H.,J.Math.Phys.,4,884(1963)
10. Ivanović,I.D.,J.Phys.A,17,2217(1984)
11. Jaynes,E.T. Phys.Rev.,106,620(1957)
12. Bloore,F.J. J.Phys.A,9,2059(1976)

THE ORIGIN OF THE NON-CLASSICAL CHARACTER OF THE QUANTUM PROBABILITY MODEL

Diederik Aerts
Vrije Universiteit, Brussels, Belgium

1. Introduction

In [1] it is shown that a lack of knowledge about the measurements of a physical system gives rize to a non classical probability calculus for this physical system. It is also shown that the non classical probability calculus of quantum mechanics can be interpreted as being the result of a lack of knowledge about the measurements. Examples are given of macroscopical real systems that have a non classical probability calculus. Also an example is given of a macroscopical real system with a quantum probability model. More specifically a model for the spin of a spin 1/2 particle is constructed. Also an example is given of a macroscopical system having neither a classical nor a quantum probability model.

What we should like to analyse in this paper, is why situations described by non classical probability models do arrive in nature.

Therefore we shall have to analyse the measuring process and we shall see that some kind of measurements give rise to non classical probability models.

2. Measurements, states and outcomes

A measurement is a happening that consists of

- a preparation of the system under consideration in a certain state
- an experiment with the system leading to an identifiable outcome

So a measurement e can in general have different outcomes $x_1, \ldots, x_n$. The set of possible outcomes can also be infinite or even a not denumerable set. We will denote measurements by symbols $e, f, g, \ldots$.

If the system is prepared, it is in a certain state. A state is a representation of the collection of properties that the system has. Hence a state represents the "reality" of the system (see [2] [3] for a detailed analysis of the concept of state). The different states in which the system can "be" will be denoted by symbols $p, q, r, \ldots$. Outcomes will be denoted by symbols $x, y, z, \ldots$. If we consider a measurement e on a system in a state p we will denote the outcome by $x(e,p)$. In general, even when an "identical" measurement is performed on the system in an "identical" state, "non identical" outcomes can be obtained. Hence the outcome is not determined by the state of the system and the measurement. What can be the origin of this lack of determinism. To be able to analyse this problem carefully, we will put it in the following form. Suppose we consider two measurements e and f and two states p and q of a system.

if $e = f$ and $p = q$, does then $x(e,p) = x(f,q)$?

By $e = f$ we want to express that the measurement e is "identical" to the measurement f. What does this mean ? Two happenings can of course never be really identical, because then there is only one happening. Hence "identical" does not refer to the happening itself, but to the conceptual model that we use to represent the happening. The same remark applies to the meaning of $p = q$ and $x(e,p) = x(f,q)$. Two systems S_1 and S_2 in two states p_1 and p_2 are "identical" systems in "identical" states, if they are so in the conceptual model that we use to represent these systems. This conceptual model is exactly what we have called an "entity" in [2].

If we have a system S in a state p and we perform a measurement e, then after the measurement, this system is generally in another state. We will denote this state by $s(e,p)$. Again in general even when an identical measurement f is performed on S in an identical state q, the state after f can be different than the state after e. Hence also the change of the state is not determined by the state of the system and the measurement.

Example 1 : Consider a system that is a classical point particle A that is located on the surface of a sphere with radius R. The particle can be in any point (θ,ϕ) of the surface of the sphere and can move on the surface of this sphere. We suppose that the particle has some property (e.g. mass, charge) represented by a positive number m_A.

A possible state of the particle A is where we indicate its place on the surface of the sphere and the value of m_A. To indicate the place it is sufficient to specify two angles θ and ϕ. Hence such a state can be represented by

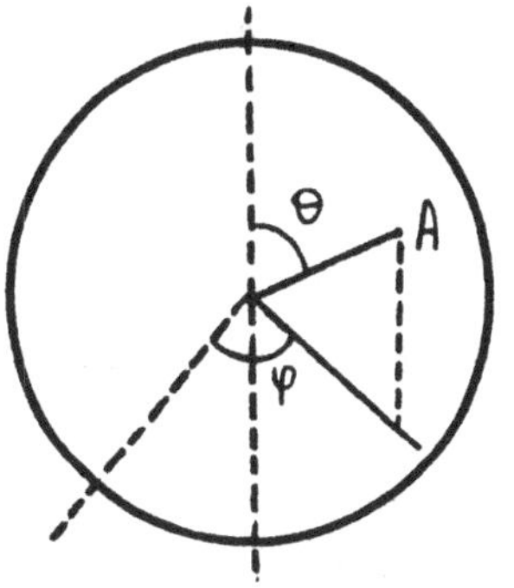

$$p_{m_A,\theta,\phi}$$

Another possible state is where we do not know its place but we do know the value of m_A. We can represent this state by p_{m_A}. Another possible state is, when we do know the place of the particle, but we do not know the value of m_A. This state can be represented by $p_{\theta,\phi}$. Sometimes we will not know the value of m_A, neither the place θ,ϕ, but we will know that the value of m_A is in the interval I of the real line and the place θ,ϕ is in the subset E of the surface of the sphere. Let us represent such a state by $p_{I,E}$.

A lot of other states are possible. But again, we want to remark that the collection of states that will be considered depend on the conceptual model that we use to describe the system.

This set of states has a natural structure. Consider a system S and two possible states p and q of S . If it is so that, whenever the system S is in the state p it is also in the state q , we will denote

$$p \subset q$$

If the system has a state, such that the system is in this state if and only if it is in the state p and in the state q , we will denote this state by $p \cap q$.
Clearly $\subset$ is partial order relation, which means that

$$p \subset p \tag{1}$$

$$p \subset q \text{ and } q \subset r \text{ then } p \subset r \tag{2}$$

$$p \subset q \text{ and } q \subset p \text{ then } p = q \tag{3}$$

For two states p and q , $p \cap q$ is an infimum for this partial order relation, which means that,

$$r \subset p \cap q \text{ iff } r \subset p \text{ and } r \subset q \tag{4}$$

If the system has a state, such that the system is in this state if and only if it is in the state p or in the state q , we will denote this state by $p \cup q$.
If the system has a state, such that it is in this state if and only if it is not in the state p , we will denote this state by $p^{\sim}$.
For two states p and q , $p \cup q$ is a supremum for the partial order relation.
The state in which the system never is will be denoted by o . The state in which the system just "is", and which in a certain sense is a trivial state, will be denoted by t . Clearly for every state p we have

$$o \subset p \subset t \tag{5}$$

If p and $p^{\sim}$ both exist we have

$$p \cap p^{\sim} = o \tag{6}$$

$$p \cup p^{\sim} = t \tag{7}$$

We also have $(p^{\sim})^{\sim} = p$, and for two states p and q if $p^{\sim}$ and $q^{\sim}$ exist we have

$$p \subset q \rightarrow q^{\sim} \subset p^{\sim} \tag{9}$$

Example 2 : Consider again the system of example 1. Then

$$p_{m_A,\theta,\phi} = p_{m_A} \cap p_{\theta,\phi}$$

And

$$p_{I_1,E_1} \cap p_{I_2,E_2} = p_{I_1 \cap I_2, E_1 \cap E_2}$$

On the other hand

$$p_{I,E}^{\sim} = p_{R^+/I, Sp/E}$$

where Sp is the surface of the sphere.

Example 3 : We consider again the system of example 1, but we will now introduce some measurements. We will consider two other point particles B and C , that have also some property represented by the numbers m_B and m_C . We suppose that $m_B + m_C = m$ and $m_B \in [0,m]$ and $m_C \in [0,m]$.

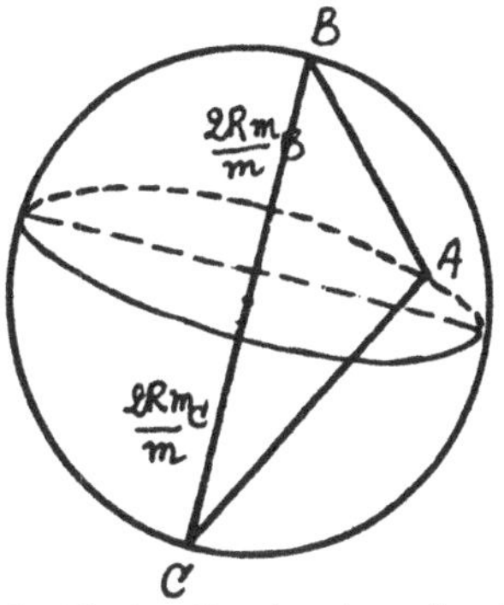

The first type of measurement we will call $e_{m_B,\alpha,\beta}$ and consists of putting B and C diametrically on the sphere such that B is in (α,β) and C is in $(\pi-\alpha, \pi+\beta)$. Consider on

the diameter between B and C the point that is located at a distance $m_B.\frac{2R}{m}$ from B and a distance $m_C.\frac{2R}{m}$ from C , and consider the plane, perpendicular to the line between B and C through this point. If the particles A and B are at the same side of this plane, we give the outcome e_1 to $e_{m_B\alpha\beta}$. If the particles A and C are at the same side of the plane we give the outcome e_2 to $e_{m_B\alpha\beta}$.
The second type of measurement we will call $f_{m_B\alpha\beta}$ and it consists again of putting B and C diametrically on the sphere such that B is in (α,β) and C is in $(\pi-\alpha, \pi+\beta)$.

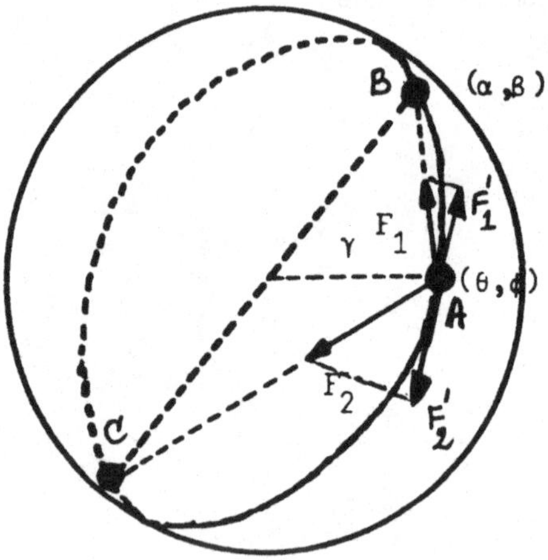

We suppose that between the particles there are attractive forces that are proportional to the values of m_A , m_B and m_C and inverse proportional to the distance between the particles.
So

$$|F_1| = G \frac{m_A m_B}{2R \sin \frac{\gamma}{2}}$$

$$|F_2| = G \frac{m_A m_B}{2R \cos \frac{\gamma}{2}}$$

where γ is the angle between (θ,ϕ) and (α,β) . The particle A will move under the influence of the forces F_1 and F_2 . If it reaches B we will give the outcome e_1 to $e_{m_B\alpha\beta}$. If it reaches C we will give the outcome e_2 to $e_{m_B\alpha\beta}$. Since the particle A moves on the surface of the sphere, the motion will be governed by the projections of $\bar{F}_1$ and $\bar{F}_2$ on the tangent plane at (θ,ϕ) . Let us call these projections F_1' and F_2' .
Then

$$|F_1'| = G \frac{m_A m_B}{2R \sin \frac{\gamma}{2}} \cdot \cos \frac{\gamma}{2}$$

$$|F_2'| = G \frac{m_A\ m_B}{2R\ \cos\frac{\gamma}{2}} \cdot \sin\frac{\gamma}{2}$$

So for the physical situation represented here we have

$$x\ (e_{m_B\alpha\beta}\ ,\ p_{m_A\theta\phi}) = e_1 \quad \text{if} \quad |F_1'| > |F_2'|$$

$$x\ (e_{m_B\alpha\beta}\ ,\ p_{m_A\theta\phi}) = e_2 \quad \text{if} \quad |F_1'| < |F_2'|$$

And for this physical situation we can say that

$$\text{if} \quad e_{m_B\alpha\beta} = e_{m_B'\alpha'\beta'}$$

$$\text{and} \quad p_{m_A\theta\phi} = p_{m_A'\theta'\phi'}$$

then $x\ (e_{m_B\alpha\beta},\ p_{m_A\theta\phi}) = x\ (e_{m_B'\alpha'\beta'},\ p_{m_A'\theta'\phi'})$

then $x\ (e_{m_B\alpha\beta}\ ,\ p_{m_A\theta\phi}) = x\ (e_{m_B'\alpha'\beta'}\ ,\ p_{m_A'\theta'\phi'}\ )$

with other words ; "identical" measurements on the system in "identical" states give rize to "identical" outcomes. We could also say : the outcome is determined by the measurement and the state of the system.
As a third type of measurement we consider the measurement $e_{m_B\alpha\beta}$ but we introduce a "lack of knowledge" on this measurement $e_{m_B\alpha\beta}$. In the sense that we suppose that m_B is not determined. We suppose that m_B is chosen at random in the interval $[0,m]$. We will call this measurement $e_{\alpha\beta}$. It is a new measurement and consists of first choosing at random the value m_B and then performing the measurement $e_{m_B\alpha\beta}$. As a fourth type of measurement we consider $f_{m_B\alpha\beta}$ and introduce the same lack of knowledge. Hence $f_{\alpha\beta}$ consists of first choosing at random the value m_B and then performing the measurement $f_{m_B\alpha\beta}$.

The set of measurements has also a natural structure. If we have two measurements e and f then we can consider the measurement that consists of performing e or performing f . This is again a measurement which we will denote by $e + f$.

Clearly

$$(e + f) + g = e + (f + g)$$

and
$$f + g = g + f$$

A measurement that cannot be performed on the system will be denoted by d .

Hence

$$e + d = e = d + e$$

If we have two measurements e and f then we can consider the measurement that consists of performing first e and then f . We will denote this measurement by $f.e$.

Clearly

$$(f.e).g = f.(e.g)$$

But in general $f.e$ does not equal $e.f$.

A measurement that does not do anything on the system will be denoted by i .

Clearly

$$e.i = i.e = e$$

We also have $e.d = d.e = d$

$$(e + f).g = e.g + f.g$$

$$e.(f + g) = e.f + e.g$$

3. Probability

A lot of physical measurement situations are not deterministic. What we mean is that, even when an identical measurement is made on an identical system in an identical state, non identical outcomes are obtained.
Physical theories that describe these kind of measurement situations are e.g. statistical mechanics and quantum mechanics.
There is a widespread belief that the non determinism represented in statistical mechanics is of a completely different nature than the non determinism represented in quantum mechanics. The non determinism in statistical mechanics is due to the lack of knowledge about the state of the system. While it is thought generally that the non determinism in quantum mechanics is of an intrinsic (ontological) nature. The main reason for this belief is the fact that the probability model of statistical mechanics is Kolmogorovian (satisfies the axioms of Kolmogorov) while the probability model of quantum mechanics is non Kolmogorovian.
And again, it is a widespread belief that, a probability model that formalizes a situation in which the origin of the probability is a "lack of knowledge", should satisfy the axioms of Kolmogorov.
As we have shown in [1], also the non determinism of quantum mechanics can be explained as due to a lack of knowledge. But not a lack of knowledge about the state of the system, but a lack of knowledge about the measurement. We have shown this in detail in [1]. What we should like to show in this paper is, why a lack of knowledge about the state of the system gives rise to a classical (Kolmogorovian) probability calculus, and why a lack of knowledge about the measurements gives rise to a non classical (non Kolmogorovian) probability model.

Therefore we first have to explain what we mean with probability.

Let us consider again the happening which we called a measurement. First there is a preparation of the system S in a certain state p. Then there is an experiment e with the system resulting in an outcome x_i. We can now "repeat" the measurement. This means that we prepare an "identical" system in an "identical" state and perform an "identical" measurement. If the measurement situation is not deterministic, we will in general find a different outcome x_j. If we call N the number of times that we "repeat" the measurement and N_i the number of times that we find outcome x_i, then we can consider the relative frequency

$$\frac{N_i}{N}$$

This ratio is a rational number between 0 and 1. For two different series of identical measurements, the ratio will not necessarely be the same. But we can construct a conceptual model where these ratio's are represented by numbers between 0 and 1, that are interpreted as probabilities.
Hence $\frac{N_i}{N}$ for any series of repeated measurements will be represented by

$$P\ (e = x_i|p)$$

which is "the probability to find an outcome x_i for the experiment e when the system is prepared in the state p ".

4. Observations

An observation is a measurement that "does not disturb" the system.
To the concept "does not disturb" the same remark as the one we made for the concept "identical" applies. Indeed, every measurement disturbs the system. So what we mean with "does not disturb" is "does not disturb" in the conceptual model that we use to represent the system. An observation is a measurement that only increases the amount of information about the state of the system, but does not change the properties of

the system. Hence for an observation e , we must have

$$s\,(e,p) \subset p \quad \text{for every state } p\ .$$

Example 4 : Consider the first type of measurement $e_{m_B\alpha\beta}$ introduced in example 3. And suppose that we describe the system of example 3 by means of the set of states $\{p_{I,E}\ ;\ I \subset [0,m]$ and E is a subset of the surface of the sphere$\}$.
Then clearly

$$s\,(e_{m_B\alpha\beta}\ ,\ p_{I,E}) \subset p_{I,E}$$

for every state $p_{I,E}$
because indeed

$$s\,(e_{m_B\alpha\beta}\ ,\ p_{I,E}) = p_{I,E\cap G}$$

where G is the zone cut by the plane of $e_{m_B\alpha\beta}$, if particle A was in this zone and

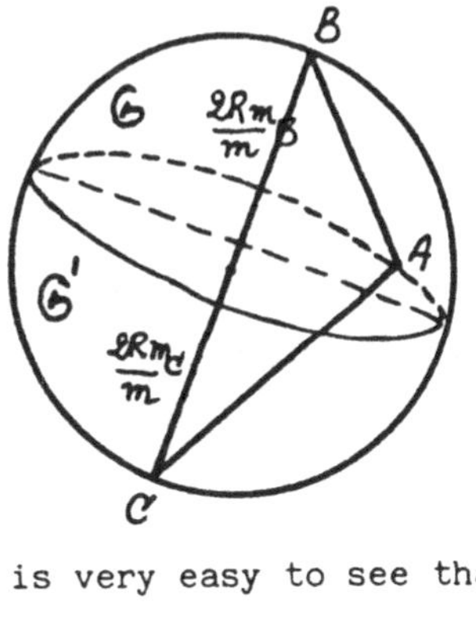

$$s\,(e_{m_B\alpha\beta}\ ,\ p_{I,E}) = p_{I,E\cap G'}$$

where G' is the zone which is the complement of the zone G' , if the particle A was in the zone G'
Hence in this conceptual model, $e_{m_B\alpha\beta}$ are observations.

It is very easy to see that $f_{m_B\alpha\beta}$ of example 3 are not observations.
Indeed

$$s\,(f_{m_B\alpha\beta}\ ,\ p_{m_A\theta\phi}) = p_{m_A\alpha\beta}$$

or

$$s\,(f_{m_B\alpha\beta}\ ,\ p_{m_A\theta\phi}) = p_{m_A,\pi-\alpha,\pi+\beta}$$

depending on whether

$$|F_1'| > |F_2'| \quad \text{or}$$

$$|F_1'| < |F_2'|$$

5. Reality and probability

As we remarked already, the reality of the system is described by its state. We have introduced a "probability model" as a conceptual model for the relative frequencies of the outcomes of measurements. There is however also a concept of probability that refers to the knowledge (or the lack of knowledge) about the reality of the system. Hence the concept of probability should describe in this case "the way" in which we lack knowledge about the reality of the system.

Example 5 :

Consider again the system of example 1. The particle is in state $p_{m_A,\theta,\phi}$ if we know the value of m_A and its "point" place (θ,ϕ) on the surface of the sphere. We have also introduced the state p_{m_A} that represents the situation where we only know the value of m_A and we also know that the particle is on the surface of the sphere, but we do not know its exact place (θ,ϕ). It is however also possible that the state is p_{m_A} but we have an extra knowledge of a different nature. Namely we know that the particle A "is" at some place $(\theta\phi)$, and we also know the probability distribution that describes the lack of knowledge we have about the exact place $(\theta\phi)$ of the particle. Such a probability distribution is for example given by the map

$$\mu : \mathcal{B}(Sp) \rightarrow [0,1]$$

$$E \rightarrow \mu(E)$$

where Sp is the surface of the sphere and $\mathcal{B}(Sp)$ is a set of subsets of Sp , E is a subset of Sp and $\mu(E) = \frac{\text{surface of } E}{\text{surface } Sp}$. Sometimes, in the litterature such a probability distributions is also called a "state" (a mixed state).

So for a general physical system S such a situation can be formalized as follows :

There are now two different kind of states. There are the states where the system "is" in. Let us call them macrostates and represent them by p, q, r, ... But every macrostate is the union of a collection of states

$$p = \cup u$$

which we will call microstates. Let us denote the set of all microstates by Σ .

Then for every macrostate p there is a probability distribution :

$$\mu_p : \mathcal{B}(\Sigma) \rightarrow [0,1]$$

and if $E \subset \Sigma$ then $\mu_p(E)$ is the probability that the microstate of the system is an element of the set E , when the system is in the macrostate p .

And all these probability measures satisfy the axioms of Kolmogorov.

So the probability model that must be used to describe the reality of the system is a Kolmogorovian.

6. The non Kolmogorovian structure of probability

We have introduced two concepts of probability. The probability describing an extra knowledge about the reality of the system. This probability satisfies the axioms of Kolmogorov. And the probability describing the relative frequencies of outcomes of repeated measurements.

As we have shown in [1] and as it is also shown in [4] for general quantum systems, this probability does not satisfy the axioms of Kolmogorov in general. Let us analyse now carefully on the hand of the four types of measurements introduced in example 3 why the probability structure describing the relative frequencies of repeated measurements is not necessarely Kolmogorovian.

Example 6 : The probability structure describing the reality of the system of example 3 is explained in example 5. Hence we suppose that the system is in the macrostate p_{m_A} . The set of microstates that we consider is $\Sigma = \{p_{m_A\theta\phi}\}$. The extra knowledge that we have is the following : we know that the particle A is in some microstate $p_{m_A\theta\phi}$ and we know the probability distribution μ that describes the lack of knowledge we have about the exact microstate $p_{m_A\theta\phi}$ where the particle is in

$$\mu : \mathcal{B}(Sp) \to [0,1]$$

$$E \to \mu(E)$$

where $\mu(E) = \dfrac{\text{surface of } E}{4\pi R^2}$

We consider now the first type of measurement $e_{m_B\alpha\beta}$ introduced in example 3. These measurements are observations.
Let us calculate

$$P(e_{m_B\alpha\beta} = e_1 \mid p_{m_A}, \mu)$$

which is the probability to find the outcome e_1 for $e_{m_B\alpha\beta}$ if the state of the system is p_{m_A}, μ . This probability is given by $\mu(E_{m_B\alpha\beta})$ where $E_{m_B\alpha\beta}$ is the zone of the sphere consisting of those points that are above the plane through the point located at a distance of $m_B.\frac{2R}{m}$ from B .

The surface of this zone is given by

$$4\pi R^2 \cdot \frac{m_B}{m}$$

Hence $$\mu(E_{m_B\alpha\beta}) = \frac{m_B}{m}$$

So $$P(e_{m_B\alpha\beta} = e_1 \mid p_{m_A}, \mu) = \frac{m_B}{m}$$

And $$P(e_{m_B\alpha\beta} = e_2 \mid p_{m_A}, \mu) = \mu(E_{m_{C\alpha\beta}}) = \frac{m_C}{m}$$

Let us consider now the second type of measurement $f_{m_B\alpha\beta}$ introduced in example 3.

Let us calculate again

$$P(f_{m_B\alpha\beta} = e_1 \mid p_{m_A}, \mu)$$

$$= P(|F_1'| > |F_2'|)$$

$$= P\left(G \frac{m_A m_B}{2R \sin \frac{\gamma}{2}} \cdot \cos \frac{\gamma}{2} > G \frac{m_A m_C}{2R \cos \frac{\gamma}{2}} \cdot \sin \frac{\gamma}{2}\right)$$

$$= P(m_B \cos^2 \frac{\gamma}{2} > m_C \sin^2 \frac{\gamma}{2})$$

$$= P(m_B \cos^2 \frac{\gamma}{2} > (m - m_B) \sin^2 \frac{\gamma}{2})$$

$$= P(m_B > m \sin^2 \frac{\gamma}{2})$$

$$= P\left(\sqrt{\frac{m_B}{m}} > \sin \frac{\gamma}{2}\right)$$

$$= P\left(2R \sqrt{\frac{m_B}{m}} > 2R \sin \frac{\gamma}{2}\right)$$

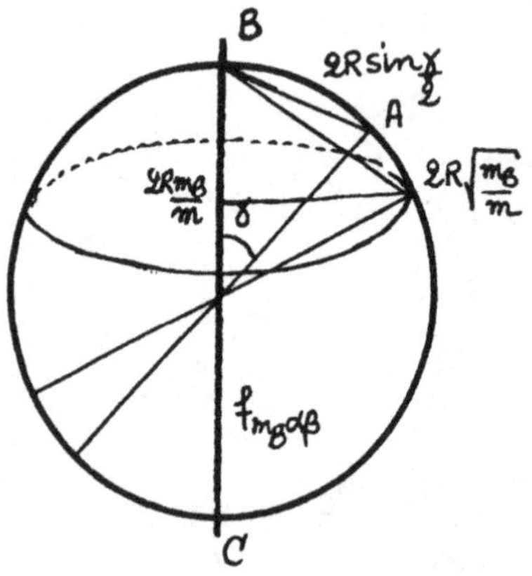

And this probability is again given by $\mu\ (E_{m_B\alpha\beta})$ and hence

$$P\ (f_{m_B\alpha\beta} = e_1 \mid p_{m_A},\ \mu) = \frac{m_B}{m}$$

$$P\ (f_{m_B\alpha\beta} = e_2 \mid p_{m_A},\ \mu) = \frac{m_C}{m}$$

The difference between $e_{m_B\alpha\beta}$ and $f_{m_B\alpha\beta}$ is that $e_{m_B\alpha\beta}$ are observations, while $f_{m_B\alpha\beta}$ are not.

Let us now consider the third and fourth type of measurement introduced in example 3.

From symmetry considerations it is easy to see that

$$P\ (e_{\alpha\beta} = e_1 \mid p_{m_A}) = \frac{1}{2}$$

$$P\ (e_{\alpha\beta} = e_2 \mid p_{m_A}) = \frac{1}{2}$$

$$P\ (f_{\alpha\beta} = e_1 \mid p_{m_A}) = \frac{1}{2}$$

$$P\ (f_{\alpha\beta} = e_2 \mid p_{m_A}) = \frac{1}{2}$$

In [1] and [4] it is shown that Bayes formula for the conditional probability is not correct for a quantum mechanical system. Let us analyse on the hand of the examples why this is the case.

So let us consider

$$P\ (e_{m_B\alpha\beta} = e_1 \mid e_{m_B'\alpha'\beta'} = e_1)$$

which is the probability that for a measurement of $e_{\alpha\beta}$ we find e_1 if the system is prepared in such a way that for a measurement of

$e_{m_B'\alpha'\beta'}$ we would find e_1 . In this conditional probability, the two concepts of probability are present

$$|e_{m_B'\alpha'\beta'} = e_1)$$

is a conditioning on the reality of the system and

$$(e_{m_B\alpha\beta} = e_1 \,|$$

is a formalization of the relative frequency of repeated measurements of $e_{m_B\alpha\beta}$.

The reason why Bayes formula was proposed and works is because the following intuitive reasoning can be made.

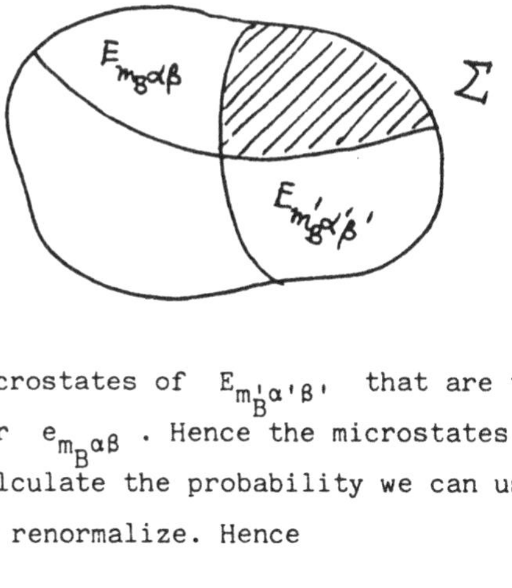

If we condition on the states of the system by asking that an eventual measurement of $e_{m'_B\alpha'\beta'}$ should give e_1 as outcome, this comes to the same as redefining our set of microstates as the set $E_{m_B'\alpha'\beta'}$. And to calculate the probability that a measurement of $e_{m_B\alpha\beta}$ would give e_1 , we just consider those microstates of $E_{m_B'\alpha'\beta'}$ that are the one's that gave an outcome e_1 for $e_{m_B\alpha\beta}$. Hence the microstates of the set $E_{m_B\alpha\beta} \cap E_{m_B'\alpha'\beta'}$. To calculate the probability we can use the old measure μ but we have to renormalize. Hence

$$P\,(e_{m_B\alpha\beta} = e_1 \mid e_{m_B'\alpha'\beta'} = e_1) = \frac{\mu\,(E_{m_B\alpha\beta} \cap E_{m_B'\alpha'\beta'})}{\mu\,(E_{m_B'\alpha'\beta'})}$$

And this is correct for the measurements $e_{m_B\alpha\beta}$. Also for the measurements $f_{m_B\alpha\beta}$ this formula is correct.

$$P\,(f_{m_B\alpha\beta} = e_1 \mid f_{m_B'\alpha'\beta'} = e_1) = \frac{\mu\,(E_{m_B\alpha\beta} \cap E_{m_B'\alpha'\beta'})}{\mu\,(E_{m_B'\alpha'\beta'})}$$

But for the measurements $e_{\alpha\beta}$ and $f_{\alpha\beta}$ the formula is not correct. This because the intuitive reasoning that leads to the formula cannot be made for these measurements. Why does this reasoning cannot be made. Because there is no set of microstates $E_{\alpha\beta}$ such that

$$P\ (e_{\alpha\beta} = e_1 \mid p_{m_A}) = \mu\ (E_{\alpha\beta})$$

The reason that there is not set of microstates is because, part of the probability comes from the measurements $e_{\alpha\beta}$ itself.
We show explicitely in [1] III, that there indeed does not exist such a set of microstates and more generally that there does not exist a Kolmogorovian model.
The same is true for the measurements $f_{\alpha\beta}$. What is very amusing is the fact that the measurements $f_{\alpha\beta}$ can be described by a quantum mechanical probability model. This result can be found in [1]. But let us make the calculation again.
Let us calculate

$$P\ (f_{\alpha\beta} = e_1 \mid f_{\alpha'\beta'} = e_1)$$

First we have to analyse the conditioning $|f_{\alpha'\beta} = e_1)$. There is only one microstate, namely the microstate $p_{m_A\alpha'\beta'}$, such that when the system is in this microstate the conditioning $|f_{\alpha'\beta'} = e_1)$ is true. Hence

$$P\ (f_{\alpha\beta} = e_1 \mid f_{\alpha'\beta'} = e_1) = P\ (f_{\alpha\beta} = e_1 \mid p_{m_A\alpha'\beta'})$$

Let us calculate $P\ (f_{\alpha\beta} = e_1 \mid p_{m_A\alpha'\beta'})$

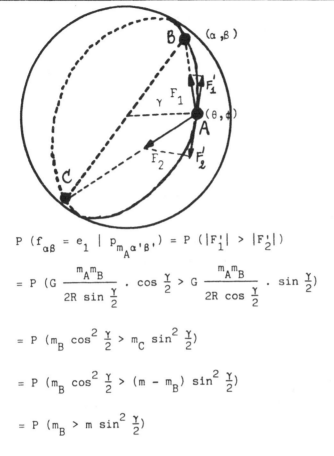

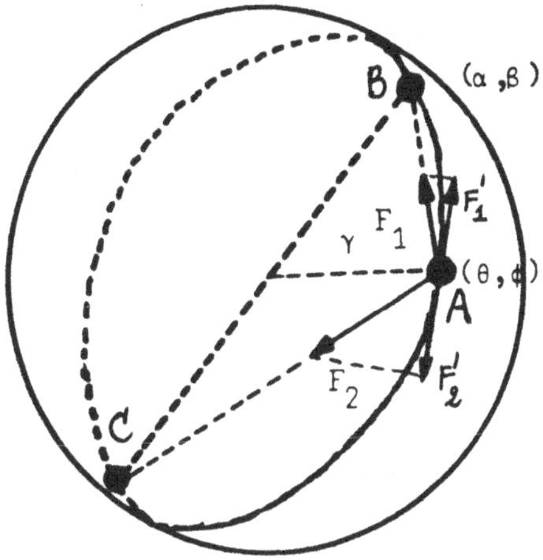

$$P\,(f_{\alpha\beta} = e_1 \mid p_{m_A\alpha'\beta'}) = P\,(|F_1'| > |F_2'|)$$

$$= P\,\left(G\,\frac{m_A m_B}{2R\sin\frac{\gamma}{2}}\cdot\cos\frac{\gamma}{2} > G\,\frac{m_A m_B}{2R\cos\frac{\gamma}{2}}\cdot\sin\frac{\gamma}{2}\right)$$

$$= P\,(m_B\cos^2\frac{\gamma}{2} > m_C\sin^2\frac{\gamma}{2})$$

$$= P\,(m_B\cos^2\frac{\gamma}{2} > (m - m_B)\sin^2\frac{\gamma}{2})$$

$$= P\,(m_B > m\sin^2\frac{\gamma}{2})$$

and m_B is chosen at random in the interval $[0,m]$.

$$= \frac{m - m\sin^2\frac{\gamma}{2}}{m} = \cos^2\frac{\gamma}{2}$$

On the other hand

$$P\,(f_{\alpha\beta} = e_2 \mid p_{m_A\alpha'\beta'}) = \sin^2\frac{\gamma}{2}$$

This is exactly the probabilities that we would find if $f_{\alpha\beta}$ would represent the measurement of the spin of a spin $\frac{1}{2}$ particle in the (α,β) direction, while the particle had spin in the (α',β') direction.

Hence we can describe this system with these measurements by means of a two dimensional complex Hilbert space. We then represent the state $p_{m_A\theta\phi}$ of the particle A by means of the vector

$$\chi_{\theta\phi} = (e^{-i\,\phi/2}\cos\frac{\theta}{2}\ ,\ e^{i\,\phi/2}\sin\frac{\theta}{2})$$

and the measurement $f_{\alpha\beta}$ by means of the self adjoint operator

$$S_{\alpha\beta} = \frac{1}{2}\begin{pmatrix} \cos\alpha & e^{-i\beta}\sin\alpha \\ e^{i\beta}\sin\alpha & -\cos\alpha \end{pmatrix}$$

We can then apply the calculus of quantum mechanics to our system with these measurements.

Let us show now explicitely that the Kolmogorovian model cannot be applied to this situation. Therefore we consider three measurements e, f and g of the type $f_{\alpha\beta}$. More specifically

$$e = f_{0,0}\quad,\quad f = f_{\frac{\pi}{3},0}\quad,\quad g = f_{\frac{2\pi}{3},0}$$

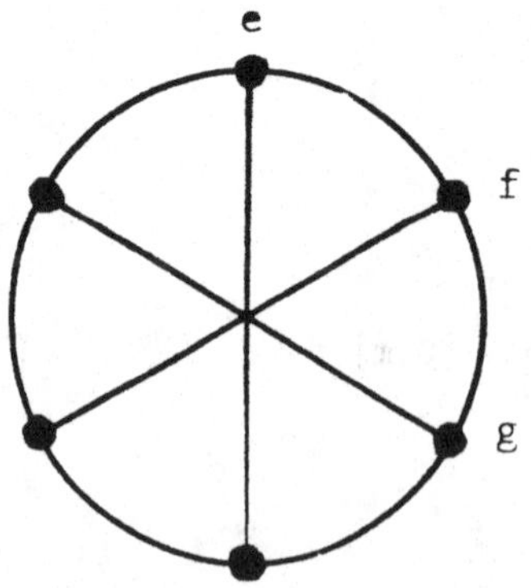

We suppose the particle A to be in the state p_{m_A} , μ . Then clearly

$$P(e=e_1) = P(f=f_1) = P(g=g_1)$$

$$= P(e=e_2) = P(f=f_2) = P(g=g_2)$$

$$= \frac{1}{2}\ .$$

If there does exist a Kolmogorovian model, we must have a probability measure μ and

$$\mu(E_1) = \mu(E_2) = \mu(F_1) = \mu(F_2) = \mu(G_1) = \mu(G_2) = \frac{1}{2}$$

Using Bayes formula and the properties of the probability measure we have :

$$\frac{1}{2} P\ (f = f_1 \mid g = g_1)$$

$$= \mu\ (F_1 \cap G_1) = \mu\ (E_1 \cap F_1 \cap G_1) + \mu\ (E_2 \cap F_1 \cap G_1)$$

$$\frac{1}{2} P\ (e = e_1 \mid g = g_1)$$

$$= \mu\ (E_1 \cap G_1) = \mu\ (E_1 \cap F_1 \cap G_1) + \mu\ (E_1 \cap F_2 \cap G_1)$$

$$P\ (f = f_1 \mid g = g_1) = \cos^2\ (\frac{\pi}{6}) = \frac{3}{4}$$

$$P\ (e = e_1 \mid g = g_1) = \cos^2\ (\frac{\pi}{3}) = \frac{1}{4}$$

Hence

$$\mu\ (E_1 \cap F_1 \cap G_1) + \mu\ (E_2 \cap F_1 \cap G_1) = \frac{3}{8}$$

$$\mu\ (E_1 \cap F_1 \cap G_1) + \mu\ (E_1 \cap F_2 \cap G_1) = \frac{1}{8}$$

From this follows that

$$\mu\ (E_2 \cap F_1 \cap G_1) = \frac{1}{4} + \mu\ (E_1 \cap F_2 \cap G_1)$$

Hence

$$\mu\ (E_2 \cap F_1 \cap G_1) > \frac{1}{4}$$

On the other hand, we have

$$\frac{1}{2} P\ (e = e_2 \mid f = f_1)$$

$$= \mu\ (E_2 \cap F_1) = \mu\ (E_2 \cap F_1 \cap G_1) + \mu\ (E_2 \cap F_1 \cap G_2)$$

and

$$P(e = e_2 \mid f = f_1) = \cos^2 \frac{\pi}{3} = \frac{1}{4}$$

Hence

$$\frac{1}{8} = \mu(E_2 \cap F_1 \cap G_1) + \mu(E_2 \cap F_1 \cap G_2)$$

which shows that

$$\mu(E_2 \cap F_1 \cap G_1) < \frac{1}{8}$$

It is not possible to have $\mu(E_2 \cap F_1 \cap G_1) > \frac{1}{4}$ and $\mu(E_2 \cap F_1 \cap G_1) < \frac{1}{8}$, and so this shows that this probability model cannot be described by a Kolmogorovian model.

It is very easy to see that also the probability model of the measurements $e_{\alpha\beta}$ is not Kolmogorovian.

7. Conclusion

The probability model describing the reality of the system is Kolmogorovian, in the sense that for every macrostate p we can describe the lack of knowledge about the microstate of the system by means of a probability measure

$$\mu_p : \Sigma_p \to [0,1]$$

where Σ_p is the set of microstates of the system when it is in the macrostate p .

If measurements are such that they are deterministic on the microstates of the system (as for example $e_{m_B\alpha\beta}$ and $f_{m_B\alpha\beta}$) , then the probability model describing the relative frequency of repeated measurements is also Kolmogorovian.
If measurements are such that they are not deterministic on the microstates of the system (as for example $e_{\alpha\beta}$ and $f_{\alpha\beta}$), then in general no Kolmogorovian model can be found for these measurements.
As we have shown, it is Bayes formula that does not work anymore. This is however not the only thing that goes wrong. New axioms have to be formulated, and a new structure in general has to be introduced to treat these situations. We shall try to do this in a following paper. We must mention that the shortcoming of the Kolmogorovian model had been remarked already quiet some time ago by Randall and Foulis, and they already proposed a generalization of the theory (see [5]). In their approach, the problems that we mention here can be avoided.

References

[1] D. Aerts ; J. Math. Phys. 27 (1), 1986.

[2] D. Aerts ; Foundations of Physics 12, 1131 (1982) and "The one and the many", Doctoral thesis, Vrije Universiteit Brussel, TENA (1981) and "The description of one and many physical systems" in "Les Fondements de la Mécanique Quantique", 25e cours de perfectionnement de l'Association Vaudoise des chercheurs en physique Ed. Christian Gruber et al.

[3] C. Piron ; Foundations of quantum physics W.A. Benjamin, Inc. 1976.

[4] L. Accardi ; Rend. Sem. Mat. Univ. Politech. Torino, 1982, 249.
L. Accardi and A. Fedullo ; Lett. Nuovo Cimento 34, 161 (1982).

[5] Foulis D., Piron C., Randall C. ; "Realism, operationalism and quantum mechanics", Foundations of Physics, Vol. 13, n° 8, 1983, 813 - 841.
Foulis D., Randall C. ; Journal Math. Phys. (1972), 1667.

ON THE UNIVERSALITY OF THE EINSTEIN-PODOLSKY-ROSEN PHENOMENON

Luigi Accardi*
Princeton University, New Jersey, U.S.A.**

* On leave of absence from Dipertimento di Matematica, University of Roma II.

In [4] (chap. 6) von Neumann deduced a canonical form for the states of a quantum system composite of two sub-systems, but he did not discuss the uniqueness of this representation. In [3] Margenau and Park discussed a generalized form of the Einstein, Podolsky, Rosen paradox. In [1], [2] Baracca, Bergia, Cannata, Ruffo and Savoia remarked that von Neumann's theorem might be interpreted as describing a generalized EPR type situation. In fact, the statement of von Neumann's theorem can be expressed by saying that any state of a composite system can be written in the form discussed by Margenau and Park in [3]. These authors also discussed how to generalize the EPR construction in the case when a Lie group is involved.

In this note, which is a comment to their paper, it is shown that a discussion of the uniqueness of the von Neumann representation shows that also the situation envisaged by Bergia et al. [2] is universal, in the sense that a Lie group (and in fact a whole sequence of them) is always involved, being canonically associated to the state of the composite system.

More precisely, the invariants of von Neumann representation of the state of a composite system are :

i) a family (σ_k) of "generalized singlet states" of one of the systems (which we will call " the measured system ").

ii) the probabilities (p_k) of these generalized singlet states.

iii) the (finite dimensional , unitary) invariance groups of these states.

** Research sponsored by the Office of Naval Research under Contract No. 00014-84-K-0421

The above result means that in a certain sense, made precise in Theorem (2), every state of every composite system can be thought of, up to isomorphism, as an orthogonal sum of generalized singlet states of the "measured system with respect to some finite dimensional group.
This suggests that even from the mathematical point of view, the fine structure of the states of a composite system is something which deserves a better comprehension.

Theorem(1) Let H_1, H_2 be two complex separable Hilbert spaces, and let ϕ be a unit vector in $H_1 \otimes H_2$. Then there exist two orthonormal bases (ψ_α), (η_α) (α = 1,2,...) of H_1, H_2 respectively, and positive numbers p_α such that :

$$\phi = \sum_\alpha \sqrt{p_\alpha}\, \psi_\alpha \otimes \eta_\alpha \tag{1}$$

$$\sum_\alpha p_\alpha = 1 \tag{2}$$

Proof For $\psi \varepsilon H_1$; $\eta \varepsilon H_2$ one has :

$$|<\phi , \psi \otimes \eta>| \leq ||\psi||\ ||\eta|| \tag{3}$$

hence, for any ψ in H_1 the linear form :

$$\eta\ \varepsilon H_2 \dashrightarrow <\phi , \psi \otimes \eta>$$

is continuous and therefore there exists a vector $F(\psi)$ in H_2 characterized by the property :

$$<F(\psi) , \eta>_{H_1} = <\phi , \psi \otimes \eta> \tag{4}$$

for every η in H_2. From (3) and (4) it follows that the map :

$$F : \psi \varepsilon H_1 \dashrightarrow F(\psi)\ \varepsilon H_2$$

is linear and continuous. The linear operators :

$$D_1 = F^+F : H_1 \dashrightarrow H_2 \tag{5}$$

$$D_2 = FF^+ : H_2 \dashrightarrow H_2 \tag{6}$$

are clearly positive, and if (ψ_m) is an orthonormal basis in H_1 and $\eta_n)$ an orthonormal basis in H_2, one has :

$$Tr(D_1) = \sum_m <\psi_m , D_1\psi_m> = \tag{7}$$

$$\sum_m <F\psi_m , F\psi_m> = \sum_{m,n} <F\psi_m , \eta_n><\eta_n , F\psi_m> =$$

$$= \sum_{m,n} <\phi , \psi_m \otimes \eta_n><\eta_n \otimes \psi_m , \phi> = 1$$

and similarly for D_2. So D_1, D_2 are both density matrices. Moreover, from :

$$FD_1 = FF^+F = D_2F \tag{8}$$

one deduces that :

$$FD_1F^+ = D_2FF^+ = (D_2)^2 \tag{9}$$

Let now :

$$D_1 = \sum_{k \geq 0} p_k E_k \tag{10}$$

$$D_2 = \sum_{k \geq 0} q_k E'_k \tag{11}$$

be the spectral decompositions of D_1, D_2 respectively. E_o (resp. E'_o) is the projection onto the null space of D_1 (resp. D_2) and p_k , q_k are > 0 for each k > 0 and $p_o = q_o = 0$. Since for each $k \geq 0$:

$$D_1E_k = p_kE_k$$

it follows that :

$$(FE_kF^+)(FE_hF^+) = FE_kD_1E_hF^+ = p_h\delta_{k,h}FE_kF^+ \tag{12}$$

hence, for h > 0 the operators $\frac{1}{\sqrt{p_k}}FE_hF^+$'s are orthogonal projections. Therefore, using (9) we find :

$$D_2^2 = \sum_k q_k^2 E'_k = FD_1F^+ = \sum_k p_k \frac{F}{\sqrt{p_k}} E_k \frac{F^+}{\sqrt{p_k}}$$

hence, by the uniqueness of the spectral decomposition :

$$p_k = q_k \quad ; \; K = 1 , 2 , \cdots \tag{13}$$

$$FE_hF^+ = E'_h \quad ; \; h = 0 , 1 ,2 , \cdots \tag{14}$$

Let now $\{ \psi_{k,v}: k \geq 0 ; v \geq 0 \}$ be an orthonormal basis of H_1 such that, for each $k \geq 0$ one has :

$$E_k = \sum_v | \psi_{k,v} >< \psi_{k,v} | \tag{15}$$

(in Dirac's notations) . Then denoting :

$$\eta_{k,v} = \frac{1}{\sqrt{p_k}} F \psi_{k,v} \tag{16}$$

one finds :

$$<\eta_{k,v} , \eta_{k',v'}> = \frac{1}{\sqrt{p_kp_{k'}}} <F\psi_{k,v} , F\psi_{k',v'}> \tag{17}$$

$$= \frac{1}{\sqrt{p_kp_{k'}}} <\psi_{k,v} , D_1\psi_{k',v'}> = \delta_{k,k'}\delta_{v,v'}$$

Therefore $\{ \eta_{k,v} : k \geq 0 ; v \geq 0 \}$ is an orthonormal basis of H_2 . Moreover :

$$\phi = \sum_{k,v,k',v'} <\phi , \psi_{k,v} \otimes \eta_{k',v'}> \psi_{k,v} \otimes \eta_{k',v'} =$$

$$= \sum_{k,k',v,v'} <F\psi_{k,v} , \eta_{k',v'}> \psi_{k,v} \otimes \eta_{k',v'} =$$

$$\sum_{k,k',v,v'} \sqrt{p_k} <\eta_{k,v} , \eta_{k',v'}> \psi_{k,v} \otimes \eta_{k',v'} =$$

$$= \sum_{k,v} \sqrt{p_k} \psi_{k,v} \otimes \eta_{k',v'}$$

and this proves the theorem.

Remark that by grouping together, in equation (1), all the vectors corresponding to the same p_k, one can write ϕ in the form :

$$\phi = \sum_k \sqrt{p_k} \sum_j \psi_{k,j} \otimes \eta_{k,j} \tag{18}$$

where, for each k, the index j runs over a (necessarily finite) set I_k.
In the following, if a vector ϕ can be written in the form (18), we will say that the right hand side of (18) is a von Neumann representation of ϕ with weights (p_k).

Theorem (2) Let (18) be a von Neumann representation of the vector ϕ in the Hilbert space $H_1 \otimes H_2$. Define :

$$H_{1,k} = \textit{closed linear span of } \psi_{k,j} ; j \in I_k \subseteq H_1 \tag{19}$$

$$\sigma_k = \sum_j \psi_{k,j} \otimes \psi_{k,j} \in H_{1,k} \otimes H_{1,k} \tag{20}$$

$$G_k = [U_k \in Un(H_{1,k}) : (U_k \otimes U_k)\sigma_k = \exp i\alpha_k \sigma_k] \tag{21}$$

for some real number α sub k , and where Un(H) denotes the group of unitaries of the Hilbert space H .
Then if

$$\phi = \sum_k \sqrt{q_k} \sum_j \psi'_{k,j} \otimes \eta'_{k,j} \tag{22}$$

is any other von Neumann representation of ϕ one has, possibly after a permutation of the indices k :

$$q_k = p_k \tag{23}$$

Moreover, possibly after a relabeling, for each k , the indices j in (22) can be supposed to belong to I_k and, for each index k, there exists a unitary $U_k \in G_k$ such that :

$$\psi'_{k,j} = U_k \psi_{k,j} \tag{24}$$

$$\eta'_{k,j} = \frac{1}{\sqrt{p_k}} F \psi'_{k,j} \tag{25}$$

where F is the operator defined by (4) in Theorem (1).
Conversely, for any $U_k \in G_k$, if q_k, $\psi'_{k,j}$, $\eta'_{k,j}$ are defined respectively by (23), (24), (25) then the identity (22) holds (i.e. the right hand side of (22) is a von Neumann representation of ϕ).

Proof. Let (22) above be another von Neumann representation of ϕ . Then, using the definition (4) of F , one has :

$$\begin{aligned}\phi &= \sum_{k,j,k',j'} \langle \phi , \psi'_{k,j} \otimes \eta'_{k',j'} \rangle \psi'_{k,j} \otimes \eta'_{k,j} \\ &= \sum_{k,j,k',j'} \langle F\psi'_{k,j} , \eta'_{k,j} \rangle \psi'_{k,j} \otimes \eta_{k',j'}\end{aligned} \tag{26}$$

Comparing (22) and (26) one immediately deduces that :

$$F\psi'_{k,j} = \sqrt{q_k}\, \eta'_{k,j} \tag{27}$$

Similarly, by considering the scalar product

$$\langle \psi'_{k,j} , F^{+}\eta'_{k',j'} \rangle$$

one obtains :

$$F^{+}\eta'_{k,j} = \sqrt{q_k}\, \psi'_{k,j} \tag{28}$$

hence also :

$$D_1\psi'_{k,j} = F^+F\,\psi'_{k,j} = q_k\psi'_{k,j} \tag{29}$$

where D_1 is the density matrix defined by (5) in Theorem (1). This implies that, up to relabeling, (23) holds and the vectors $\psi'_{k,j}$ belong, for k -fixed and any j , to the space $H_{1,k}$ defined by (19) . But then one must also have :

$$\sum_j \psi_{k,j} \otimes \eta_{k,j} = \sum_j \psi'_{k,j} \otimes \eta'_{k,j} \tag{30}$$

for each k . Therefore the j-indices in the two sides of (30) must run through sets of equal cardinality.

Keeping into account (16) and (27) , we can write (30) in the equivalent form :

$$\sum_j \psi_{k,j} \otimes F\,\psi_{k,j} = \sum_j \psi'_{k,j} \otimes F\,\psi'_{k,j} \tag{31}$$

and, since F is invertible on $H_{1,k}$, (31) is equivalent to :

$$\sum_j \psi_{k,j} \otimes \psi_{k,j} = \sum_j \psi'_{k,j} \otimes \psi'_{k,j} = \sigma_k \tag{32}$$

with σ_k defined by (20) .

Thus, choosing any unitary U_k on $H_{1,k}$ satisfying :

$$U_k\psi_{k,j} = \exp i\,\alpha_k\;\psi'_{k,j} \tag{33}$$

(for some real number α), one easily sees that $U_k \in G_k$ and :

$$\psi'_{k,j} = \exp - i\frac{\alpha_k}{2}\,U_k\psi'_{k,j} \tag{34}$$

$$\eta'_{k,j} = \frac{1}{\sqrt{p_k}}\,F\,\psi'_{k,j} \tag{35}$$

Conversely, if U_k is any element of G_k , then defining $\psi'_{k,j}$, $\eta'_{k,j}$ by the right hand sides of (34), (35) respectively, one finds that :

$$\begin{aligned}\sum_k \sqrt{p_k}\,\sum_j \psi'_{k,j} \otimes \eta'_{k,j} &= (1 \otimes F)\sum_k \exp i\,\alpha_k \sum_j \psi_{k,j} \otimes \psi_{k,j}\\ &= (1 \otimes F)\sum_k \sum_j (\psi_{k,j} \otimes \psi_{k,j})\\ &= \sum_k \sqrt{p_k}\sum_j \psi_{k,j} \otimes \psi_{k,j} = \phi\end{aligned}$$

Bibliography

Baracca A., Bergia S., Cannata F., Ruffo S., Savoia M. Int. J. Theor. Phys. 16(1977)491

Bergia S., Cannata F., Russo S., Savoia M. Group theoretical interpretation of von Neumann' s theorem on composite systems. Am. Journ. of Phys. 47(1979)548-552

Margenau H., Park J.L. The logic of noncommutability of quantum mechanical operators and its empirical consequences. in: "Perspectives in quantum theory" W Yourgrau and A. van der Merwe (ed.) Dover (1971)

von Neumann J. Mathematical foundations of quantum mechanics. Princeton University Press 1955

THE SYMMETRY BETWEEN ELECTRICITY AND MAGNETISM AND THE WAVE EQUATION OF A SPIN $\frac{1}{2}$ MAGNETIC MONOPOLE

Georges Lochak
C.N.R.S. Fondation Louis de Broglie, Paris, France

1. INTRODUCTION

Symmetry laws are introduced in physical theories in two different ways which may be called the constructive way and the abstract (or a priori) way.

An example of the first one is given by the development of electromagnetism : step by step, the theory was built in order to explain the observed phenomena, and symmetry laws were expressed only afterwards as general properties of the physical laws already discovered.

The second way is now dominating in quantum physics. The most famous example is Dirac's discovery of the equation of the electron, which was suggested by general geometrical conditions which were miraculously rewarded by celebrated physical results.

The present paper is inspired by both methods of introduction of symmetry laws. It is based on the constructive Curie laws [1] on the symmetry of magnetism, but it aims to express them in quantum terms and to show that the Dirac abstract way is rewarded once again : it will be shown indeed, that not only Dirac's equation describes an electron, but moreover it can describe a magnetic monopole. More precisely, it happens that the Dirac equation admits, not only one local gauge invariance, but two. And only two. The first one corresponds to the well known electromagnetic interaction with an electric charge (which leads to the theory of the electron). The second one will be shown to involve another electromagnetic interaction which corresponds not to an electric charge but to a magnetic monopole.

The properties of this monopole are in full accordance with the Curie symmetry laws and even give a better account of these symmetry laws than the classical expression given by Curie himself. The reason is that the chirality properties are related to the helicity, i.e. to a wave-like property which is therefore better described in quantum physics than in classical physics.

As could be predicted, some limits will appear in the symmetry between electricity and magnetism. The origin of the difference between the two kinds of charges lies obviously in the difference between the symmetry laws of electric and magnetic fields. But the consequences are magnified by quantum properties. It occurs indeed that the symmetry laws of a monopole imply that it is necessarily massless, at least when it is described by a linear equation. The mass term is then non linear and its general form will be given. This non linear term will be interpreted geometrically, namely as expressing a torsion of physical space. Besides, this mass term implies the possibility of three kinds of states for the monopole : bradyon, luxon and tachyon states. It seems that this is even the first example of a wave equation which gives these three categories of particles.

The second part of the paper will be devoted to the motion of a massless monopole in a central electric field. Our equation gives the Poincaré integral of motion (the total angular momentum)[2]. The famous Dirac relation [3],[4],[5] $eg/\hbar c = n/2$ is shown to be simply a consequence of the quantization of the projection of the total angular momentum on the symmetry axis of the system formed by the monopole and the central field. This projection of the total angular momentum is equal to eg/c. These properties are in a strong analogy with the properties of a quantum symmetrical top and this analogy allows us to get the angular part of the wave-functions (the so called "monopole-harmonics") without any calculation, only as a consequence of an elementary property of the rotation group.

This interpretation leads to the idea that the neutrino may be considered as a special case of our monopole : namely the one which occurs when a monopole is produced in such a way that the angular momentum of the system constituted by itself and a certain electric charge, remains orthogonal to the symmetry axis of the system. This hypothesis will be developed in connection with the problem of weak interactions and of solar neutrinos.

2. CURIE'S SYMMETRY LAWS, CLASSICAL EQUATION OF MOTION OF A MAGNETIC MONOPOLE

The essential difference between an electric and a magnetic charge comes from the difference between an electric field $\vec{E}$ and a magnetic field $\vec{H}$: $\vec{E}$ is a *polar* vector, i.e. a vector the image of which in a mirror is opposite to $\vec{E}$, while $\vec{H}$ is an *axial* vector the image of which is equal to $\vec{H}$ (Fig. 1). $\vec{E}$ has the symmetry of a radial vector : $\vec{r}$, a velocity : $\vec{v}$, a linear momentum : $\vec{p}$, or a force : $\vec{F}$, while $\vec{H}$ has the symmetry of the external product of two radial vectors : $\vec{r} \times \vec{r}'$ or of an angular momentum $\vec{r} \times \vec{p}$.

An immediate consequence is that equations which give the force exerted by a field on a charge :

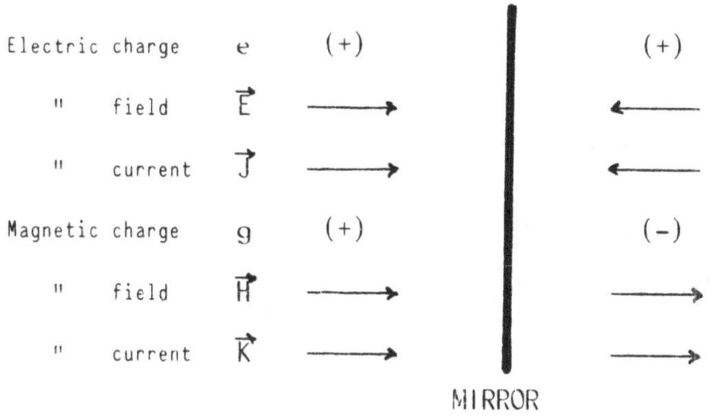

Fig. 1

Symmetry laws of Electricity and Magnetism.

$$\vec{F} = e\vec{E}\ , \qquad \vec{F} = g\vec{H} \tag{2.1}$$

are quite different : the electric charge e is a scalar because $\vec{F}$ and $\vec{E}$ are both polar, while the magnetic charge g must be a pseudo-scalar because $\vec{F}$ is polar, but $\vec{H}$ is axial. This means that the image of an electric charge in a mirror is the charge itself while the image of a magnetic charge will be the opposite one or, in other words : the image of a north pole will be a south pole ! (Fig. 1). We shall see that this strange conclusion of classical physics is not exactly the one of quantum physics and we shall find a quite different interpretation of this fundamental *Curie law* [1].

An immediate consequence of this law is that the following relations :

$$e\vec{v} = \vec{J}, \qquad g\vec{v} = \vec{K} \tag{2.2}$$

lead to different conclusions concerning $\vec{J}$ and $\vec{K}$. The electric current $\vec{J}$ is a polar vector like the velocity $\vec{v}$ (and like the electric field $\vec{E}$), while a magnetic current $\vec{K}$ must be an axial vector contrary to $\vec{v}$ (and like the magnetic field $\vec{H}$).

Just like it occurs for the magnetic charge, quantum mechanics will not exactly agree with classical physics, on the properties of the magnetic current. Some axial current will appear indeed, but it will be the *total* current of two opposite charges and the respective currents of these charges will not be axial, but *chiral* currents.

In conclusion of this section, it is interesting to complete the second formula (2.3) and to give the Lorentz force for a monopole and the corresponding equation of motion [6]

$$\frac{d\vec{p}}{dt} = g\vec{H} - \frac{g}{c}\vec{v} \times \vec{E}. \tag{2.3}$$

Note the minus sign before the second term.

3. CHIRAL GAUGE INVARIANCE OF THE DIRAC EQUATION

Now we come to the quantum problems, with a fundamental gauge property. Let us first consider the Dirac equation in absence of field :

$$\gamma_\mu \partial_\mu \psi + \frac{m_0 c}{h} \psi = 0, \tag{3.1}$$

where :

$$x_\mu = \{x,y,z,ict\}, \quad \gamma_k = i\begin{pmatrix} 0 & s_k \\ -s_k & 0 \end{pmatrix}, \quad \gamma_4 = \begin{pmatrix} I & 0 \\ 0 & -I \end{pmatrix} \tag{3.2}$$

s_k are the Pauli matrices and I the unit matrix.

One can prove the following theorem [8] :

The Dirac equation is invariant with respect to 2 (only 2) global gauge transformations :

$$\psi \to e^{i\Gamma\theta}\psi \tag{3.3}$$

where Γ is a 4×4 matrix. The two possible Γ matrices are :
$\Gamma = I$: ordinary *phase invariance* valid for all values of m
$\Gamma = \gamma_5 = \gamma_1\gamma_2\gamma_3\gamma_4$: *chiral gauge invariance* only valid when $m_0 = 0$

The associated *local gauges* define two different electromagnetic couplings :

1) $\Gamma = I$ gives the ordinary Dirac equation of the electron :

$$\gamma_\mu \nabla_\mu \psi + \frac{m_0 c}{\hbar} \psi = 0, \tag{3.4}$$

where ∇_μ is the covariant derivative :

$$\nabla_\mu = \partial_\mu - i\frac{e}{\hbar c} A_\mu, \tag{3.5}$$

and the local gauge transformation ($\phi = \phi(x,y,z,t)$) :

$$\psi \to e^{i\frac{e}{\hbar c}\phi}\psi, \quad A_\mu \to A_\mu + \partial_\mu \phi. \tag{3.6}$$

The constant e is the electric charge and the polar four vector A_μ :

$$A_\mu = \{\vec{A}, iV\} \tag{3.7}$$

is the classical Lorentz potential. Let us remind that the electromagnetic field is defined as :

$$\vec{H} = \text{rot}\,\vec{A}, \qquad \vec{E} = -\vec{\nabla}V + \frac{1}{c}\frac{\partial A}{\partial t} \tag{3.8}$$

and that the principal experimental test of the equation (3.4) is given by the properties of the hydrogen atom.

2) $\Gamma = \gamma_5$ defines in an analogous way the new equation [7][8] :

$$\gamma_\mu \nabla_\mu \psi = 0 \qquad (m_0 = 0) \tag{3.9}$$

where ∇_μ is the covariant derivative :

$$\nabla_\mu = \partial_\mu - \frac{g}{\hbar c}\gamma_5 B_\mu \tag{3.10}$$

and the local gauge invariance of (3.9) is now defined by the *chiral gauge* transformation :

$$\psi \to e^{i\frac{g}{\hbar c}\gamma_5 \phi}\psi, \qquad B_\mu \to B_\mu + i\partial_\mu \phi. \tag{3.11}$$

The constant g will be shown to be a *magnetic* charge and we shall prove that (3.9) is the equation of a massless *magnetic monopole*. In the same way as the hydrogen atom was a test for the equation (3.4), the new experimental test for (3.9) will be now the Birkeland-Poincaré effect, i.e. the interaction between an electric charge and a magnetic pole. g will be a *scalar* (like every physical constant) but we see on (3.10) and (3.11) that the coupling with the potential B_μ is defined by a *pseudo-scalar charge operator* :

$$G = g\gamma_5 \tag{3.12}$$

ϕ will be a *pseudo-scalar* phase and B_μ an *axial* four-vector, i.e. the dual of an antisymmetric tensor of rank 3. We have :

$$i\,B_\mu = \{\vec{B}, iW\}. \tag{3.13}$$

B_μ is the Cabibbo and Ferrari potential [9]. In $\mathbb{R}^3$, $\vec{B}$ is axial and W a pseudo-scalar and the electromagnetic field is given by :

$$\vec{H} = \vec{\nabla}W + \frac{1}{c}\frac{\partial \vec{B}}{\partial t}, \qquad \vec{E} = \text{rot}\,\vec{B} \tag{3.14}$$

We know that the invariance of (3.4) by the transformation (3.6) involves the conservation of a polar vector, a electric density current :

$$\partial_\mu J_\mu = 0, \qquad J_\mu = -ie\,\bar{\psi}\gamma_\mu\psi. \tag{3.15}$$

One can prove that the invariance of (3.9) by the *chiral* gauge transformation (3.11) is associated likewise with the conservation of an *axial* vector, a magnetic density current ; see [7],[8] and [10](where K_μ was suggested on the basis of a priori symmetry arguments) :

$$\partial_\mu K_\mu = 0, \quad K_\mu = -ig\,\bar{\psi}\gamma_\mu\gamma_5\psi. \tag{3.16}$$

The fact that K_μ is an axial vector is in accordance with the Curie law (see (2.2) and Fig. 1) but this property occurs in a quite different way than in the classical case, because our g is a scalar instead of a pseudo-scalar and K_μ cannot be written in the form (2.2). Making use of (3.12), K_μ can be written, rather as :

$$K_\mu = -i\bar{\psi}\gamma_\mu G\psi, \tag{3.17}$$

where G is a magnetic charge-operator which is a q-number.

But another problem arises from the Darwin-de Broglie identities [11]:

$$J_\mu J_\mu = -e^2(\Omega_1^2 + \Omega_2^2), \quad K_\mu K_\mu = +g^2(\Omega_1^2 + \Omega_2^2), \tag{3.18}$$

where Ω_1, Ω_2 are respectively the Dirac scalar and pseudo-scalar quantities :

$$\Omega_1 = \bar{\psi}\psi, \quad \Omega_2 = -i\bar{\psi}\gamma_5\psi. \tag{3.19}$$

The difficulty lies in the fact that (3.18) means that J_μ is a *time-like* vector but conversely, K_μ will be thus *space-like* which seems inadmissible for a current density. This objection will be removed a little further.

4. THE SYMMETRY LAWS OF A MONOPOLE

It is easy to show that the equation (3.9) is P, T, C invariant, i.e. invariant under the three transformations :

$$(P)\quad \vec{x} \to -\vec{x}, \quad B_4 \to -B_4, \quad B_k \to B_k, \quad \psi \to \gamma_4\psi \tag{4.1}$$

$$(T)\quad t \to -t, \quad B_4 \to B_4, \quad B \to -B_k, \quad \psi \to \gamma_1\gamma_2\gamma_3\psi \tag{4.2}$$

$$(C)\quad g \to g \quad , \quad \psi \to \gamma_2\psi^*. \tag{4.3}$$

It could seem in contradiction with the Curie law, that in (4.1) the sign of the magnetic charge g does not change. But actually the Curie law is satisfied because when $\vec{x} \to -\vec{x}$, we have $\psi \to \gamma_4\psi$ and therefore the sign of the magnetic charge density will change :

$$K_4 = -ig\,\bar{\psi}\gamma_4\gamma_5\psi \to -ig\,\bar{\psi}\gamma_4\gamma_4\gamma_5\gamma_4\psi = ig\psi\gamma_4\gamma_5\psi = -K_4 \tag{4.4}$$

At first sight the transformation (4.3) seems anomalous too, because we find that a monopole and its charge conjugated particle have both the same magnetic charge constant g, contrary to what occurs with an electric charge. A consequence which can be easily guessed is that, if we quantize the monopole field, namely if we describe the creation of monopole pairs, we shall face the strange problem of a creation of monopoles with the same sign of magnetic charge, which signifies that the conservation law (3.16) will be lost. And this is exactly what happens : this phenomenon is nothing but the well known "anomalies" which are associated with an axial current like (3.16). But an important point is that despite the strange property (4.3), the *magnetic charge densities* of a monopole and of its charge conjugated have opposite signs, as could be expected. Using (4.3), we have indeed :

$$K_4 = -ig\,\bar{\psi}\gamma_4\gamma_5\psi \rightarrow -ig\,\tilde{\psi}\gamma_2\gamma_5\gamma_2\psi^* = -K_4. \tag{4.5}$$

Nevertheless, the question remains : *what is an antimonopole* if g does not change its sign ? This is the subject of the next section.

5. THE 2-COMPONENT THEORY AND THE QUESTION OF CHIRALITY

Consider the 2-component spinors ξ and η and let us apply to (3.9), (3.10) the unitary transformation :

$$\psi = \frac{1}{\sqrt{2}}(\gamma_4 + \gamma_5)\begin{pmatrix}\xi\\ \eta\end{pmatrix}. \tag{5.1}$$

The equation (3.9) splits into two independent equations. Droping the relativistic notations and using (3.13), we find [7][8] :

$$\begin{aligned}
&(\mathrm{L})\quad \left\{\frac{1}{c}\frac{\partial}{\partial t} - \vec{s}.\vec{\nabla} - i\,\frac{g}{\hbar c}(W + \vec{s}.\vec{B})\right\}\xi = 0\\
&(\mathrm{R})\quad \left\{\frac{1}{c}\frac{\partial}{\partial t} + \vec{s}.\vec{\nabla} + i\,\frac{g}{\hbar c}(W - \vec{s}.\vec{B})\right\}\eta = 0
\end{aligned} \tag{5.2}$$

We recognize in (L) and (R) a generalization of the equations of a left and a right neutrino. The chiral gauge invariance (3.11) takes now the form of a phase invariance of (L) and (R), but with opposite phases (do not forget : Φ is a *pseudo* scalar) :

$$\xi \rightarrow e^{i\frac{g}{\hbar c}\Phi}\xi, \quad \eta \rightarrow e^{-i\frac{g}{\hbar c}\Phi}\eta, \quad W \rightarrow W + \frac{1}{c}\frac{\partial\Phi}{\partial t}, \quad \vec{B} \rightarrow \vec{B} - \vec{\nabla}\Phi \tag{5.3}$$

The P, T, C transformations are now :

$$(\mathrm{P})\quad \vec{x} \rightarrow -\vec{x}, \quad W \rightarrow -W, \quad \vec{B} \rightarrow \vec{B}, \ \xi \leftrightarrow \eta \tag{5.4}$$

$$(\mathrm{T})\quad t \rightarrow -t, \quad W \rightarrow W, \quad \vec{B} \rightarrow -\vec{B}, \ \xi \leftrightarrow \xi \tag{5.5}$$

$$(\mathrm{C})\quad g \rightarrow g, \quad -is_2\xi^* \rightarrow \eta, \quad is_2\eta^* \rightarrow \xi \tag{5.6}$$

(5.4) shows that the left and the right monopole are each the image of the other in a mirror. (5.5) shows that they are exchanged by a time reversal, and (5.6) that they are charge conjugated (but obviously, g remains unchanged by the transformation). Therefore, *the antimonopole is the image of the monopole in a mirror* : one is Right, the other is Left.

But why g does not change ? For the following reason : the transformation (5.1) diagonalizes γ_5 and thus diagonalizes the charge operator G defined by (3.12) :

$$\frac{1}{\sqrt{2}}(\gamma_4 + \gamma_5)G\,\frac{1}{\sqrt{2}}(\gamma_4 + \gamma_5) = g\,\gamma_4 = g\begin{pmatrix} I & 0 \\ 0 & -I \end{pmatrix} \tag{5.7}$$

This means that ξ (and therefore the left monopole) is the +g state of charge, while η (and therefore the right monopole) is the -g state of charge. This is the quantum expression of the Curie law. In other words, the monopole and the antimonopole have opposite helicities, just like the neutrino and the antineutrino, and opposite eigenvalues of the charge operator G.

6. CHIRAL CURRENTS. CHIRAL INVARIANTS

Consider the two currents :

$$X_\mu = \{\xi^+\xi,\ -\xi^+\vec{s}\xi\},\quad Y_\mu = \{\eta^+\eta,\ \eta^+\vec{s}\eta\}. \tag{6.1}$$

They are isotropic : it is easily found that :

$$X_\mu X_\mu = Y_\mu Y_\mu = 0. \tag{6.2}$$

They are the respective currents of the left and right monopole and they are conserved by the corresponding equations (L) and (R) in (5.2). We have :

$$\frac{1}{c}\frac{\partial}{\partial t}(\xi^+\xi) - \vec{\nabla}(\xi^+\vec{s}\xi) = 0,\quad \frac{1}{c}\frac{\partial}{\partial t}(\eta^+\eta) + \vec{\nabla}(\eta^+\vec{s}\eta) = 0 \tag{6.3}$$

The currents X_μ and Y_μ will be called *chiral currents*. Using the expression (3.16) of K_μ and (5.1), we find the decomposition :

$$K_\mu = g(X_\mu - Y_\mu) = g\,X_\mu + (-g)Y_\mu. \tag{6.4}$$

Therefore, K_μ is not the current density of a monopole but the sum of the current densities $g\,X_\mu$ and $(-g)Y_\mu$ of two monopoles (respectively left and right) which are *independent* from each other because the splitting of the equations. These currents are both isotropic as it must be for massless particles, but the fact that their *sum* is space-like or time-like has not any physical meaning. This is the answer to the question asked at the end of the section 3.

It is interesting to note that the electric current density J defined in (3.15) may be written as :

$$J_\mu = e(X_\mu + Y_\mu). \tag{6.5}$$

But in the Dirac equation we have : $m_0 \neq 0$, so that X_μ and Y_μ are no more separately conservative and are not independent. Only J_μ is an electric current density and the fact that it is time-like is a fundamental property : if it had been space-like, the whole theory had failed because this would be a violation of the principle of causality.

Now, in the following, we shall need *chiral invariants*, i.e. tensorial bilinear forms of ξ and η which are invariant by the chiral gauge (3.11) or (5.3). More conveniently we shall write here :

$$\psi \to e^{i\gamma_5\frac{\theta}{2}}\psi, \qquad \xi \to e^{i\frac{\theta}{2}}\xi, \qquad \eta \to e^{-i\frac{\theta}{2}}\eta. \tag{6.6}$$

X_μ, Y_μ and thus K_μ and J_μ are trivially chiral invariants. But Ω_1 and Ω_2 are not. Using (3.19) we find :

$$\begin{pmatrix}\Omega_1\\ \Omega_2\end{pmatrix} \to \begin{pmatrix}\cos\theta & -\sin\theta\\ \sin\theta & \cos\theta\end{pmatrix}\begin{pmatrix}\Omega_1\\ \Omega_2\end{pmatrix} \tag{6.7}$$

We call this relation a *chiral rotation* in the *chiral plane* $\{\Omega_1,\Omega_2\}$. Obviously,

$$\rho^2 = \Omega_1^2 + \Omega_2^2 \tag{6.8}$$

is a scalar chiral-invariant and it may be shown, owing to classical identities that all scalar invariants are functions of ρ [8].

7. NON LINEAR GENERALIZATION OF THE MONOPOLE THEORY

Let us add to the lagrangian density of the equation (3.9), (3.10) a function of ρ^2 :

$$L = \frac{1}{2}\bar{\psi}\gamma_\mu[\partial_\mu]\psi - \frac{g}{\hbar c}\bar{\psi}\gamma_\mu\gamma_5 B_\mu\psi + \frac{c}{4\hbar}\mathfrak{M}(\rho^2) \tag{7.1}$$

It is the most general lagrangian which is invariant under chiral transformations (3.11) provided that we add no differential terms.

The equation will be $\left\{m(\rho^2) = \frac{\partial\mathfrak{M}(\rho^2)}{\partial\rho^2}\right\}$ [8] :

$$\left\{\gamma_\mu(\partial_\mu - \frac{g}{\hbar c}\gamma_5 B_\mu) + \frac{m(\rho^2)c}{2\hbar}(\Omega_1 - i\Omega_2\gamma_5)\right\}\psi = 0 \tag{7.2}$$

Equivalently, using (3.18), we could write :

$$\left\{\gamma_\mu(\partial_\mu - \frac{g}{\hbar c}\gamma_5 B_\mu) - \frac{m(\rho^2)c}{2\hbar}(\bar{\psi}\gamma_\mu\psi)\gamma_\mu\right\}\psi = 0 \tag{7.3}$$

$$\left\{\gamma_\mu(\partial_\mu - \frac{g}{\hbar c}\gamma_5 B_\mu) + \frac{m(\rho^2)c}{2\hbar}(\bar{\psi}\gamma_\mu\gamma_5\psi)\gamma_\mu\gamma_5\right\}\psi = 0 \tag{7.4}$$

The equation (7.2) is P-invariant and CT-invariant (CT is the "weak" time reversal), but it is neither T nor C invariant. In terms of the 2-component spinors ξ and η, (7.2) becomes.

$$\left.\begin{aligned}&\frac{1}{c}\frac{\partial\xi}{\partial t} - \vec{s}.\vec{\nabla}\xi - i\,\frac{g}{\hbar c}(W + \vec{s}.\vec{B})\xi + i\,\frac{mc}{\hbar}(\eta^{+}\xi)\eta = 0,\\&\frac{1}{c}\frac{\partial\eta}{\partial t} + \vec{s}.\vec{\nabla}\eta + i\,\frac{g}{\hbar c}(W + \vec{s}.\vec{B})\eta + i\,\frac{mc}{\hbar}(\xi^{+}\eta)\xi = 0,\end{aligned}\right\} \tag{7.5}$$

where $m = m(\rho^2) = m(4|\xi^{+}\eta|^2)$.

An important consequence of the invariance of (7.2), or (7.5) under the chiral transformations (3.11) or (5.3) is that *separate conservation* (6.3) *of the chiral currents* X_μ and Y_μ still holds, despite the non linear coupling between the ξ and η equations.

An important case occurs when the non linear term cancels i.e. :

$$\xi^{+}\eta = \eta^{+}\xi = 0. \tag{7.6}$$

Hence, the system (7.5) reduces to the previous linear system (5.2). One can prove what follows :

(7.6) occurs only if :

a) $\eta = 0$: only a left monopole remains.

b) $\xi = 0$: only a right monopole remains.

c) $\xi = is_2\eta^*$: ξ and η constitute then a *pair monopole-antimonopole*. This is the most interesting case because one can show that $X_\mu = Y_\mu$, which involves $K_\mu = 0$ and the total magnetic current thus cancels. Furthermore, *all* the tensorial densities cancel, except the electric current density which is now : $J_\mu = 2eX_\mu$. But actually our monopole is supposed to be not electrically charged, J_μ finally cancels as the other densities. Consequently, although the wave functions and the chiral currents of such a solution do not vanish, all the observable quantities disappear. If we are surrounded by an ether made of such pairs, not any electro-magnetic phenomenon could allow us to observe it. Only another kind of field could break such a homogenity.

8. NON LINEAR PLANE WAVES. TACHYON STATES

It is very interesting to examine the plane solutions of the non linear equation (7.5) :

$$\xi = a\,e^{i(\omega t-\vec{k}.\vec{r})}, \qquad \eta = b\,e^{i(\omega' t-\vec{k}'.\vec{r})} \tag{8.1}$$

where a and b are constant spinors. The phases of ξ and η may be different, owing to the properties of the non linear terms. We find the system :

$$\left.\begin{aligned}&\left(\frac{\omega}{c} + \vec{s}.\vec{k}\right)a + \frac{mc}{\hbar}(b^{+}a)b = 0\\&\left(\frac{\omega'}{c} - \vec{s}.\vec{k}'\right)b + \frac{mc}{\hbar}(a^{+}b)a = 0.\end{aligned}\right\} \tag{8.2}$$

The compatibility condition provides us with the following dispersion relation [8] :

$$\left(\frac{\omega^2}{c} - k^2\right)\left(\frac{\omega'^2}{c^2} - k'^2\right) - 2\left(\frac{\omega\omega'}{c^2} - \vec{k}.\vec{k}'\right)\frac{M^2c^2}{\hbar^2} + \frac{M^4c^4}{\hbar^4} = 0, \tag{8.3}$$

$$M = m(4|a^+b|^2)|a^+b|. \tag{8.4}$$

Numerous types of plane waves are compatible with this relation, among which we find the following important types :

1) Bradyons, that is ordinary particles. For example :

$$\omega = \omega', \quad \vec{k} = \vec{k}', \qquad \frac{\omega^2}{c^2} = k^2 + \frac{M^2c^2}{\hbar^2}, \text{ or :}$$

$$\omega = \omega', \ |\vec{k}| = |\vec{k}'|, \quad \vec{k}\,\vec{k}' = 0, \quad \frac{\omega^2}{c^2} = \left(|k| + \frac{1}{\sqrt{2}}\frac{Mc}{\hbar}\right)^2 + \frac{1}{2}\frac{M^2c^2}{\hbar^2}.$$

2) Luxons, i.e. luminal particles as the monopole-antimonopole pairs already considered :

$$a = is_2b^*, \quad \frac{\omega^2}{c^2} = k^2, \quad \frac{\omega'^2}{c^2} = k'^2$$

3) Tachyons, i.e. supraluminal particles. For example :

$$\omega = -\omega', \quad \vec{k} = -\vec{k}', \quad \frac{\omega^2}{c^2} = k^2 - \frac{M^2c^2}{\hbar^2}.$$

It is noteworthy that, on the basis of pure relativistic arguments, the possibility of tachyon-monopoles was suggested for a long time [12], [13]. The problem of a tachyon-neutrino was also investigated [14],[15] and this is also related to the present theory because the neutrino is obviously a special case of our monopole (we shall examine this point more extensively in the last section. Let us add that the possible evidence of a tachyon related to a light monopole has been put forward quite recently in the interpretation of experiments involving the absorption of γ-rays by copper [16] and that the presence of light monopoles was also suspected in the observation of ferromagnetic aerosols (iron dust) irradiated by a laser beam [17].

9. CHIRAL GAUGE AND TWISTED SPACE

The chiral gauge and especially the chiral invariant (6.8), may be linked up to a torsion of the physical space, which is induced by the electromagnetic coupling of a monopole.

Let us consider a space with an affine connection, that is a space where covariant derivatives of a vector are defined [18] :

$$\nabla_k T^i = \partial_k T^i + \Gamma^i_{rk} T^r, \qquad \nabla_k T_i = \partial_k T_i - \Gamma^r_{ik}\, T_r \tag{9.1}$$

In a transformation of coordinates, the connection coefficients obey the law :

$$\Gamma^{i'}_{k'j'} = \Gamma^i_{kj} \frac{\partial x^{i'}}{\partial x^i} \frac{\partial x^k}{\partial x^{k'}} \frac{\partial x^j}{\partial x^{j'}} + \frac{\partial x^{i'}}{\partial x^i} \frac{\partial^2 x^i}{\partial x^{k'}\partial x^{j'}}. \tag{9.2}$$

Although the Γ do not behave like a tensor, they define two tensors :

The *curvature* :

$$-R^i_{qk\ell} = \frac{\partial \Gamma^i_{q\ell}}{\partial x^k} - \frac{\partial \Gamma^i_{qk}}{\partial x^\ell} + \Gamma^i_{pk}\,\Gamma^p_{q\ell} - \Gamma^i_{p\ell}\,\Gamma^p_{qk} \qquad (9.3)$$

and the *torsion* :

$$S^i_{[kj]} = \Gamma^i_{kj} - \Gamma^i_{jk} \qquad (9.4)$$

When the latter is ≠ 0, the space is said to be twisted for the following reason:

Consider the parallel transport of a tensor T along a direction :

$$\nabla_\xi T = \xi^k \nabla_k T = \frac{dx^k}{dt}\,\nabla_k T = 0 \qquad (9.5)$$

A transported vector depends on the curve along which it is transported (except in a euclidean space). Let us try to construct a geodesic parallelogram* going successively from a point P, to the points Q, R, S, T (Fig. 2).

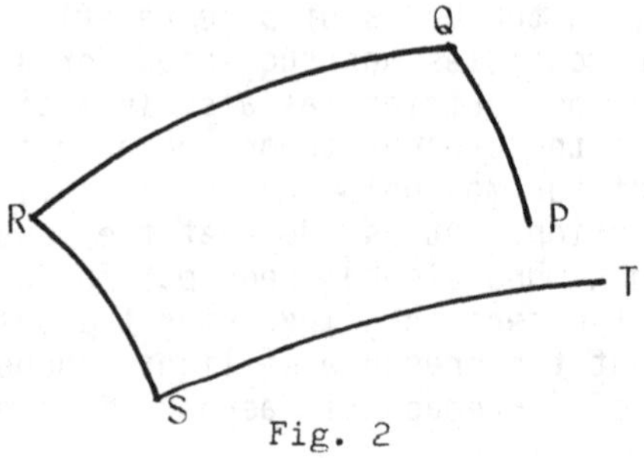

Fig. 2

A geodesic parallelogram generallly remains open

Except in the trivial case of a euclidean space the point T does not coïncides with P. Hence the parallelogram remains open and we have for infinitesimal dimensions :

$$x^k_P - x^k_T = S^k_{ij}\left(\frac{dx^i}{dt}\right)_T\left(\frac{dx^j}{dt}\right)_T dt^2 + dt^3 \text{ (termes depending on curvature).} \qquad (9.6)$$

(*)Remind that a geodesic line is enveloped by the parallel transport of its tangent.

The principal part of the gap between P and T thus depends on the torsion S^k_{ij} of the space.

Following an old work of Rodichev [19] we shall consider a space which is *flat* (the geodesics will thus be straight lines) but twisted. Hence :

$$\Gamma^i_{kj} + \Gamma^i_{jk} = 0, \quad \Gamma^i_{kj} - \Gamma^i_{jk} = S^i_{kj} \neq 0. \tag{9.6}$$

We can choose Γ completely antisymmetric :

$$\Gamma_{\lambda[\mu\nu]} = \phi_{[\lambda\mu\nu]}. \tag{9.7}$$

The spinor covariant derivative will be [19] :

$$\nabla_\mu\psi = \partial_\mu\psi - \frac{i}{4}\phi_{\mu\nu\lambda}\gamma_\nu\gamma_\lambda\psi$$

and we shall consider the lagrangian density :

$$L = \frac{1}{2}\{\bar{\psi}\gamma_\mu(\nabla_\mu\psi) - (\nabla_\mu\bar{\psi})\gamma_\mu\psi\}. \tag{9.9}$$

Introducing the dual tensor :

$$\Phi_\mu = \frac{i}{3!}\varepsilon_{\mu\nu\lambda\sigma}\phi_{\nu\lambda\sigma} \tag{9.10}$$

the lagrangian becomes :

$$L = \frac{1}{2}\{\bar{\psi}\gamma_\mu[\partial_\mu]\psi - \frac{1}{2}\Phi_\mu\bar{\psi}\gamma_\mu\gamma_5\psi\} \tag{9.11}$$

and the field equation is :

$$\gamma_\mu[\partial_\mu - \frac{1}{2}\Phi_\mu\gamma_5]\psi = 0 \tag{9.12}$$

This is exactly our equation (3.9), (3.10) provided that :

$$\Phi_\mu = \frac{2g}{\hbar c}B_\mu \tag{9.13}$$

(this Φ_μ is not to be confused with $\partial_\mu\Phi$ in (3.11), which has nothing to do with it !)

Therefore, we may assert that a monopole in an electromagnetic field "sees" a torsion in the space, which is defined as the dual tensor of the pseudo electromagnetic potential B_μ (up to the factor 2g/hc).

Then, with the help of (9.3) and (9.7), the total curvature of the space will be :

$$R = R^i_{\ell i \ell} = -6\,\Phi_\mu\Phi_\mu. \tag{9.14}$$

Following Rodichev once again, we write the following Einstein-like lagrangian :

$$L = \frac{1}{2}\{\bar{\psi}\gamma_\mu(\nabla_\mu\psi) - (\nabla_\mu\bar{\psi})\gamma_\mu\psi - bR\} \tag{9.15}$$

where ∇_μ is defined as in (9.12), R as in (9.14) and b is a constant.

The variation of L with respect to Φ_μ gives :

$$\Phi_\mu = \frac{1}{24b}\,\bar{\psi}\gamma_\mu\gamma_5\psi \tag{9.16}$$

and the variation with respect to ψ then leads to the equation :

$$\gamma_\mu\partial_\mu\psi - \frac{1}{48b}(\bar{\psi}\gamma_\mu\gamma_5\psi)\gamma_\mu\gamma_5\psi = 0 \tag{9.17}$$

which is a special case of (7.4) and thus of our equation (7.2). This equation was precedingly considered by H. Weyl [20] with a linear term of mass and later on by other authors (see bibliography in [8]). But none of these works has never related this equation to the magnetic monopole.

Using (9.14) and (9.16) and taking (3.16) and (3.17) into account, we find that :

$$R = -\frac{1}{4b}(\Omega_1^2 + \Omega_2^2). \tag{9.18}$$

We see that, in the lagrangian density (7.1), the chiral invariant ρ defined by (6.8) is the total curvature (up to a constant factor).

10. THE PROBLEM OF A MASSLESS MONOPOLE IN A COULOMBIAN ELECTRIC FIELD

In the equations (5.2) and according to the formulae (3.14), the potentials W and $\vec{B}$ are now defined by the conditions :

$$\vec{H} = 0, \quad W = 0, \quad \vec{E} = \mathrm{rot}\,\vec{B} = \frac{e\vec{r}}{r^3}\ . \tag{10.1}$$

Here arises Dirac's famous problem. Because the Stokes theorem the solution $\vec{B}$ cannot be at once continuous and univocal. Dirac abandoned continuity and considered solutions $\vec{B}$ with a discontinuous string [3] ; conversely, Wu and Yang abandoned univocity and considered the wave functions as sections, in terms of fiber bundle theory [21]. In both cases considerations on the observability of physical properties of the phase factor* :

$$e^{i\frac{e}{\hbar c}\oint A_\mu dx^\mu} \tag{10.2}$$

lead to the Dirac relation between electric and magnetic charges :

$$\frac{eg}{\hbar c} = \frac{n}{2} \quad (n : \text{integer}). \tag{10.3}$$

(*)The factor is written here as it were in the original papers, i.e. for a fixed monopole and a moving electric charge, whereas only the converse has a physical meaning in the case of a massless monopole. But the problem remains the same one for the pseudo-potentials B_μ.

This relation will be find here in a different-and in author's opinion simpler-way. It will be physically interpreted by an analogy with the problem of symmetrical top.

At first we shall accept, with Dirac, a discontinuous solution of (10.1) :

$$\vec{B} = e\mathcal{B}, \quad \mathcal{B}_x = \frac{1}{r}\,\frac{yz}{x^2+y^2}, \quad \mathcal{B}_y = \frac{1}{r}\,\frac{-xz}{x^2+y^2}, \quad \mathcal{B}_z = W = 0. \tag{10.4}$$

Hence, in polar coordinates, $x = r\sin\theta\cos\phi$, $y = r\sin\theta\sin\phi$, $z = r\cos\theta$:

$$\mathcal{B}_x = \frac{1}{r}\,\frac{\sin\phi}{\mathrm{tg}\theta}, \quad \mathcal{B} = \frac{1}{r}\,\frac{-\cos\phi}{\mathrm{tg}\theta}, \quad \mathcal{B}_z = W = 0. \tag{10.5}$$

This solution has actually the variance of an axial vector and is more convenient for the calculations than the one chosen by Dirac. The latter differs from the former by a gauge transformation. More generally, the string is gauge dependent : one can, in principle, choose any curve without laps, going to infinity [22]. On another side, such a singular line is absolutely irrelevant to the symmetry of the physical system. For these reasons, the string is physically meaningless and must be excluded from the solution. We shall formulate this condition in a simple way : *we shall postulate that spinors* ξ *or* η *in (5.2), with* $\vec{B}$, W *given by (10.4) must be continuous functions of the angles*. We need not any phase condition to be added to this condition of continuity which was already shown to be sufficient to integrate the equation of a quantum symmetrical top [23].

11. ANGULAR MOMENTUM. ANGULAR FUNCTIONS

With the conditions (10.4), the L and R equations admit the constant of motion

$$\vec{J} = \hbar(\vec{\Lambda} + \frac{1}{2}\vec{s}) \tag{11.1}$$

where $\vec{\Lambda}$ is respectively equal (for L and R) to :

$$\vec{\Lambda} = \vec{r} \times (-i\vec{\nabla} \pm D\mathcal{B}) \pm D\,\frac{\vec{r}}{r}\,. \tag{11.2}$$

The *plus* sign is for L and the *minus* sign for R. We shall confine ourselves to L and thus keep only the *plus* sign. $\vec{J}$ is the total angular momentum. $\hbar\vec{\Lambda}$ corresponds to the Poincaré first integral [2] : it is the sum of the orbital momentum and of the angular momentum of the electromagneti field [24][25]. We have :

$$[J_1, J_2] \underset{0}{=} i\hbar\, J_3, \qquad [\Lambda_1, \Lambda_2] \underset{0}{=} i\,\Lambda_3. \tag{11.3}$$

The coefficient D is the Dirac number :

$$D = \frac{eg}{\hbar c}\,. \tag{11.4}$$

In terms of polar angles, we find :

$$\left.\begin{aligned} \Lambda^{+} &= \Lambda_1 + \Lambda_2 = e^{i\phi}\left\{i\cot\theta\frac{\partial}{\partial\phi} + \frac{\partial}{\partial\theta} + \frac{D}{\sin\theta}\right\} \\ \Lambda^{-} &= \Lambda_1 - \Lambda_2 = e^{-i\phi}\left\{i\cot\theta\frac{\partial}{\partial\phi} - \frac{\partial}{\partial\theta} + \frac{D}{\sin\theta}\right\} \\ \Lambda_3 &= -i\frac{\partial}{\partial\phi}\,. \end{aligned}\right\} \qquad (11.5)$$

These expressions are simpler than those which are ordinary found, owing to the choice (10.5) for the gauge of $\vec{B}$. Now the clue remark is the following. Let us consider the infinitesimal operators R_1, R_2, R_3 of the rotation group expressed in the fixed axis (instead of the rotating frame which is usually chosen). We have [23][26] :

$$\left.\begin{aligned} R^{+} &= R_1 + iR_2 = e^{i\phi}\left\{i\cot\theta\frac{\partial}{\partial\phi} + \frac{\partial}{\partial\theta} - \frac{i}{\sin\theta}\frac{\partial}{\partial\chi}\right\} \\ R^{-} &= R_1 - iR_2 = e^{-i\phi}\left\{i\cot\theta\frac{\partial}{\partial\phi} - \frac{\partial}{\partial\theta} - \frac{i}{\sin\theta}\frac{\partial}{\partial\chi}\right\} \\ R_3 &= -i\frac{\partial}{\partial\phi}\,, \end{aligned}\right\} \qquad (11.6)$$

These operators are also, up to a constant factor, the components of the angular momentum of a *symmetrical top* [23]. Comparing (11.5) and (11.6), we see that, if $Z(\theta,\phi)$ is an eigenfunction of $\vec{\Lambda}^2$ and Λ_3, i.e. a *monopole harmonic*, and $\mathcal{D}(\theta,\phi,\chi)$ an eigenfunction of $\vec{\Lambda}^2$ and Λ_3, i.e. a matrix element of a unitary representation of the rotation group (the so called generalized spherical harmonics) we have the simple relation :

$$Z(\theta,\phi) = e^{-iD\chi}\mathcal{D}(\theta,\phi,\chi) = \mathcal{D}(\theta,\phi,0). \qquad (11.7)$$

As the $\mathcal{D}$ functions are known, we have nothing more to do to find the monopole harmonics. The generalized spherical harmonics are :

$$\mathcal{D}_j^{m'm}(\theta,\phi,\chi) = e^{im'\chi}e^{im\phi}d_j^{m',m}(\theta) \qquad (11.8)$$

$$d_j^{m',m}(\theta) = N(1-u)^{-\frac{m-m'}{2}}(1+u)^{-\frac{m+m'}{2}}\left(\frac{d}{du}\right)^{j-m}\left[(j-u)^{j-m'}(j+u)^{j+m'}\right] \qquad (11.9)$$

$$u = \cos\theta, \quad N = \frac{(-1)^{j-m}i^{m-m'}}{2^j}\sqrt{\frac{(j+m)!(2j+1)}{(j-m)!(j-m')!(j+m')!}} \qquad (11.10)$$

where the $(2j+1)$ factor in the root normalizes $\mathcal{D}_j^{m'm}$ to unity. The values of j, m', m are :

$$j = \frac{n}{2}\ (n = \text{integer}),\ m',m = -j,-j+1,\ldots j-1,j. \qquad (11.11)$$

Although it may seem to be not evident on the formula (11.9) $D_j^{m',m}(\theta,\phi,\chi)$ is a *continuous* function of the angles θ, ϕ, χ *for all values* (11.11) of j, m', m. This condition was assumed in the integration of the differential equation $\vec{\Lambda}^2 D = \lambda D$ [23] and let us remind that $D_j^{m'm}$ is a *polynomial function* of the Cayley-Klein parameters [23][27][28]:

$$a = \cos\frac{\theta}{2}\, e^{i(\chi+\phi)/2}, \qquad b = i\, \sin\frac{\theta}{2}\, e^{i(\chi-\phi)/2}. \tag{11.12}$$

Now, introducing (11.8) in (11.7), we readily find the monopole harmonics :

$$Z_j^{m',m}(\theta,\phi) = D_j^{m',m}(\theta,\phi,0) \tag{11.13}$$

and we see that the continuity condition which gave us the expression of $D_j^{m',m}$, now implies that in the formulae (11.5) and hence in the angular momentum (11.1), and finally in the equations (5.2) themselves, the number D must be equal to m'. So that, in accordance with (11.4) and (11.11):

$$D = \frac{eg}{\hbar c} = m' = \frac{n}{2} \quad (n = \text{integer}) \tag{11.14}$$

This is Dirac's condition. It is a little laborious but not difficult to find now the wave functions of a massless monopole in a coulombian electric field. The solution was given in our previous papers [7][8].

12. SOME RESULTS ON THE QUANTUM TOP

The kinetic energy of a classical top is :

$$T = \frac{1}{2}\left(\frac{L_1^2}{I_1} + \frac{L_2^2}{I_2} + \frac{L_3^2}{I_3}\right) \tag{12.1}$$

where I_k are the inertia momenta and L_k the projections of the *angular momentum* on the principal axis of the top i.e. in a *rotating frame* linked to the top. In quantum mechanics the corresponding operators are :

$$L_k = \hbar\, M_k \quad (k=1,2,3) \tag{12.2}$$

where the M_k are the *infinitesimal operators of the rotation group*, i.e. exactly the same operators as the R_k which were defined in (11.6) for the magnetic monopole. The only difference is that in the theory of the monopole these operators are expressed in the laboratory, i.e. a fixed frame instead of a rotating one. M_k and R_k are related by the formula [23] :

$$M_k(\theta,\phi,\chi) = R_k(\theta,\pi-\chi,\pi-\phi). \tag{12.3}$$

The M_k and R_k commute and therefore there is a system of eigenfunctions which is common to $\vec{M}^2$, M_3, $\vec{R}^2$, R_3 : it is of course the $D_j^{m'm}(\theta,\phi,\chi)$ defined in the preceding section.

Using (12.1) and (12.2), we can write the Schrödinger equation of the quantum top :

$$i\hbar \frac{\partial\psi}{\partial t} = \frac{\hbar^2}{2}\left(\frac{M_1^2}{I_1} + \frac{M_2^2}{I_2} + \frac{M_3^2}{I_3}\right) \tag{12.4}$$

which becomes, in the symmetric case $I_1 = I_2 = I$, $I_3 = K$:

$$i\hbar \frac{\partial\psi}{\partial t} = \frac{\hbar^2}{2}\left(\frac{\vec{M}^2}{I} + \left(\frac{1}{K} - \frac{1}{I}\right)M_3^2\right)\psi. \tag{12.5}$$

The structure of the hamiltonian shows that the stationary (*continuous*) wave functions are obviously $D_j^{m',m}(\theta,\phi,\chi)$. The fact that the angular parts of the wave functions of a monopole in a central electric field* are stationary functions of the quantum symmetrical top was observed for a long time [29] but it seems to have never been clearly explained. We see that the explanation simply lies in the identity of symmetry laws which involves a trivial correspondance (12.3) between the angular momenta operators of the monopole and the top. Finally, both systems differ only in two specificities : 1) The radial properties are different because the top is a rigid body while the system monopole-electric charge is not. 2) The last system has only one dimension because both particles are supposed dimensionless, which implies that the proper rotation angle may be chosen arbitrarily in the wave functions.

13. THE INTERPRETATION OF THE DIRAC RELATION

As long as the Dirac relation (10.3) was deduced from a condition of uniformity, it could hardly be considered as something more than an algebraic formula. But now, the relation is deduced from (11.14) and it becomes easy to give it a simple physical meaning.

Let us go back to the expressions (11.1) and (11.2) of $\vec{J}$ and of the orbital part $\hbar\vec{\Lambda}$ which is itself the sum of the orbital momentum of the monopole :

$$\hbar\, \vec{r} \times \vec{v} = \hbar\, \vec{r} \times (-i\vec{\nabla} + D\vec{B}) \tag{13.1}$$

and of the angular momentum of the field [24][25]

$$\hbar\, D\, \frac{\vec{r}}{r} = \frac{eg}{c}\,\frac{\vec{r}}{r}. \tag{13.2}$$

*Or conversely of an electric charge in a central magnetic field, which is the equivalent more usual problem.

These two components are trivially *orthogonal*. Then, we know from the theory of the symmetrical top, and it is easily verified using (15.5), (12.3) and (11.8), that $\hbar j(j+1)$ are the eigenvalues of $\hbar\vec{\Lambda}^2$, $\hbar m$ the eigenvalues of the projection of $\hbar\Lambda$ on the z axis and $\hbar m'$ *the eigenvalues of the projection of* $\hbar\vec{\Lambda}$ *on the symmetry axis of the system*. But in the present case we know this axis : it the one which joins the electric center to the monopole, i.e. the vector $\vec{r}$ itself and (13.2) shows that the projection of $\hbar\vec{\Lambda}$ on the symmetry axis has the amplitude $\frac{eg}{c}$. Moreover, the orthogonality of the components (13.1) and (13.2) means that this projection of $\hbar\vec{\Lambda}$ is nothing else than the angular momentum of the field.

Therefore the Dirac relation in the form (11.14) simply expresses the quantization law :

$$\frac{eg}{c} = \hbar m' = \hbar \frac{n}{2} \qquad (n = \text{integer}) \tag{13.3}$$

of the projection of the angular momentum $\hbar\vec{\Lambda}$ on the symmetry axis, or of the angular momentum of the electromagnetic field.

Let us recall that this is not a postulate but a *deduction from the postulate of continuity of the angular wave functions which is equivalent to the isotropy of the system.*

It is of interest to note that the continuity of angular wave functions implies that numbers j, m and m' must be *integers or half integers*. The possibility of half integers is independent of the fact that the monopole is assumed to be a spin 1/2 particle, because to get the total angular momentum, we must still add this spin 1/2 to j, so that the total momentum itself may be an integer or a half integer, according to the charge of the monopole.

14. ANALOGIES BETWEEN A MONOPOLE AND A SYMMETRICAL TOP IN CLASSICAL MECHANICS. THE BIRKELAND-POINCARE EFFECT

We shall now investigate the geometrical-optics limit of the monopole. It will be very useful in order to confirm that it is truely a monopole, and then to understand some geometrical properties and to show *a contrario* how some essential properties disappear at the classical limit.

Let us introduce in the equation (5.2) (L) the following decomposition of the spinor ξ :

$$\xi = a\, e^{\frac{i}{\hbar} S} , \tag{14.1}$$

puting a common phase S in evidence.

At the $\hbar = 0$ order, i.e. the zero W.K.B. limit, we find :

$$\left\{\frac{1}{c}\left(\frac{\partial S}{\partial t} - gW\right) - \left(\vec{\nabla} S + \frac{g}{c}\vec{B}\right).\vec{s}\right\} a = 0 \tag{14.2}$$

and in order to have $a \neq 0$, we must satisfy the condition :

$$\frac{1}{c^2}\left(\frac{\partial S}{\partial t} - gW\right)^2 - \left(\vec{\nabla} S + \frac{g}{c}\vec{B}\right)^2 = 0. \tag{14.3}$$

This is a Hamilton-Jacobi condition with $m_0 = 0$. Let us now define :

$$\left.\begin{aligned} &\text{- the energy : } E = -\frac{\partial S}{\partial t} + gW, \\ &\text{- the linear momentum : } \vec{p} = \vec{\nabla} S + \frac{g}{c}\vec{B}, \\ &\text{- the Lagrange momentum : } \vec{P} = \vec{\nabla} S. \end{aligned}\right\} \qquad (14.4)$$

Using (14.3), the Hamilton function is thus :

$$H = c\sqrt{(\vec{P} + \frac{g}{c}\vec{B})^2} - gW. \qquad (14.5)$$

Hence, owing to some simple algebra, the equation of motion will be :

$$\frac{d\vec{p}}{dt} = g\left(\vec{\nabla} W + \frac{1}{c}\frac{\partial \vec{B}}{\partial t}\right) - \frac{g}{c}(\vec{v} \times \text{rot}\,\vec{B}) \qquad (14.6)$$

where the luminal velocity $\vec{v}$ is defined as :

$$\vec{v} = \frac{c^2}{E}\vec{p}. \qquad (14.7)$$

Introducing in (14.6) the expressions (3.14) of the fields, we immediately find the classical equation of motion (2.3) with the Lorentz force for a monopole.

It is interesting, in an analogous manner, to introduce in the second equation (5.2), that is (R), a spinor η similar to (14.1). Let us write it, at first, with the same phase S :

$$\eta = b\, e^{\frac{i}{\hbar}S}. \qquad (14.8)$$

This gives, instead of (14.3) :

$$\frac{1}{c}\left(\frac{\partial S}{\partial t} + gW\right)^2 - (\vec{\nabla} S - \frac{g}{c}\vec{B})^2 = 0, \qquad (14.9)$$

which means that, at the classical limit, the antimonopole has a charge which is opposite to the charge of the monopole (remember : in quantum mechanics, only the helicities were opposite, but classical mechanics does not know helicity). But we may, in place of (14.8), put :

$$\eta = b\, e^{-\frac{i}{\hbar}S}, \qquad (14.10)$$

which seems to be better because the charge conjugation implies opposite phases. But infortunately, we find (14.9) again, i.e. the same result as with the spinor (14.8). Nevertheless, in order to make use of the opposite phase in (14.10), we may replace (14.4) by the definitions :

$$E = \frac{\partial S}{\partial t} + gW, \quad \vec{p} = -\vec{\nabla} S + \frac{g}{c}\vec{B}, \quad \vec{P} = -\vec{\nabla} S. \qquad (14.11)$$

But if we introduce these definitions in (14.9), we find (14.5) and (14.6) and there is no more difference between a monopole and an antimonopole !

There is no good solution because the essential quantum property of a monopole is its *helicity*, and the latter is lost at the classical limit : *a monopole can never be considered as a classical particule.*

Now, a special case of the equation (14.6) is the one when W = 0 and $\vec{B}$ is given by (10.4), i.e. the coulombian case (10.1). We get :

$$\frac{d\vec{p}}{dt} = -\frac{eg}{c}\vec{v}\times\frac{\vec{r}}{r^3} \qquad (\vec{p} = \frac{E}{c^2}\vec{v}), \tag{14.12}$$

but, as $\vec{A} = 0$, we have E = const, hence :

$$\frac{d^2\vec{r}}{dt^2} = \frac{\lambda}{r^3}\vec{r}\times\frac{d\vec{r}}{dt} \qquad (\lambda = \frac{egc}{E}) \tag{14.13}$$

This is the Poincaré equation [2] which he found, not in order to describe a monopole in a coulombian field, but conversely, to describe the motion of an electron beam (cathodic rays) in the vicinity of a pole of a rectilinear magnet. The experiment shows, in accordance with (14.13) and thus in accordance with the classical limit (14.6) of the equation (5.2), a focusing of the electron beam : it is the Birkeland-Poincaré effect [30] which definitely shows that the equations (5.2) have really a clear connection with the monopole problem.

15. THE ANALOGY BETWEEN A MONOPOLE IN A CENTRAL FIELD AND A SYMMETRICAL TOP IN CLASSICAL MECHANICS

From (14.13), it is easy to find the following constants of motion :

$$|\dot{\vec{r}}| = \text{Const}, \qquad |\vec{L}| = \text{Const}, \qquad \vec{\Lambda} = \text{Const} \tag{15.1}$$

where :

$$\vec{L} = \vec{r}\times\dot{\vec{r}}, \qquad \vec{\Lambda} = \vec{r}\times\dot{\vec{r}} + \lambda\frac{\vec{r}}{r}. \tag{15.2}$$

$\vec{\Lambda}$ is the Poincaré integral. (11.2) is the quantum expression of this first integral.

We have :

$$\vec{L}.\vec{r} = 0, \qquad \vec{\Lambda}.\frac{\vec{r}}{r} = \lambda = \text{Const}, \tag{15.3}$$

so that we have the Fig. 3.

$\vec{\Lambda}$ has a fixed direction in space (it is a first integral) and owing to (15.3) the angle Θ' is fixed. Therefore, $\vec{r}$ rotates around $\vec{\Lambda}$ on a cone with a vertex angle Θ' : it is the Poincaré cone [2]. Note that $\vec{r}$ is the radial vector going from the electric charge to the monopole. Then, (14.13) implies :

$$\ddot{\vec{r}}.\vec{r} = \ddot{\vec{r}}.\dot{\vec{r}} = 0. \tag{15.4}$$

This means that the monopole trajectory is a geodesic line on the cone. But we may look at this figure in another way an only consider the

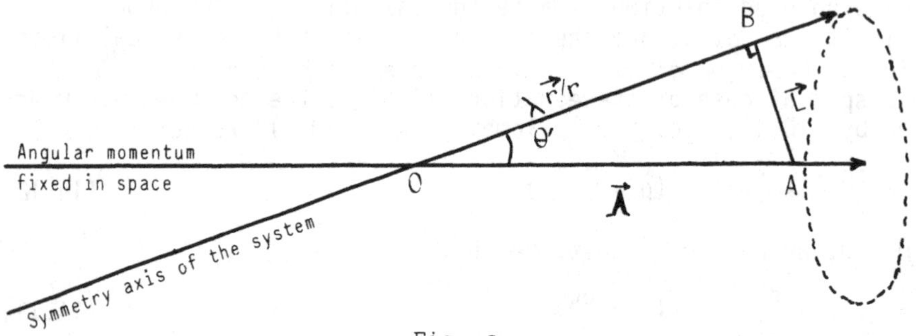

Fig. 3

The generation of the cone of Poincaré and Poinsot and the decomposition of the angular momentum of a monopole in a central electric field.

vector $\vec{r}$ as defining the *symmetry axis* of the physical system formed by the electric charge and the monopole and so we see that the envelope of this axis is the Poincaré cone. But we already know such a motion. It is exactly the one of a symmetrical top, the axis of which passes through a fixed point.

It is known that the symmetry axis of the cone rotates on a revolution cone around the angular momentum vector which remains fixed in space : this is a particular case of the Poinsot construction [31]. The Poincaré cone and the Poinsot cone are therefore the same ones and are only defined by the rotation group invariance properties.

Finally, let us remember that in quantum mechanics (for a monopole in a coulombian electric field just as well for a symmetrical top) the total angular momentum is defined by the quantum number j, while the number m' is the projection of the angular momentum on the symmetry axis. At the geometrical optics limit, the vertex angle of the cone is thus defined by [23] :

$$\cos \theta' = \frac{m'}{j} . \qquad (15.5)$$

16. GENERAL CONCLUSION AND EXPERIMENTAL SUGGESTIONS. IS THE MONOPOLE AN EXCITED STATE OF THE NEUTRINO ?

First of all it is necessary to underline that the symmetry between electricity and magnetism is not so strong as it could seem when one "symmetrizes" both groups of Maxwell's equations by additional terms corresponding to magnetic charges and magnetic currents.

Our equation shows that there are important differences between electric charges and magnetic poles (in the mass terms, charge conjugation, chiral properties and so on), which are essentially due to the differences between the symmetries of electric and magnetic fields.

An important point is that, although quantum theory fundamentally confirms the symmetry laws enounced by Pierre Curie and admits the Poincaré equation as a classical limit, it may be said that the *classical* theory of magnetic poles does not exist because the essential feature of this particle lies in the helicity of monopoles, which is a wave property excluded from classical mecanics.

Of course (and despite some experimental expectations) the problem of the existence of monopoles remains open, but it seems difficult to be indifferent to the fact enounced in the present paper that Dirac's equation has *two and only two gauge laws* and that one of them corresponds to an electric charge and the other one to a magnetic monopole. The miraculous heuristic qualities of Dirac's equation allows us to hope than such a property is not an illusion.

In connection with the observability of monopoles, another difference with electrons must be pointed out : *the conservation of magnetism is certainly not so strong and so universal as the conservation of electricity*. The reason is that the chiral gauge invariance, which involves the conservation of magnetic charges, is obviously subject to violations, while a violation of the phase invariance which is responsible for the conservation of electric charges is absolutely precluded in quantum mechanics. Therefore, it is possible that monopoles are created in many cases but unoberved until now for different reasons among which a possible one could be the very high values of the elementary magnetic charges predicted by the Dirac relation. The latter can be written indeed :

$$g = ng_0 \ ; \ n=0,1,2,3,\ldots \ ; \quad g_0 = \frac{e}{2\alpha} \ ; \ \alpha = \frac{e^2}{\hbar c}, \qquad (16.1)$$

which means that $g_0 = 68{,}5\ e$. This implies a strong ionization ability (or at least important interactions) in a condensed medium and therefore a big loss of impulse and energy of a monopole. But, since our monopole is massless (or very light), that does not signifie a braking of the motion, but a series of deviations and probably a chaotic trajectory wound on itself, which may be not so easily observable.

As a last conclusion we would like to make the following suggestion. It was already noted that the equations (5.2) of the monopole reduce to the equations of the left and of the right neutrino when $g = 0$. It could then happen that the *neutrino* is only a kind of low level or a *ground state of the monopole* and corresponds to the case $n = 0$ in the formula (16.1) giving the magnetic charge. Since the monopole has exactly the chiral properties of a neutrino, it could in principle be produced in the *weak reactions*. For instance, a reaction like :

$$p + p \to D + e^+ + \nu_e, \qquad (16.2)$$

could be generalized as :

$$p + p \to D + e^+ + m, \qquad (16.3)$$

where m would be a massless monopole obeing our equation.

Such a reaction could constitute a certain proportion of weak reactions and the difference with the ordinary reaction (16.2) could be observed owing to the Bremsstrahlung and the Čerenkov effect. The latter must exist in refringent media because the monopole has the velocity of light. It is easy to show that the Čerenkov effect is not the same as the one of an electron because the field component in the direction of motion will be a *magnetic component* instead of the electric component which appears in the case of an electron.

The generation of monopoles in weak interactionscould play a role in the *solar activity*, in particular in the *magnetic properties of the sunspots* which could be due to sources of *monopole flows*. A production of monopoles in a part of weak interactions, could also explain the well known *lack of solar neutrinos registred on the earth*. These monopoles would undergo an important energy loss and strong disturbances in the sun atmosphere, so that they could be unable to reach the earth. Therefore, *only a part of these monopoles could be registred* : the ones which are in the charge ground-state, i.e. the ones which have a magnetic charge g = 0, namely the "true" neutrinos

REFERENCES

[1] P. Curie, J. Phys. 3e Série, 3, 415, 1894.

[2] H. Poincaré, Comptes rendus, 123, 530, 1896.

[3] P.A.M. Dirac, Proc. Roy. Soc. A 133, 60, 1931.

[4] P.A.M. Dirac, Phys. Rev. 74, 817, 1948.

[5] P.A.M. Dirac, Directions in Physics, J. Wiley & Sons, N.Y. 1978.

[6] A. Sommerfeld, Elektrodynamik, Akad. Verlag, Leipzig, 1949.

[7] G. Lochak, Annales Fond. L. de Broglie, 8, 345, 1983 (I), 9, 1, 1984 (II).

[8] G. Lochak, Int. J. Theor. Phys. 24, 1019, 1985.

[9] N. Cabibbo, G. Ferrari, Nuovo Cim. 23, 1147, 1962.

[10] A. Salam, Phys. Letters 22, 683, 1966.

[11] T. Takabayasi, Prog. Theor. Phys. Suppl. 4, 1957

[12] R. Mignani, E. Recami, Nuovo Cim. 30A, 533, 1975.

[13] E. Recami, R. Mignani, Physics Letters, 62B, 41, 1976.

[14] A. Chodos, A. Hauser, A. Kostelecky, Physics Letters, 150B, 431, 1985.

[15] E. Recami, W. Rodriges, in "Progress in Particle and Nuclear Physics", vol. 15 A. Faessler ed., Pergamon Press, 1985.

[16] J. Steyaert, Université Catbolique de Louvain, Belgium (private communication), 1986.

[17] V. Mikhailov, J. Phys. A Math. Gen. 18, L 903, 1985.

[18] D. Doubrovine, S. Novikov, A. Fomenko, Géométrie Contemporaine, Mir, Moscow, 1982.

[19] V. Rodichev, Soviet Physics JETP, 13, 1029, 1961.

[20] H. Weyl, Phys. Rev., 77, 699, 1950.

[21] T.T. Wu, C.N. Yang, Phys. Rev. D12, 3845, 1975.

[22] A. Frenkel, P. Hrasho, Ann. of Phys. (U.S.A.), 105, 288, 1977.

[23] G. Lochak, Cahiers de Physique, 13, 41, 1959.

[24] J.J. Thomson, Elements of the mathematical theory of Electricity and Magnetism, Camb. Univ. Press, Oxford, 1904.

[25] A.S. Goldhaber, Phys. Rev. B140, 1407, 1965.

[26] I.M. Gelfand, R.A. Minlos, Z.Y. Shapiro, Representations of the rotation and Lorentz groups and their applications, Pergamon, N.Y., 1963.

[27] E. Cartan, Leçons sur la théorie des spineurs, Hermann, Paris, 1938.

[28] E. Cartan, The Theory of Spinors, Dover, N.Y.

[29] I. Tamm, Zeitschrift für Physik, 71, 141, 1986.

[30] M. Birkeland, Archives des Sciences Physiques et Naturelles de Genève, 1, June 1896.

[31] H. Goldstein, Classical Mechanics, 2nd ed., Addison-Wesley, 1980.

PART II

STOCHASTIC CONTROL AND QUANTUM MECHANICS

NOTE ON CLOSED-LOOP CONTROLLED RANDOM PROCESSES IN QUANTUM MECHANICS

Augustin Blaquière
Université de Paris VII, Paris, France

1. INTRODUCTION.

The idea according to which the motion of a quantum particle can be modelized by a stochastic process appears more or less explicitely in the work of Louis de Broglie [1]. Other investigations in that direction were carried on from time to time [2]-[4], and their number rapidly increased since the early and mid sixties, in connection with a stream of renewed interest in the conceptual foundations of quantum mechanics. Also in the early and mid sixties the impact of optimal control and differential games theories on the scientific community was intense. This phenomenon, apparently, passed unheeded in Physics, as most of the methods newly introduced did not enter the curriculum of physicists. Nevertheless, it gave rise in 1966 to a new approach of quantum mechanics based on the concept of closed-loop controlled random process (Blaquière [5] [6]). This approach was followed, in the same year, by one based on the theory of Markov processes (Nelson [7]). The latter had a large audience among the physicists, to such an extent that it almost occulted the former.

However we feel that Nelson's stochastic approach may have introduced some confusion among the physicists by letting them believe that the motion of a quantum particle can be modelized by a Markov process. We will see that, in one of the simplest cases, a Markovian model does not fit with experimental results.

On the other hand a closed-loop controlled random process provides a nice model for the motion of a quantum particle, in agreement with experience.

As this question apparently deserves some attention, we shall devote some space to it in the present chapter.

2. NOTATIONS.

Let Ω be the set of all functions defined on $[t_o,t_1]$ with range in a 3-dimensional Euclidean space E. The members ω of Ω,

$$\omega : \begin{cases} [t_o,t_1] \to E \\ t \mapsto \omega(t) \end{cases}$$

are *sample paths*.

For $t \in [t_o,t_1]$, function x_t

$$x_t : \begin{cases} \Omega \to E \\ \omega \mapsto x_t(\omega), \qquad x_t(\omega) \triangleq \omega(t) \end{cases}$$

is a *random variable*.

For $\omega \in \Omega$, $x_t(\omega)$ is the position at time $t \in [t_o,t_1]$ of a point in E, moving randomly, as time evolves. The set $\{x_t(\omega) : t \in [t_o,t_1]\}$ is its *trajectory*.

We shall also make use of the following notation, namely

$\mathcal{B}$, σ-algebra (or Borel field) on E generated by the class $\mathcal{C}$ of open subsets of E ;

$\mathcal{M}_t^o$, $t \in [t_o,t_1]$, σ-algebra on Ω generated by the sets

$\{\omega : \omega \in \Omega, \quad x_t(\omega) \in B\}, \quad B \in \mathcal{B}$.

More concisely we shall write

$\{x_t \in B\}, \quad B \in \mathcal{B}$.

$\mathcal{M}$, σ-algebra on Ω generated by the sets

$\{x_t \in B\}, \quad B \in \mathcal{B}, \quad t \in [t_o,t_1]$.

$\mathcal{M}^s$, σ-algebra on Ω generated by the sets

$\{x_t \in B\}, \quad B \in \mathcal{B}, \quad t \in [s,t_1] \subseteq [t_o,t_1]$.

Indeed, we have $\mathcal{M} \supset \mathcal{M}_t^o$ and $\mathcal{M} \supseteq \mathcal{M}^s$.

3. FIRST DEFINITION OF A MARKOV PROCESS.

Let us provisionally introduce

Definition 1. A Markov process on $[t_o,t_1]$ is a triplet $(E,\mathcal{B},Q)$, where Q is a *transition probability function* relative to $[t_o,t_1]$.

By transition probability function relative to $[t_o,t_1]$, we mean a function

$$Q : \begin{cases} [t_o,t_1]\times E\times[t_o,t_1]\times\mathcal{B} \to R_+ \\ (t,x,u,B) \mapsto Q(t,x;u,B) \end{cases}$$

where R_+ denotes the set of non-negative numbers, such that

(i) for $t_o \leq t \leq u \leq t_1$ and $B \in \mathcal{B}$, Q is $\mathcal{B}$-measurable with respect to x ; and

(ii) for $t_o \leq t \leq u \leq t_1$ and $x \in E$, Q is a probability measure on $\mathcal{B}$; and

(iii) for $t_o \leq t \leq t_1$ and $x \in E$, $Q(t,x;t,E-x) = 0$; and

(iv) for $t_o \leq s \leq t \leq u \leq t_1$ and $B \in \mathcal{B}$,

$$Q(s,x;u,B) = \int_E Q(t,y;u,B)Q(s,x;t,dy) .$$

We shall denote by Q_s a transition probability function relative to $[s,t_1] \subseteq [t_o,t_1]$.

Dynkin (8) starts with a more general definition, then shows how a Markov process in a σ-compact measurable topological space can be constructed starting from a complete transition probability function ; that is, one satisfying (i)-(iv) above.

4. RANDOM PROCESS.

Otherwise, one finds in the literature (see Kunio Yasue (9) for instance) the definition of a *random process*.

Definition 2. A random process is a triplet (Ω, M, P) where P is a probability measure on M.

For $B \in \mathcal{B}$ and $t \in [t_o,t_1]$, let

$$\mathrm{Prob}(B,t) \triangleq P\{x_t \in B\} .$$

The function

$$\mathrm{Prob}(\cdot,t) : \begin{cases} \mathcal{B} \to R_+ \\ B \mapsto \mathrm{Prob}(B,t) \end{cases}$$

is a probability measure on $\mathcal{B}$. It is called the *position probability function* of the process at time t.

5. SECOND DEFINITION OF A MARKOV PROCESS.

For given transition probability function Q, and $s \in [t_o,t_1]$, and $x \in E$, one can prove as in Dynkin that there exists a unique probability measure $P_{s,x}$ on M^s such that for all $s_1, s_2,\ldots,s_n$ with $s \leqslant s_1 \leqslant s_2 \leqslant \ldots s_n \leqslant t_1$, and for all $B_1, B_2,\ldots B_n \in \mathcal{B}$

$$P_{s,x}\{ x_{s_1} \in B_1,\ldots x_{s_n} \in B_n \} = Q(s,x;s_1,B_1,\ldots s_n,B_n)$$

where

$$Q(s,x;s_1,B_1,\dots s_n,B_n) \triangleq$$

$$\int_{B_1}\dots\int_{B_{n-1}} Q(s,x;s_1,dy_1)Q(s_1,y_1;s_2,dy_2)\dots Q(s_{n-1},y_{n-1};s_n,B_n) \ .$$

Let

$$Q(s,x;A) \triangleq P_{s,x}(A) \quad \text{for} \quad A \in \mathcal{M}^s \ .$$

Now we have

Definition 1'. A Markov process in measurable space $(E,\mathcal{B})$, with transition probability function Q is the *family* of all random processes $(\Omega,\mathcal{M}^s,P_{s,\mu})$, where $P_{s,\mu}$ is the probability measure on $\mathcal{M}^s$ defined by

$$P_{s,\mu}(A) \triangleq \int_E Q(s,x;A)\mu(dx)$$

for $A \in \mathcal{M}^s$, $s \in [t_o,t_1]$, μ probability measure on $\mathcal{B}$.

In particular we have

$$P_{s,\mu}\{x_u \in B\} = \int_E Q(s,x;u,B)\mu(dx) \tag{1}$$

for $u \in [s,t_1]$ and $B \in \mathcal{B}$.

μ is called the *initial state* (at time s) of the random process $(\Omega,\mathcal{M}^s,P_{s,\mu})$.

Among the initial states is the one, say μ_x, such that x is given in E and

$$\text{for } \Delta x \in \mathcal{B} \begin{cases} x \in \Delta x \Rightarrow \mu_x(\Delta x) = 1 \\ x \in E - \Delta x \Rightarrow \mu_x(\Delta x) = 0 \ . \end{cases}$$

Then

$$P_{s,\mu_x}\{x_u \in B\} = Q(s,x;u,B)$$

for $t_o \leqslant s \leqslant u \leqslant t_1$ and $B \in \mathcal{B}$.

By varying s in $[t_o,t_1]$ and x in E, one generates a family of random processes $(\Omega, M^s, P_{s,\mu_x})$. Such a family is called by Dynkin a *Markovian family*.

With a Markov process, there is associated a Markovian family. Conversely, the definition of a Markov process corresponding to a given Markovian family is easy.

In these paragraphs we have (perhaps too heavily) insisted on equivalent definitions of a Markov process for a few reasons. First, we suspect, though it is not always clear, that different authors refer to different and not equivalent definitions. We believe that the definition given by Dynkin is both general enough and widely accepted. Secondly, we note that a Markov process is *not* a random process ; that is, it is a *family* of random processes indexed by the initial state μ.

6. CLOSED-LOOP CONTROLLED RANDOM PROCESS.

Let $\{\mu\}$ denote the set of all probability measures on $\mathcal{B}$, and let there be given a non-empty subset M of $\{\mu\}$, termed the *state-space of the system.*

For any $s \in [t_0, t_1]$, let $\{Q_s\}$ be the set of all transition probability functions relative to $[s, t_1]$; let $\{P_s\}$ be the set of all probability measures on M^s and let C^s

$$C^s \subseteq \{P_s\} \times \{Q_s\} , \qquad C^s \neq \phi$$

be a given relation which we call *closed-loop (or feedback) control relation at time* s .

The set of all pairs (s, C^s), for all $s \in [t_0, t_1]$, is the graph of a mapping

$$C : s \mapsto C(s) \triangleq C^s .$$

This mapping C we call *closed-loop (or feedback) control function of the system.*

Let there be associated with any $\mu \in M$ and any $s \in [t_0, t_1]$ the family

$F_{s,\mu}(C^s)$ of all random processes satisfying the following property :

Any member $(\Omega, M^s, P_{s,\mu})$ of $F_{s,\mu}(C^s)$ is such that there exists a transition probability function $Q_{s,\mu}$ such that

$$P_{s,\mu}(A) = \int_E Q_{s,\mu}(s,x;A)\mu(dx) \quad \text{for all} \quad A \in M^s \text{ ; and} \tag{a}$$

$$(P_{s,\mu}, Q_{s,\mu}) \in C^s \ . \tag{b}$$

In particular we have

$$P_{s,\mu}\{ x_u \in B \} = \int_E Q_{s,\mu}(s,x;u,B)\mu(dx) \tag{2}$$

for all $u \in [s,t_1]$ and for all $B \in \mathcal{B}$.

Note that $F_{s,\mu}(C^s)$ may be empty, and if it is non-empty it may have more than one member.

Definition 3. A closed-loop controlled random process with closed-loop control function C is the family F(C) of all random processes

$$(\Omega, M^s, P_{s,\mu}) \in F_{s,\mu}(C^s), \quad s \in [t_o,t_1] \ , \quad \mu \in M \ .$$

In other words, it is the union of the $F_{s,\mu}(C^s)$ for all $s \in [t_o,t_1]$ and all $\mu \in M$. Note that

$$P_{s,\mu}\{x_u \in B\} \triangleq \text{Prob}(s,\mu;u,B)$$

for $B \in \mathcal{B}$, $s \in [t_o,t_1]$, $u \in [s,t_1]$, $\mu \in M$, defines a probability measure on $\mathcal{B}$, namely Prob$(s,\mu;u,\cdot)$.

1. Contrary to a Markov process, in a closed-loop controlled random process there may be more than one process $(\Omega, M^s, P_{s,\mu})$ associated with any given initial state μ at time s .
2. Measures μ_x, $x \in E$, need not belong to M. In other words, with a closed-loop controlled random process there need not be associated a Markovian family.
3. Even in the case where $M = \{\mu\}$, and C is such that there is associated with each $\mu \in M$ and each $s \in [t_o,t_1]$ a unique random

process $(\Omega, M^s, P_{s,\mu})$, a closed-loop controlled random process needs not be a Markov process, since

$\forall\mu, \ \forall s, \ \exists Q_{s,\mu} \ldots$

does not imply

$\exists Q$ such that $\forall\mu, \ \forall s, \ \ldots$

4. With any member $(\Omega, M^s, P_{s,\mu})$ of a closed-loop controlled random process F(C) there is associated a Markov process $(E, \mathcal{B}, Q_{s,\mu})$, that is, a family of random processes generated by the transition probability function $Q_{s,\mu}$, to which $(\Omega, M^s, P_{s,\mu})$ belongs.
Of course, the latter property 4 *does not imply* that a closed-loop controlled random process is a Markov process.

7. EXAMPLE OF INADEQUACY OF A MARKOVIAN MODEL IN PHYSICS.

Now, concerning a description of the behavior of a single quantum particle in the framework of random processes, let us consider a modified version of the well-known experiment of "Young holes". Suppose that, at each instant of time t in the interval of time $[t_o, t_1]$, the particle has a position in E, namely $x_t(\omega)$, with $\omega \in \Omega$. x_t is a random variable. *Suppose that there is a transition probability function* Q.

We modelize the two Young holes by giving two bounded subsets of E, namely, $B_1, B_2 \in \mathcal{B}$, such that $B_1 \cap B_2 = \phi$ and $B_1 \cup B_2 = B_3$.

We consider 3 experiments :

1. in the first one, the hole corresponding to B_1 is open and the other one is closed ;
2. in the second one, the hole corresponding to B_2 is open and the other one is closed ;
3. in the third one, both holes are open.

In both cases, the position of the particle is detected after the time of flight $t_1 - t_o$.

Concerning experiment 3, we suppose that $\mu = \mu_3$ with $\mu_3(B_1 \cup B_2) = 1$;

and $\mu_3(B_1) = \mu_3(B_2)$, so that $\mu_3(B_1) = \mu_3(B_2) = 1/2$. Concerning experiments 1 and 2, we suppose that $\mu = \mu_1$ and $\mu = \mu_2$, respectively, with

$$\mu_1(\Delta x) = \mu_3(\Delta x)/\mu_3(B_1) = 2\mu_3(\Delta x), \quad \text{for} \quad \Delta x \in \mathcal{B}[B_1],$$

$$\mu_2(\Delta x) = \mu_3(\Delta x)/\mu_3(B_2) = 2\mu_3(\Delta x), \quad \text{for} \quad \Delta x \in \mathcal{B}[B_2].$$

From formula (1) we deduce

$$\mathrm{Prob}(t_o,\mu_1;t_1,B) = 2\int_{B_1} Q(t_o,x;t_1,B)\mu_3(dx),$$

$$\mathrm{Prob}(t_o,\mu_2;t_1,B) = 2\int_{B_2} Q(t_o,x;t_1,B)\mu_3(dx),$$

$$\mathrm{Prob}(t_o,\mu_3;t_1,B) = \int_{B_1} Q(t_o,x;t_1,B)\mu_3(dx) + \int_{B_2} Q(t_o,x;t_1,B)\mu_3(dx),$$

for all $B \in \mathcal{B}$.

It follows that

$$\mathrm{Prob}(t_o,\mu_3;t_1,B) = \frac{1}{2}[\mathrm{Prob}(t_o,\mu_1;t_1,B) + \mathrm{Prob}(t_o,\mu_2;t_1,B)] \tag{3}$$

for all $B \in \mathcal{B}$.

It is well known that this relation is in contradiction with experiment. This is not so because of the elementary rules of the classical calculus of probabilities which are what they are. This is so because we have supposed that there is a transition probability function Q ; that is, because our model is Markovian.

Now, if instead of modelizing the Young holes experiment by a Markov process, we modelize it by a closed-loop controlled random process, we deduce from formula (2)

$$\mathrm{Prob}(t_o,\mu_1;t_1,B) = 2\int_{B_1} Q_{\mu_1}(t_o,x;t_1,B)\mu_3(dx),$$

$$\mathrm{Prob}(t_o,\mu_2;t_1,B) = 2\int_{B_2} Q_{\mu_2}(t_o,x;t_1,B)\mu_3(dx),$$

$$\mathrm{Prob}(t_o,\mu_3;t_1,B) = \int_{B_1} Q_{\mu_3}(t_o,x;t_1,B)\mu_3(dx) + \int_{B_2} Q_{\mu_3}(t_o,x;t_1,B)\mu_3(dx),$$

for all $B \in \mathcal{B}$.

Hence, relation (3) needs not hold, which agrees with experiment.

That comes from the fact that the transition probability function $Q_\mu(t_0,x;t_1,B)$ is not the same when the first hole is closed, or when the second hole is closed, or when the two holes are both open.

REFERENCES.

1. de Broglie, L., Une tentative d'interprétation causale nonlinéaire de la Mécanique ondulatoire, Gauthier-Villars, Paris, France, 1956.
2. Fürth, R., Zeits. Phys., 81, 143 (1933).
3. Fényes, I., Zeits. Phys., 132, 81 (1952).
4. Braffort, P., and Tzara, C., Compt. Rend. Acad. Sc. 239, 157 (1954).
5. Blaquière, A., Compt. Rend. Acad. Sc., Série A, 262, 593 (1966) ; 262, 721 (1966).
6. Blaquière, A. : "On the Geometry of Optimal Processes with Applications in Physics", AM-66-6, Univ. of California, Berkeley, Report Nonr-3565 (31), (1966).
7. Nelson, E., Phys. Rev., 150, 1079 (1966).
8. Dynkin, E.B., Théorie des processus markoviens, Dunod, Paris, France, 1963.
9. Kunio Yasue, J. Funct. Anal., 41, 327 (1981).
10. Blaquière, A. : "Modèle stochastique non markovien en Mécanique ondulatoire non relativiste", Compt. Rend. Acad. Sc., Série A, 285, 145 (1977).
11. Blaquière, A., Journal of Optimization Theory and Applications, Vol.32, N°4, 1980.
12. Blaquière, A. : "More on Stochastic Processes in Wave Mechanics", to be published in "Dynamical Systems a Renewal of Mechanism", Eds. S. Diner, D. Fargue, G. Lochak, Springer-Verlag, Series Synergetics.

AN ALTERNATIVE APPROACH TO WAVE MECHANICS OF A PARTICLE AT THE NON-RELATIVISTIC APPROXIMATION

Augustin Blaquière, Angelo Marzollo
Université de Paris VII, Paris, France

1. INTRODUCTION.

The mathematical theory of stochastic dynamic programming has cast a new light on the conceptual foundation of wave mechanics (1). In this chapter we shall return to an illustration of that point of view, and more specifically to section 5 of the paper by Blaquière and Marzollo, published in (2). Recently we have been aware, thanks to J.C. Zambrini, of an article by Schrödinger published in 1932 (3), in which a model identical with ours was presented as a possible starting point for an alternative approach to wave mechanics.

This paper comforts our choice of that direction and invites us to go further in the study of the model of Ref.(2). First, let us briefly recall what this model is. As in Ref.(2), we shall introduce it through three seemingly separate optimal control problems.

2. THREE OPTIMAL STOCHASTIC CONTROL PROBLEMS.

2.1. Problem 1.

A point in a 3-dimensional Euclidean space E is moving randomly, as time evolves on the interval $[t_o,t_1]$. For given *control function* in a proper class $v(\cdot) : E \times [t_o,t_1] \to R^3$, and given initial position $\xi^i \in E$ at time $t_i \in [t_o,t_1)$, its trajectory is

$$\{\xi(t) : \quad t \in [t_i,t_1]\}$$

where $\xi(\cdot) : [t_i,t_1] \to E$ is the stochastic process such that for all $t \in [t_i,t_1]$

$$\xi(t) = \xi^i + \int_{t_i}^{t} v(\xi(s),s)ds + z(t) - z(t_i) \tag{1}$$

where $z(\cdot) : [t_o,t_1] \to E$ is a Wiener process with infinitesimal generator

$$\sum_{k=1}^{3} D \frac{\partial^2}{\partial \xi_k^2}, \qquad D = \text{constant} > 0 .$$

We allow a criterion, or as we say a *payoff*, that involves a combination of a definite integral and a function evaluated at time t_1. For that purpose, let there be given the function $V_1(\cdot) : E \to R^1$.

For given control function $v(\cdot)$ and given initial state ξ^i at arbitrary time t_i, the payoff is

$$E\left[\int_{t_i}^{t_1} (1/2)\, m\, (v(\xi(t),t))^2 dt + V_1(\xi(t_1))\right] \tag{2}$$

where m is a constant, and the expectation $E(\cdot)$ is defined on the set of random trajectories emanating from ξ^i at time t_i and ending at time t_1.

The purpose is to find a control function $v^*(\cdot)$ *for which the payoff is minimum.*

Under proper conditions, which the reader can find in classical books (4), a necessary condition of optimality (i.e., minimality) of a control is provided by the following equation of stochastic dynamic programming, namely

$$\min_{v}\left[(1/2)mv^2 + (\partial V_f^*/\partial t) + \sum_k (\partial V_f^*/\partial \xi_k)v_k + D\sum_k (\partial^2 V_f^*/\partial \xi_k^2)\right] =$$

$$= (1/2)mv^{*2} + (\partial V_f^*/\partial t) + \sum_k(\partial V_f^*/\partial \xi_k)v_k^* + D\sum_k(\partial^2 V_f^*/\partial \xi_k^2) = 0 \tag{3}$$

$V_f^* \triangleq V_f^*(\xi,t)$ is the minimal value of the payoff for an arbitrary initial condition $(\xi,t) \in E\times[t_o,t_1]$. Function $V_f^*(\cdot)$ must satisfy the end-condition

$$V_f^*(\xi,t_1) = V_1(\xi) \quad \text{for all} \quad \xi \in E .$$

We deduce from (3)

$$mv^* = -\operatorname{grad} V_f^*(\xi,t) \ ; \text{ and} \tag{4}$$

$$(\partial V_f^*/\partial t) - \frac{1}{2m}(\operatorname{grad} V_f^*)^2 + D\nabla^2 V_f^* = 0 \tag{5}$$

Equation (5) is a generalized Hamilton-Jacobi equation.

Now let

$$\rho_f(\xi,t) \triangleq K' \exp(-V_f^*(\xi,t)/2mD), \tag{6}$$

$$K' = \text{constant} > 0.$$

Then Equation (5) is written

$$\frac{\partial \rho_f(\xi,t)}{\partial t} = -D\sum_k \frac{\partial^2 \rho_f(\xi,t)}{\partial \xi_k^2} . \tag{7}$$

This is the adjoint of the equation of heat transfer. It has a unique solution satisfying the given *terminal condition* at time t_1 : $\rho_f(\xi,t_1) = K' \exp(-V_1(\xi)/2mD)$ for all $\xi \in E$.

At any $(\xi,t) \in E\times[t_o,t_1)$, $v^*(\xi,t)$ is the *drift forward* of our stochastic process, and it follows from (4) and (6) that

$$v_k^*(\xi,t) = \frac{2D}{\rho_f(\xi,t)}\frac{\partial \rho_f(\xi,t)}{\partial \xi_k}, \quad k = 1,2,3 . \tag{8}$$

One can easily prove that, if $\rho_f(\cdot,t_o) : E \to R_+^\dagger$ is taken for the *position probability density function* of our stochastic process at time t_o, then, for all $(\xi,t) \in E\times[t_o,t_1]$, $\rho_f(\xi,t)$ is its position probability density so that, in view of (8) and of a well-known relation between the drift forward and the *drift backward*, the drift backward of our process is identically zero and Equation (7) is its *backward Fokker-Planck equation*.

It is interesting to note that, if $m > 0$,
$V_f^*(\xi,t) = -2mD \,\mathrm{Log}\, \rho_f(\xi,t) + \text{const.}$, $(\xi,t) \in E\times[t_o,t_1]$, is a *negentropy*.

2.2. Problem 2.

As a second step we consider the random motion of a point which is controlled, so as to say, backward in time. Precisely, for given control function $\overline{v}(\cdot) : E\times(t_o,t_1] \to R^3$, in a proper class and given end position $\eta^f \in E$ at time $t_f \in (t_o,t_1]$, its trajectory is

$$\{\eta(t) : \quad t \in [t_o,t_f]\}$$

where $\eta(\cdot) : [t_o,t_f] \to E$ is the stochastic process such that for all $t \in [t_o,t_f]$

$$\eta(t) = \eta^f - \int_t^{t_f} \overline{v}(\eta(s),s)ds + z(t) - z(t_f) . \tag{9}$$

In other words, $\overline{v}(\eta,t)$ is the drift backward of the process at (η,t).

The payoff is

$$E\left[\int_{t_o}^{t_f} (1/2)\, m\, (\overline{v}(\xi(t),t))^2 dt + V_o(\xi(t_o))\right], \tag{10}$$

where $V_o(\cdot) : E \to R^1$ is a given initial cost function, and $E(\cdot)$ is defined on the set of random trajectories ending at η^f at time t_f and emanating at time t_o.

† R_+ is the set of non-negative real numbers.

We wish to find a control function $\overline{v}*(\cdot)$ for which the payoff is minimum.

Now, under proper conditions, the equation of stochastic dynamic programming is

$$\min_{\overline{v}} \left[(1/2)m\overline{v}^2 - (\partial V_b^*/\partial t) - \sum_k (\partial V_b^*/\partial \eta_k)\overline{v}_k + D \sum_k (\partial^2 V_b^*/\partial \eta_k^2) \right] =$$

$$= (1/2)m\overline{v}^{*2} - (\partial V_b^*/\partial t) - \sum_k (\partial V_b^*/\partial \eta_k)\overline{v}_k^* + D \sum_k (\partial^2 V_b^*/\partial \eta_k^2) = 0. \qquad (11)$$

$V_b^* \triangleq V_b^*(\eta,t)$ is the minimal value of the payoff for an arbitrary end condition $(\eta,t) \in E\times[t_o,t_1]$. Function $V_b^*(\cdot)$ must satisfy the initial condition

$$V_b^*(\eta,t_o) = V_o(\eta) \quad \text{for all} \quad \eta \in E .$$

We deduce from (11)

$$m\overline{v}^* = \text{grad } V_b^*(\eta,t) \ ; \text{ and} \qquad (12)$$

$$(\partial V_b^*/\partial t) + \frac{1}{2m} (\text{grad } V_b^*)^2 - D\nabla^2 V_b^* = 0 \qquad (13)$$

Putting

$$\rho_b(\eta,t) \triangleq K'' \exp(-V_b^*(\eta,t)/2mD) \qquad (14)$$

$$K'' = \text{constant} > 0,$$

Equation (13) is written

$$\frac{\partial \eta_b(\eta,t)}{\partial t} = D \sum_k \frac{\partial^2 \rho_b(\eta,t)}{\partial \eta_k^2} . \qquad (15)$$

This is the equation of heat transfer. It has a unique solution satisfying the given *initial condition* at time t_o :

$$\rho_b(\eta,t_o) = K'' \exp(-V_o(\eta)/2mD) \quad \text{for all} \quad \xi \in E .$$

It follows from (12) and (14) that

$$\overline{v}_k^*(\eta,t) = - \frac{2D}{\rho_b(\eta,t)} \frac{\partial \rho_b(\eta,t)}{\partial \eta_k} , \quad k = 1,2,3 . \qquad (16)$$

Here one can prove that if $\rho_b(\cdot,t_1) : E \to R_+$ is taken for the position probability density function of our stochastic process at time t_1, then, for all $(\eta,t) \in E\times[t_o,t_1]$, $\rho_b(\eta,t)$ is its position probability density so that, in view of (16) and of a well-known relation between the drift forward and the drift backward, the drift forward of our process is identically zero and Equation (15) is its *forward Fokker-Planck equation.*

Again, if $m > 0$,

$V_b^*(\eta,t) = -2mD \text{ Log } \rho_b(\eta,t) + \text{const.}$, $(\eta,t) \in E\times[t_o,t_1]$, is a negentropy.

2.3. Problem 3.

In problems 1 and 2 we have defined two classes of random trajectories, say Class 1 and Class 2. Members of Class 1 are optimal trajectories for the stochastic control problem 1 in the sense that their drift forward function is $\overline{v}^*(\cdot) : E\times[t_o,t_1) \to R^3$. Members of Class 2 are optimal trajectories for the stochastic control problem 2 in the sense that their drift backward function is $\overline{v}^*(\cdot) : E\times(t_o,t_1] \to R^3$. Now let us define a third class of random trajectories, namely the one whose members belong to both Class 1 and 2. A trajectory of Class 3 namely

$$\{\zeta(t) : \quad t \in [t_o,t_1]\}$$

must be such that function $\zeta(\cdot) : [t_o,t_1] \to E$ be a solution of both stochastic differential equations

$$d\zeta(t) = v^*(\zeta(t),t)dt + dz(t) \text{ ; and} \tag{17}$$

$$d\zeta(t) = -\overline{v}^*(\zeta(t),t)dt + dz(t), \tag{18}$$

with the provisio that, as concerns Equation (17), the integration is performed for t increasing from t_o to t_1 whereas, as concerns Equation (18), the integration is performed for t decreasing from t_1 to t_o. In other words $v^*(\cdot) : E\times[t_o,t_1) \to R^3$ and $\overline{v}^*(\cdot) : E\times(t_o,t_1] \to R^3$ are the drift forward function *and* the drift backward function, respectively, for the latter process. Indeed, $v^*(\cdot)$ and $\overline{v}^*(\cdot)$ depend on the payoffs we have chosen for problems 1 and 2 ; in particular they depend on the arbitrarily given functions $V_1(\cdot)$ and $V_o(\cdot)$, respectively. Letting

$$\frac{1}{2}(V_b^* - V_f^*) = V, \qquad \frac{1}{2}(V_b^* + V_f^*) = W$$

$$\frac{1}{2}(\overline{v} + v) = u, \qquad \frac{1}{2}(\overline{v} - v) = w,$$

we deduce from Equations (5) and (13), by addition and by substraction, the following pair of equations

$$(\partial V/\partial t) + \frac{1}{2m}(\text{grad } V)^2 + \frac{1}{2m}(\text{grad } W)^2 - D\nabla^2 W = 0, \tag{19}$$

$$(\partial W/\partial t) + \frac{1}{m} \text{ grad } V \text{ grad } W - D\nabla^2 V = 0, \tag{20}$$

with $V = V(\zeta,t)$ and $W = W(\zeta,t)$, and from (4) and (12) we deduce

$$mu^* = \text{grad } V, \qquad mw^* = \text{grad } W . \tag{21}$$

Further let

$$\rho^c(\zeta,t) \triangleq \rho_b(\zeta,t)\rho_f(\zeta,t) = K'K'' \exp(-W(\zeta,t)/mD) \tag{22}$$

for all $(\zeta,t) \in E\times[t_o,t_1]$.

One can readily verify that function $\rho^c(\cdot) : E\times[t_o,t_1] \to R_+$ satisfies the continuity equation of hydrodynamics

$$\frac{\partial\rho^c}{\partial t} = -\text{ div } (\rho^c u^*), \qquad u^* \triangleq u^*(\zeta,t), \tag{23}$$

and the forward Fokker-Planck equation

$$\frac{\partial\rho^c}{\partial t} = -\text{ div } (\rho^c v^*) + D\nabla^2\rho^c, \qquad v^* \triangleq v^*(\zeta,t), \tag{24}$$

and the backward Fokker-Planck equation

$$\frac{\partial\rho^c}{\partial t} = -\text{ div } (\rho^c \overline{v}^*) - D\nabla^2\rho^c, \qquad \overline{v}^* \triangleq \overline{v}^*(\zeta,t), \tag{25}$$

with $\rho^c \triangleq \rho^c(\zeta,t)$, $\quad (\zeta,t) \in E\times[t_o,t_1]$.

Indeed, function $\rho^c(\cdot)$ satisfies both end conditions

$$\rho^c(\zeta,t_o) = \rho_b(\zeta,t_o)\rho_f(\zeta,t_o) = K'K'' \exp\,[-(V_o(\zeta) + V_f^*(\zeta,t_o))/2mD]$$

$$\rho^c(\zeta,t_1) = \rho_b(\zeta,t_1)\rho_f(\zeta,t_1) = K'K'' \exp\,[-(V_b^*(\zeta,t_1) + V_1(\zeta))/2mD]$$

for all $\zeta \in E$.

One can easily prove that if $\rho^c(\cdot,t_o) : E \to R_+$ is taken for the position probability density function of our stochastic process at time t_o, then $\rho^c(\cdot,t) : E \to R_+$ is its position probability density function at time t, and, indeed, it also satisfies the preassigned terminal condition at time t_1.

A problem of this kind in which the position probability density function of a stochastic process is asked to satisfy both given initial and terminal conditions is deeply different from classical problems in which a stochastic process is asked to satisfy a single initial or terminal given condition.

When introducing Problem 3 in Ref.(2) we have drawn the attention of the reader on striking similarities between some of its features and some features which are specific of quantum mechanics. We shall not return to that discussion here. This point is stressed by Schrödinger in Ref.(3).

3. ANOTHER APPROACH TO PROBLEM 3.

Instead of focusing our attention on trajectories which belong to both Class 1 and Class 2, let us consider two points ξ and η moving randomly in E, or equivalently a point (ξ,η) moving randomly in E×E, as time evolves on the interval $[t_o,t_1]$. Let us suppose that the position probability density function

$$\sigma(\cdot) : \begin{cases} E\times E\times[t_o,t_1] \to R_+ \\ (\xi,\eta,t) \mapsto \sigma(\xi,\eta,t) \end{cases}$$

is defined and of class $C^{2,2,1}$, and that the random motion of the pair $(\xi,\eta) \in E\times E$ is controlled by a pair of control functions

$$(v^1(\circ),v^2(\circ)) : \begin{cases} E\times E\times[t_o,t_1] \to R^3 \times R^3 \\ (\xi,\eta,t) \mapsto (v^1(\xi,\eta,t),\ v^2(\xi,\eta,t)) \end{cases}$$

of class $C^{1,1,0}$, through the forward stochastic differential equations

$$d\xi(t) = v^1(\xi(t),\eta(t),t)dt + dz^1(t)$$

$$d\eta(t) = v^2(\xi(t),\eta(t),t)dt + dz^2(t)$$

where $z^1(\cdot) : [t_o,t_1] \to E$ and $z^2(\cdot) : [t_o,t_1] \to E$ are Wiener processes with infinitesimal generators

$$\sum_{k=1}^{3} D \frac{\partial^2}{\partial\xi_k^2} \quad \text{and} \quad \sum_{k=1}^{3} D \frac{\partial^2}{\partial\eta_k^2} \quad , \text{ respectively .}$$

According to a well-known relation between the drift forward $(v^1(\xi,\eta,t),v^2(\xi,\eta,t))$, the drift backward $(\overline{v}^1(\xi,\eta,t),\overline{v}^2(\xi,\eta,t))$, and the position probability density $\sigma(\xi,\eta,t)$, at any $(\xi,\eta,t) \in E\times E\times(t_o,t_1)$, we have

$$\overline{v}_k^1(\xi,\eta,t) - v_k^1(\xi,\eta,t) = - \frac{2D}{\sigma(\xi,\eta,t)} \frac{\partial\sigma(\xi,\eta,t)}{\partial\xi_k} \tag{26}$$

$$\overline{v}_k^2(\xi,\eta,t) - v_k^2(\xi,\eta,t) = - \frac{2D}{\sigma(\xi,\eta,t)} \frac{\partial\sigma(\xi,\eta,t)}{\partial\eta_k} \tag{27}$$

$$k = 1,2,3 \ .$$

Now, let us choose $v^2(\cdot)$ such that

$$v^2(\xi,\eta,t) = 0 \quad \text{for all} \quad (\xi,\eta,t) \in E\times E\times[t_o,t_1], \tag{28}$$

and $v^1(\cdot)$ such that

$$\overline{v}^1(\xi,\eta,t) = 0 \quad \text{for all} \quad (\xi,\eta,t) \in E\times E\times[t_o,t_1]. \tag{29}$$

These are the conditions met in Problems 2 and 1, respectively.

From now on, we shall rename $v^1(\cdot)$ and $\overline{v}^2(\cdot)$ and call them $v(\cdot)$ and $\overline{v}(\cdot)$, respectively.

From (26) - (29) it follows that

$$v_k(\xi,\eta,t) \equiv \frac{2D}{\sigma(\xi,\eta,t)} \frac{\partial\sigma(\xi,\eta,t)}{\partial\xi_k} \quad ; \text{ and} \tag{30}$$

$$\overline{v}_k(\xi,\eta,t) \equiv -\frac{2D}{\sigma(\xi,\eta,t)} \frac{\partial\sigma(\xi,\eta,t)}{\partial\eta_k} \quad , \; k = 1,2,3 \; . \tag{31}$$

Further, let us suppose that the continuity equation of hydrodynamics holds. When our assumptions are satisfied, it takes the form

$$\frac{\partial\sigma}{\partial t} = -\sum_k \left[\frac{1}{2} \frac{\partial(\sigma v_k)}{\partial\xi_k} + \frac{1}{2} \frac{\partial(\sigma\overline{v}_k)}{\partial\eta_k}\right] , \tag{32}$$

so that, on account of (30) and (31), we have

$$\frac{\partial\sigma}{\partial t} = -D\sum_k \frac{\partial^2\sigma}{\partial\xi_k^2} + D\sum_k \frac{\partial^2\sigma}{\partial\eta_k^2} \quad . \tag{33}$$

Also it will be interesting to translate the above formulas in terms of new variables

$$x \triangleq \frac{\xi+\eta}{2} \quad , \qquad y \triangleq \frac{\xi-\eta}{2} \quad .$$

For $\xi = \xi(x,y) = x+y$ and $\eta = \eta(x,y) = x-y$, let

$$\rho(x,y,t) \triangleq \sigma(\xi,\eta,t),$$
$$u(x,y,t) \triangleq (1/2)\,[\,\overline{v}(\xi,\eta,t) + v(\xi,\eta,t)],$$
$$w(x,y,t) \triangleq (1/2)\,[\,\overline{v}(\xi,\eta,t) - v(\xi,\eta,t)].$$

Then, we deduce from (30)-(33)

$$u_k(x,y,t) \equiv \frac{D}{\rho(x,y,t)} \frac{\partial\rho(x,y,t)}{\partial y_k} , \tag{34}$$

$$w_k(x,y,t) \equiv -\frac{D}{\rho(x,y,t)} \frac{\partial\rho(x,y,t)}{\partial x_k} , \qquad k = 1,2,3 \; ; \tag{35}$$

$$\frac{\partial\rho}{\partial t} = -\sum_k \left[\frac{1}{2} \frac{\partial(\rho u_k)}{\partial x_k} - \frac{1}{2} \frac{\partial(\rho w_k)}{\partial y_k}\right] ; \tag{36}$$

$$\frac{\partial\rho}{\partial t} = -D\sum_k \frac{\partial^2\rho}{\partial x_k \partial y_k} \quad . \tag{37}$$

Note that, in view of (34) and (35), (36) can also be written

$$\frac{\partial \rho}{\partial t} = - \sum_k \frac{\partial(\rho u_k)}{\partial x_k} = \sum_k \frac{\partial(\rho w_k)}{\partial y_k} \quad . \tag{38}$$

Now, if we let

$$\rho(x,y,t) = K \exp(-\mathcal{W}(x,y,t)/mD), \tag{39}$$

$$K = \text{constant} > 0,$$

formulas (34), (35) and (37) are transformed into

$$m u_k(x,y,t) \equiv - \frac{\partial \mathcal{W}(x,y,t)}{\partial y_k} \quad ; \tag{40}$$

$$m w_k(x,y,t) \equiv \frac{\partial \mathcal{W}(x,y,t)}{\partial x_k} \quad ; \qquad k = 1,2,3, \tag{41}$$

and

$$\frac{\partial \mathcal{W}(x,y,t)}{\partial t} - \frac{1}{m} \sum_k \frac{\partial \mathcal{W}(x,y,t)}{\partial x_k} \frac{\partial \mathcal{W}(x,y,t)}{\partial y_k} + D \sum_k \frac{\partial^2 \mathcal{W}(x,y,t)}{\partial x_k \partial y_k} = 0 \tag{42}$$

respectively.

If $m > 0$,
$\mathcal{W}(x,y,t) = - mD \operatorname{Log} \rho(x,y,t) + \text{const.}$, $(x,y,t) \in E\times E\times[t_o,t_1]$, is a negentropy.

One can readily verify that, in terms of variables ξ and η, relations (40)-(42) are written

$$\frac{1}{2} m v_k(\xi,\eta,t) \equiv - \frac{\partial \widetilde{W}(\xi,\eta,t)}{\partial \xi_k} \quad ; \tag{40'}$$

$$\frac{1}{2} m \bar{v}_k(\xi,\eta,t) \equiv \frac{\partial \widetilde{W}(\xi,\eta,t)}{\partial \eta_k} , \qquad k = 1,2,3, \tag{41'}$$

$$\frac{\partial \widetilde{W}}{\partial t} - \frac{1}{m}\left(\frac{\partial \widetilde{W}}{\partial \xi}\right)^2 + \frac{1}{m}\left(\frac{\partial \widetilde{W}}{\partial \eta}\right)^2 + D \frac{\partial^2 \widetilde{W}}{\partial \xi^2} - D \frac{\partial^2 \widetilde{W}}{\partial \eta^2} = 0 \tag{42'}$$

with

$$\widetilde{W} = \widetilde{W}(\xi,\eta,t) \triangleq \mathcal{W}(x(\xi,\eta),y(\xi,\eta),t) \ .$$

If the random trajectories of points ξ and η are ones of Problems 1 and 2, controlled by $v^*(\cdot)$ and $\bar{v}^*(\cdot)$, with the corresponding position probability density functions $\rho_f(\cdot)$ and $\rho_b(\cdot)$, respectively, then

$$\rho(x,y,t) = \rho_f(x+y,t)\rho_b(x-y,t) \ ,$$

as such trajectories are statistically independent. It follows that

$$W(x,y,t) = (1/2)[V_f^*(x+y,t) + V_b^*(x-y,t)].$$

If, on the other hand, we let

$$V(x,y,t) \triangleq (1/2)[V_b^*(x-y,t) - V_f^*(x+y,t)]$$

we have

$$\frac{\partial V(x,y,t)}{\partial x_k} = -\frac{\partial W(x,y,t)}{\partial y_k} \quad ; \text{ and} \tag{43}$$

$$\frac{\partial V(x,y,t)}{\partial y_k} = -\frac{\partial W(x,y,t)}{\partial x_k} \quad , \quad k = 1,2,3, \tag{44}$$

so that function $W(\cdot)$ (as well as function $V(\cdot)$) is *anti-harmonic* ; that is

$$\frac{\partial^2 W(x,y,t)}{\partial x_k^2} - \frac{\partial^2 W(x,y,t)}{\partial y_k^2} = 0, \qquad k = 1,2,3 \ .$$

Hence relation (42) becomes

$$\frac{\partial W(x,y,t)}{\partial t} + \frac{1}{m} \sum_k \frac{\partial V(x,y,t)}{\partial x_k} \frac{\partial W(x,y,t)}{\partial x_k} - D \sum_k \frac{\partial^2 V(x,y,t)}{\partial x_k^2} = 0 \tag{45}$$

By differentiating relation (45) with respect to y_j , then with respect to x_j, for $j = 1,2,3$, and taking account of (43) and (44) one obtains

$$\frac{\partial V(x,y,t)}{\partial t} + \frac{1}{2m} \sum_k \left(\frac{\partial V(x,y,t)}{\partial x_k}\right)^2 + \frac{1}{2m} \sum_k \left(\frac{\partial W(x,y,t)}{\partial x_k}\right)^2 - D \sum_k \frac{\partial^2 W(x,y,t)}{\partial x_k^2} = \varphi(t) \tag{46}$$

where $\varphi(t)$, which depends on t only, will be set identically to zero by a slight change in the definition of function $V(\cdot)$; that is, by adding to it a function of time only†.

† Note that conditions (43) and (44) still hold when an arbitrary function of time is added to function $V(\cdot)$.

In fact, what we have done in the present paragraph is to treat in the same setting trajectories of Class 1 and trajectories of Class 2. Trajectories belonging to both Class 1 and Class 2 are characterized by the simple condition $y \equiv 0$. Then relations (45) and (46) reduce to (20) and (19) respectively, with

$$\mathcal{V}(x,0,t) = V(x,t) \; ; \text{ and} \tag{47}$$

$$\mathcal{W}(x,0,t) = W(x,t) \; . \tag{48}$$

The latter method is more flexible than the one of the previous sections, as far as we have no need of explicit expressions for the payoffs. This gives us more freedom in the problem statement and sweep the difficulties met in Ref.[2] and by Schrödinger in Ref.[3] in the search for a new way to quantum mechanics.

4. THE CASE OF QUANTUM MECHANICS.

Let us return to relation (42) wich we have obtained under pretty general conditions. Now let us suppose that the random variables ξ and η are *not* statistically independent, and that there exists a function $\mathcal{V}(\cdot) : E\times E\times[t_o,t_1] \to R^1$ such that (43) is satisfied but (44) is replaced by

$$\frac{\partial \mathcal{V}(x,y,t)}{\partial y_k} = \frac{\partial \mathcal{W}(x,y,t)}{\partial x_k} \; , \qquad k = 1,2,3 \; . \tag{49}$$

(43) and (49) are Cauchy conditions, and consequently function $\mathcal{W}(\cdot)$ (as well as function $\mathcal{V}(\cdot)$) is *harmonic*.

Because relation (43) still holds, relation (45) is deduced from (42) like in paragraph 3. Now, by differentiating relation (45) with respect to y_j, then with respect to x_j, for $j = 1,2,3$, and taking account of (43) and (49), one obtains

$$\frac{\partial \mathcal{V}(x,y,t)}{\partial t} + \frac{1}{2m} \sum_k \left(\frac{\partial \mathcal{V}(x,y,t)}{\partial x_k}\right)^2 - \frac{1}{2m} \sum_k \left(\frac{\partial \mathcal{W}(x,y,t)}{\partial x_k}\right)^2 + D \sum_k \frac{\partial^2 \mathcal{W}(x,y,t)}{\partial x_k^2} = g(t) \tag{50}$$

in place of relation (46), where like in (46) we shall let $g(t) \equiv 0$ by a proper definition of function $V(\circ)$.

Like in paragraph 3, we are dealing with a class of random trajectory-pairs, each trajectory-pair being associated with the random motion of a pair (ξ,η) as time evolves on $[t_0,t_1]$. However, unlike in the statement of paragraph 3, this class cannot be separated into two independent classes of random trajectories.

In our class of random trajectory-pairs, of special interest are "degenerated" trajectory-pairs ; that is, those for which $y \equiv 0$. Then, relations (45) and (50) reduce to (20) and

$$(\partial V/\partial t) + \frac{1}{2m}(\text{grad } V)^2 - \frac{1}{2m}(\text{grad } W)^2 + D\nabla^2 W = 0, \tag{51}$$

respectively, where $V = V(x,t)$ and $W = W(x,t)$ are defined by (47) and (48), respectively.

The most important point is that relations (20) and (51) are now the equations of Wave mechanics which Louis de Broglie has introduced under the names Equations (C) and (J), respectively [5].

It is well-known that this pair of equations is equivalent to the Schrödinger's equation, as concerns the motion of a quantum particle, in the non-relativistic approximation. Here we have treated the case where the potential function is identically zero (or equal to an arbitrary constant). One can easily introduce in our theory any potential function.

The fact that our model leads to the Schrödinger's equation is a necessary condition for its physical validity, however this is not its essential interest. What is essential is that we have given a description of the motion of a quantum particle without introducing a wave function whose meaning is not clear after about 60 years of quantum mechanics. The search for a model more close to classical mechanics than the usually accepted one, which motivated our research, also seems to be the aim of Schrödinger's paper in Ref.[3].

One can summarize our discussion by saying that, *associated with the observable motion of a physical particle in our 3-dimensional physical space, is a set of "varied" trajectory-pairs*†. This picture has a striking similarity with the one of classical mechanics in which the observable trajectory of a mass-point is compared with "varied" trajectories, in the theoretical framework of stationarity principles. In both cases, the classical one and the quantum mechanical one, "varied" trajectories are not observable ones.

In our stochastic model, no one of the two directions of the time axis is a priviledged one, as concerns the conditions we have imposed to the forward and backward drifts of ξ and η. Their role is obviously symmetrical, so symmetrical that it would seem reasonable to describe the motion of this pair of points (ξ,η) along a "varied" trajectory-pair by saying that one of these points follows its trajectory in one of the directions of the time-axis - say as time increases - whereas the other one follows its trajectory in the opposite direction of the time-axis - say as time decreases. This remark is made more precise if we attach with ξ and η *proper times* τ' and τ'', respectively, such that, for instance, $d\tau' = dt$ and $d\tau'' = -dt$. Now, if $\xi(t)$ and $\eta(t)$ are the positions of ξ and η at a time t, and $\xi(t+\Delta t)$ and $\eta(t+\Delta t)$ are their positions at a later time $t+\Delta t$, respectively, then in terms of the proper times $\xi(t+\Delta t) = \xi(t+\tau')$ lies *in the future* of $\xi(t)$, whereas $\eta(t+\Delta t) = \eta(t-\Delta\tau'')$ lies *in the past* of $\eta(t)$.

The physical trajectory of an observable particle, being a "degenerated" pair (one for which $y \equiv 0$), appears as one followed in one direction of time by a point ξ and in the opposite direction of time by a point η.

5. END CONDITIONS.

In paragraphs 3 and 4 we have introduced two classes of functions

† They correspond to a kind of *polarization* of the physical particles, which is a degree of freedom which classical mass-points do not exhibit.

$\rho(\cdot) : E\times E\times[t_o,t_1] \to R_+$ satisfying Equation (37). In both cases, for any $(x,y,t) \in E\times E\times[t_o,t_1]$, $\rho(x,y,t)$ is the position probability density at time t of a pair (ξ,η), with $\xi = x+y$ and $\eta = x-y$; and in both cases, in view of Equation (38), the *conditional* position probability density at time t of a pair (ξ,η), *given that* $\xi = \eta$ *or equivalently* $y = 0$, can be easily shown to be $\rho(x,0,t) = \sigma(\xi,\eta,t|\xi = \eta)$ (multiplied by *a constant* normalization factor).

In the first class, considered in paragraph 3, the function $W(\cdot) \triangleq - mD \operatorname{Log} \rho(\cdot)$ is *anti-harmonic* with respect to x and y, for any $t \in [t_o,t_1]$. It must be so at time t_o and at time t_1 ; that is, as concerns the given initial or terminal condition. This special feature, as a consequence of the problem statement, is due to the fact that the random variables ξ and η are statistically independent. This is the case studied in Ref.(2) and by Schrödinger in Ref.(3).

The second class, considered in paragraph 4, is more interesting as it corresponds to an actual problem in quantum mechanics. In that case, the conditional position probability density at time t of a pair (ξ,η), given that $\xi = \eta$ or equivalently $y = 0$, is the probability density of finding a physical particle at the position $x = \xi = \eta$ at time t. In that class, the function $W(\cdot) \triangleq - mD \operatorname{Log} \rho(\cdot)$ is *harmonic* with respect to x and y, for any $t \in [t_o,t_1]$. It must be so at time t_o and at time t_1 ; that is, as concerns the given initial or terminal condition.

It is interesting to remark that Equation (37) (or its version (33) when variables ξ and η are used) is general enough to apply to Problem 3 as well as to the case of quantum mechanics. Also we see that the usual end (initial or terminal) conditions of quantum mechanics, concerning the complex function ψ, are replaced in our model by end (initial or terminal) conditions concerning a real function of two variables x and y (or ξ and η) with the simple provisio that this function is harmonic.

REFERENCES.

1. Blaquière, A., *System Theory : A New Approach to Wave Mechanics*, Journal of Optimization Theory and Applications, Vol.32, N°4, 1980.
2. Blaquière, A., Marzollo, A., *Introduction à la théorie moderne de l'optimisation et à certains de ses aspects fondamentaux en physique, in* La pensée physique contemporaine, Edts., S. Diner, D. Fargue, G. Lochak, Editions Augustin Fresnel, 1982.
3. Schrödinger, E., *Une analogie entre la mécanique ondulatoire et quelques problèmes de probabilités en physique classique*, Annales de l'Institut Henri Poincaré, 1932.
4. Dreyfus, S.E., *Dynamic Programming and the Calculus of Variations*, Academic Press, New York, 1965.
5. De Broglie, L., *Une tentative d'interprétation causale et non-linéaire de la mécanique ondulatoire*, Gauthier-Villars, Paris, 1956.

SCHRÖDINGER'S STOCHASTIC VARIATIONAL DYNAMICS

Jean Claude Zambrini
University of Bielefeld, Federal Republic of Germany

ABSTRACT

We summarize the results of a program, initiated by E. Schrödinger in 1931, and whose aim is to construct some unconventional diffusion processes associated to the classical Heat equation, in such a way that their properties are as close as possible to the ones of the probabilistic concepts involved in Quantum Mechanics. It is shown, in particular, that Nelson's stochastic Mechanics can be reinterpreted in this frame.

CONTENT

1. Variational Principle in Classical Mechanics

We consider a system of n points in the Euclidean space $\mathbb{R}^3$, i.e. in configuration space $\mathbb{R}^{3n} \equiv \mathbb{R}^N$, or more generally, in a differentiable manifold M. The trajectories of the system are applications $X: I = [-\frac{T}{2}, \frac{T}{2}] \to M : t \to X(t) \equiv X_t$. In general (for $n \neq 1$) such a trajectory has no resemblance to the path in physical space of an actual particle.

The hypothesis of Classical Kinematics is that there exists a differentiable $v: M \times I \to M$ such that $dx(t) = v(x(t),t)dt$. Due to the time-symmetry of the theory, the boundary conditions may be chosen in the past or in the future. Lagrangian Mechanics is the relevant version of Classical Mechanics for our purpose. Let $L_c : M \times M \times \mathbb{R} \to \mathbb{R}$ be the given Lagrangian. On $\Omega_T = \{X: I \to M$, differentiable, and s.t. $X_{-T/2}$ and $X_{T/2}$ are fixed$\}$, one defines the Action functional

$$A[X] \equiv \int_{-T/2}^{T/2} L_c(X(t), \dot{X}(t), t)dt . \tag{1.1}$$

The variation of A in the direction δX is $\delta A[X](\delta X)$ s.t. $A[X+\delta X] - A[X] = \delta A[X](\delta X) + O(\|\delta X\|)$. $\bar{X}$ is, by definition, extremal for A if $\delta A[\bar{X}](\delta X) = 0, \forall \delta X$ st. $\delta X(-\frac{T}{2}) = \delta X(\frac{T}{2}) = 0$. The Hamilton Principle says that the trajectories of the mechanical system described by L_c are extremal points of the Action A. An equivalent characterization of such an extremum $\bar{X}$ is that is solves the Euler-Lagrange equation

$$\frac{d}{dt}\left(\frac{\partial L_c}{\partial \dot{\bar{X}}}\right) - \frac{\partial L_c}{\partial \bar{X}} \equiv \frac{\partial A}{\partial \bar{X}} = 0 . \tag{1.2}$$

For the "Natural" physical system with Lagrangian $L_c = \sum_i \frac{1}{2} \dot{X}_i^2 - V(X)$ ($M_i = 1$), this is nothing but the Newton equation $\ddot{X} = -\nabla V$ for a (scalar) potential V.

Another concept of Action is also used in Classical Mechanics in order to derive the Hamilton-Jacobi equation. Let S be the function of the future position x,

$$S_{X_{-T/2},-T/2}(x,t) = \int_\gamma L_c(\bar{X},\dot{\bar{X}},\tau)d\tau + S_{-T/2}(\bar{X}(-\tfrac{T}{2})) \tag{1.3}$$

where $\gamma:\tau \to \bar{X}(\tau)$, such that $d\bar{X}(t) = v(\bar{X}(t),t)dt$, is an extremal between $\bar{X}_{-T/2}$ and $\bar{X}(t) = x$. Some precautions are necessary here because, for $[-\frac{T}{2}, t]$ arbitrary large, we may have more than one extremal between the two given points. But for $[-\frac{T}{2}, t]$ short enough, one shows that $S_{X_{-T/2},-T/2} = S(x,t)$ solves the Hamilton-Jacobi equation (for the natural Lagrangian L_c),

$$\frac{\partial S}{\partial t} + \frac{1}{2}(\nabla S)^2 + V = 0 . \tag{1.4}$$

This system is equivalently described by a solution of the Newton equation with $\dot{\bar{X}}(-\frac{T}{2}) = \nabla S_{-T/2}(X(-\frac{T}{2})) \equiv v_{-T/2}$, $\bar{X}(t) = x$.

Remarks:

1) This second kind of Action appears naturally in a variational context if the paths and the final time t are varied and the domain of the functional is $\{X(t);\ X_{-T/2}$ fixed and $\int_\gamma L_c(X,\dot{X},\tau)d\tau + S_{-T/2}(X(-\frac{T}{2})) = \ell$, a constant$\}$. This set of level defines the "wave front in t from the point $X_{-T/2}$."

2) An analytical continuation in t: $t \to \tau = -it$ does not change the nature of classical dynamics, since $X(t) \to X(-it) \equiv Z(\tau)$, $\dot{X}(t) \to -i\dot{X}(-it) \equiv -i\dot{Z}(\tau)$, $\ddot{X}(t) = -\nabla V(X(t)) \to \ddot{Z}(\tau) = +\nabla V(Z(\tau))$. Also $L_c = \frac{1}{2}|\dot{X}|^2 - V \to \bar{L}_c = -\frac{1}{2}|\dot{Z}|^2 - V$.

2. A Forgotten Idea of Schrödinger

My references for this Chapter are a pair of papers of E. Schrödinger [1] (1931-32) and a paper of S. Bernstein [2], a probabilist who proposed a program to realize rigorously Schrödinger's idea. In Schrödinger's description, many assumptions are implicit. We summarize here, in a somewhat axiomatic way, the key points of the resulting program. We call $Z(t) \equiv Z_t$ the stochastic process we are looking for, $(M,\mathcal{B})$ denotes its State space.

a) The hypothesis of Classical Kinematics is generalized by

$$d_*Z(t) = B_*(Z(t),t)dt + \hbar^{1/2}\, \mathbb{I}\, dW_*(t), \quad Z(\tfrac{T}{2}) = c(\omega) \qquad (2.1)$$

for $-\frac{T}{2} \le t < \frac{T}{2}$.

This is an (Itô's) stochastic differential equation with respect to a decreasing filtration F_t (decreasing because it contains the future information on Z_t). The $*$ denotes the "backward" nature of this equation, i.e. $d_*Z(t) = Z(t) - Z(t-dt), dt > 0$. W_* is a Brownian Motion with respect to F_t (an "F_t martingale") and $\hbar$ is the Planck constant over 2π. Here, $M = \mathbb{R}^N$, $\mathbb{I} = N \times N$ identity matrix, and all the particles have unit mass. At the classical limit $\hbar = 0$, this indeed reduces to the hypothesis of Classical Kinematics.

Remarks: 1) The choice of the diffusion constant $\hbar$ is not arbitrary. According to conventional Quantum Mechanics the concept of trajectory is meaningless in this theory. As a matter of fact, it is known that the quantum "paths", whatever their interpretation is, have the irregularities of the Brownian paths. This was the crucial element of Feynman's path integral approach. The fact that the increment $(\Delta Z)^2 \sim \hbar \Delta t$ is a version of the Heisenberg principle [3].

2) The process $Z(t)$ of (2.1) is Markovian and, usually, associated to physical irreversible evolutions. The class of Markov processes is not specific enough for our purpose:

b) $Z(t)$ satisfies the Bernstein property, i.e. $\forall\ -\frac{T}{2} < s < t < u < \frac{T}{2}$,

$$E[f(Z_t) | P_s \cup F_u] = E[f(Z_t) | Z_s, Z_u] \qquad (2.2)$$

for any bounded f, where P_s is the past sigma-algebra $\sigma\{Z_r, r \le s\}$ and F_u the future σ-algebra $\sigma\{Z_v, v \ge u\}$.

Remark: The Bernstein processes are intrinsically time symmetric, in contrast with Markov processes. The concept of Markovian transition probability also has to be replaced by the natural one for a time symmetric theory.

c) The Bernstein transition H of the Bernstein process $Z(t)$ is such that

$\forall(x,y)$ in $M\times M$, $-\frac{T}{2}\le s<t<u\le\frac{T}{2}$ and $A\in\mathcal{B}$, $A\longrightarrow\int_A h(s,x;t,\xi;u,y)d\xi\equiv H(s,x;t,A;u,y)$ is a probability on $\mathcal{B}$, with density h.

d) The Markovian data of an initial probability is replaced by the data of the joint probability, with density m, for $Z_{-T/2}$ and $Z_{T/2}$.

Remark: This last property is the probabilistic generalization of the classical two fixed end points conditions used for the derivation of the Euler-Lagarange equation. The points b) and c) are natural for a construction involving the joint density m .

The mathematical consistency of this program has been shown by Jamison, Beurling and Fortet [4], but the resulting Bernstein processes have, curiously, never been used in Theoretical Physics. The next chapter briefly summarizes the key steps of this construction and of one of its realizations in a dynamical context. The resulting processes are called variational processes. The proofs can be found in [5].

3. The Realization of Schrödinger's Program

It can be described in three steps:

A) Construction of the finite-dimensional distributions of any Bernstein processes.

B) Characterization of an unique Markovian Bernstein process in a dynamical context.

C) Proof of existence and uniqueness of this Markovian representative for a pair of given end points probabilities.

Only this Markovian representative in the class of Bernstein processes will be associated to a Physical Dynamics (for a given potential V). This aspect, and the explicit form of the Dynamics will be proved in terms of Stochastic Variational Principles univocally associated to these Bernstein processes.

A) Finite-dimensional distributions of Bernstein processes.

For a given Bernstein transition $H=H(s,x;t,A;u,y)$ and $m=m(x,y)$

a (density of) probability on $\mathcal{B}\times\mathcal{B}$, there is an unique probability measure P_m such that with respect to the underlying probability space (Ω, σ_T, P_m) Z_t is a Bernstein process and

$$P_m(Z_{-T/2} \in B_S, Z_{T/2} \in B_F) = \int_{B_S\times B_F} m(x,y)dxdy, \quad B_S, B_F \in \mathcal{B} \tag{A1}$$

$$P_m(Z_t \in B|Z_s, Z_u) = H(s,Z_s;t,B;u,Z_u), \quad -\frac{T}{2}\le s<t<u\le\frac{T}{2}, \quad B\in\mathcal{B}. \tag{A2}$$

Furthermore the finite-dimensional distributions densities of Z_t are given by

$$pm(x_1,t_1,x_2,t_2,\dots,x_n,t_n) =$$
$$= \int_{B_S\times B_F} dxdy\; m(x,y)\; h(-\tfrac{T}{2},x,t_1,x_1;\tfrac{T}{2},y)\dots h(t_{n-1},x_{n-1};t_n,x_n,\tfrac{T}{2},y). \tag{A3}$$

Remarks: 1) The relation A1) is what one expects from a joint density m, and A2) shows that a Bernstein transition is nothing but a conditional probability, given a position in the past and another one in the future.

2) In A3) the final position y of the Bernstein transition is fixed. The analogous formula is valid for a fixed initial position x.

Our generalization of Lagrangian Mechanics will use the following way to construct a Bernstein transition:

Let $h = h(s,x,t,y) = h(x,t-s,y)$ be the fundamental solution of the Heat equation

$$-\hbar\frac{\partial\bar{\theta}}{\partial t} = H\bar{\theta} \tag{3.1}$$

where H is the Hamiltonian $H = -\frac{\hbar^2}{2}\Delta + V$. Under weak technical assumptions on H and, for example, V continuous and bounded below on $M=\mathbb{R}^n$, h is jointly continuous and strictly positive. Then, one easily verifies that

$$h(s,x;t,y;u,z) \equiv \frac{h(s,x,t,y)h(t,y;u,z)}{h(s,x;u,z)} \tag{3.2}$$

is the density of a Bernstein transition.

Notice that if $t = i\tau$, for $\tau > 0$, Eq. (3.1) is a Schrödinger equation. Using the terminology of Field Theory, Eq. (3.1) described therefore the Euclidean version of a Quantum Dynamics.

Let us underline that this is not the only dynamical realization of Bernstein processes. For an alternative approach, see [5].

B) Characterization of an unique Markovian Bernstein process.

We assume that the (positive) fundamental solution h of Eq. (3.1) is given. Let Z_t be the Bernstein process corresponding to a given joint density of probability m. Then this process Z_t is Markovian if and only if it is possible to find $\bar{\theta}_{-T/2}$ and $\theta_{T/2}$ positive (in fact of same signs on M) such that

$$\int_{B_S \times B_F} m(x,y)dxdy = \int_{B_S \times B_F} \bar{\theta}_{-T/2}(x)\; h(-\tfrac{T}{2}, x, \tfrac{T}{2}, y)\; \theta_{T/2}(y)dxdy. \tag{3.3}$$

Remarks: 1) It will be shown in C) that it is always possible to determine such a pair of functions $\bar{\theta}_{-T/2}(x)$ and $\theta_{T/2}(y)$.

2) For the choice of the joint probability (3.3), the finite dimensional distribution A3) reduces to

$$p_m(x_1,t_1,x_2,t_2,\ldots,x_n,t_n) =$$

$$= \int_M dxdy\; \bar{\theta}_{-T/2}(x)h(-\tfrac{T}{2}, x,t_1,x_1)\ldots h(t_n,x_n,\tfrac{T}{2},y)\; \theta_{T/2}(y). \tag{3.4}$$

Let us define the backward evolution of the given $\theta_{T/2}$ by

$$\theta(x,s) = \int_M h(s,x,\tfrac{T}{2},y)\; \theta_{T/2}(y)dy \equiv e^{-(\frac{T}{2}-s)H/\hbar}\; \theta_{T/2},\; s \in I. \tag{3.5}$$

Then one shows that

$$q(s,x,t,y) \equiv h(s,x,t,y)\,\frac{\theta(y,t)}{\theta(x,s)} \tag{3.6}$$

is the density of a Markovian transition probability, and that Z_t is a diffusion process with forward drift given by

$$B(x,t) \equiv \lim_{\Delta t\downarrow 0} \frac{1}{\Delta t} \int_{S_\varepsilon(x)} (y-x)\, q(t,x,t+\Delta t,y)dy = \hbar\, \frac{\nabla\theta}{\theta}(x,t) \tag{3.7}$$

where $S_\varepsilon(x) = \{y; |y-x| \le \varepsilon\}$, and diffusion matrix $C = \hbar\,\mathrm{I\!I}$.

The same process Z_t may also be described in terms of its backward transition probability

$$\bar{q}(s,x,t,y) = \frac{\bar{\theta}(x,s)}{\bar{\theta}(y,t)}\, h(s,x,t,y) \tag{3.8}$$

involving the forward evolution of the given $\bar{\theta}_{-T/2}$,

$$\bar{\theta}(y,t) = \int_M \bar{\theta}_{-T/2}(x)\, h(-\tfrac{T}{2}, x,t,y)dx \equiv \bar{\theta}_{-T/2}\, e^{-(t+\frac{T}{2})H/h}, \quad t\in I. \tag{3.9}$$

This yields the explicit form of the backward drift (cf. 2.1),

$$B_*(y,t) = \lim_{\Delta t\downarrow 0} \frac{1}{\Delta t} \int_{S_\varepsilon(y)} (y-x)\bar{q}(t-\Delta t,x,t,y)dx = -h\, \frac{\nabla\bar{\theta}}{\bar{\theta}}(y,t) \tag{3.10}$$

with the same diffusion matrix $C_* = C$.

Using (3.4) for the one-dimensional distribution, we obtain (up to a normalization) the probability of the Markovian Bernstein Z_t,

$$p(x,t)dx = \bar{\theta}(x,t)\theta(x,t)dx \quad \forall t \in I. \tag{3.11}$$

This is the key property of this Markovian Bernstein process: since, under time reversal, $\theta \longleftrightarrow \bar{\theta}$, the probability density is reversible in time. In 1931, this analogy with the quantum probability

$$\rho(x,t)dx = \psi^*(x,t)\,\psi(x,t)\,dx$$

was the starting point of Schrödinger's investigation [1].

C) Existence and uniqueness of the Markovian Bernstein process.

The construction of the Markovian Bernstein process Z_t is complete as soon as the functions $\bar{\theta}_{-T/2}$ and $\theta_{T/2}$ are known since, from these, all the finite distributions (3.4) of this process are given. By hypothesis, our data are the probabilities $p(x,-\frac{T}{2})dx$ and $p(y,\frac{T}{2})dy$. Using the marginals of the markovian point probability (3.3) we obtain the system of equations for $\bar{\theta}_{-T/2}$ and $\theta_{T/2}$

$$\begin{cases} \bar{\theta}_{-T/2}(x) \int_M h(x,-\frac{T}{2},y,\frac{T}{2})\,\theta_{T/2}(y)dy = p(x,-\frac{T}{2}) \\ \\ \theta_{T/2}(y) \int_M \bar{\theta}_{-T/2}(x)\,h(x,-\frac{T}{2},y,\frac{T}{2})dx = p(y,\frac{T}{2}) \end{cases} \quad . \tag{3.12}$$

I call this the Schrödinger system [1] for $\bar{\theta}_{-T/2}$ and $\theta_{T/2}$. The existence and uniqueness of its solutions has been investigated by Bernstein, Fortet, Beurling and Jamison. If $p(x,-\frac{T}{2})$ and $p(y,\frac{T}{2})$ are strictly positive, we have existence and uniqueness of solutions of same signs on M [4].

4. Stochastic Variational Principles

We begin by a probabilistic generalization of the Variational Principle involving the Action depending on the future configuration.
The backward drifts $\hat{B}_*$ of the admissible processes $\hat{Z}_\tau$ for this principle will vary in a class as large as possible. The diffusion matrix will remain constant ($= \hbar \mathbb{I}$). Any $\hat{B}_*$ adapted to a decreasing filtration F_τ, and such that there is an F_τ-Brownian motion $\hat{W}_*(\tau)$ so that

$$d_*\hat{Z}(\tau) = \hat{B}_*(\tau)d\tau + \hbar^{1/2}\,\mathbb{I}\,d\hat{W}_*(\tau), \quad -\frac{T}{2} \le \tau < t < \frac{T}{2} \tag{4.1}$$

and

$$\hat{Z}(t) = z\,, \quad \int_{-T/2}^{t} |\hat{B}_*(\tau)|d\tau < \infty \quad \text{a.s.}$$

is admissible. Such a $\hat{Z}(\tau)$ is generally not Markovian.

Now, we use the forward dynamics associated to a solution $\bar{\theta}_{-T/2}$ of the Schrödinger system (3.12). We denote $\bar{\theta}_{-T/2}(x)$ by $e^{(\bar{R}_{-T/2}-\bar{S}_{-T/2})(x)/h}$ and its forward evolution under (3.9) by $\bar{\theta}(z,t)$. The stochastic Action for any admissible $\hat{Z}(\tau)$ is defined by the function of the future position z

$$\bar{I}_{h,\hat{B}_*}(z,t) = E_{z,t}\int_{-T/2}^{t}\left\{\frac{1}{2}\,|D_*\hat{Z}(\tau)|^2 + V(\hat{Z}(\tau))\right\}d\tau + E_{z,t}(\bar{S}_{-T/2}-\bar{R}_{-T/2})(\hat{Z}(-\tfrac{T}{2})) \tag{4.2}$$

where $E_{z,t}$ denotes the conditional expectation given $\hat{Z}(t) = z$ and $D_*\hat{Z}$ is the backward velocity

$$D_*\hat{Z}(\tau) = \lim_{\Delta\tau\downarrow 0} E\left[\frac{\hat{Z}(\tau)-\hat{Z}(\tau-\Delta\tau)}{\Delta\tau}\,\middle|\,F_\tau\right] = \hat{B}_*(\tau)\ .$$

Notice that this action is nothing but that (cf. Remark 2, Chapt. 1)

$$-E_{z,t}\int_{-T/2}^{t}\bar{L}_c(\hat{Z},D_*\hat{Z},\tau)d\tau + E_{z,t}(\bar{S}_{-T/2}-\bar{R}_{-T/2})(\hat{Z}(-\tfrac{T}{2})).$$

The Least Action Principle results of the comparison between $\bar{I}_{\hbar,\hat{B}*}(z,t)$ and the function $\bar{A}_{\hbar}(z,t) = -\hbar\log\bar{\theta}(z,t)$:

<u>Least Action Principle</u>

For any admissible $\hat{Z}(\tau)$, $\bar{I}_{h,\hat{B}_*}(z,t) \geq \bar{A}_{\hbar}(z,t)$ (*) . Moreover, the Markovian Bernstein process $Z(\tau)$ is admissible, with drift $\hat{B}_*(\tau) = B_*(Z(\tau),\tau) = \nabla\bar{A}_{\hbar}(z(\tau),\ )$ (cf. 3.10) and reduces (*) to an equality. Finally, this $Z(\tau)$ satisfies the Euclidean Newton equation

$$\frac{1}{2}\,(DDZ + D_*D_*Z)(\tau) = \nabla V\ ,\quad -\frac{T}{2} \leq \tau < t \tag{4.3}$$

for $D_*Z(-\frac{T}{2}) = (\nabla\bar{S}_{-T/2} - \nabla\bar{R}_{-T/2})(Z(-\frac{T}{2}))$ and $Z(t) = z$.

Remarks

1) Here $DZ(\tau) = \lim_{\Delta\tau\downarrow 0} E\left[\frac{Z(\tau+\Delta\tau)-Z(\tau)}{\Delta\tau} \,|P_\tau\right] = B(Z(\tau),\tau)$, the forward analog of the velocity used before.

2) At the classical limit $h = 0$, the Action reduces to the classical expression, just as the Newton equation with force $+\nabla V$ (since we are in "imaginary time").

3) This is a reversible dynamics associated to the Heat equation, i.e. a new probabilistic interpretation of it (cf. Chapter 5). We shall call this new theory "(Schrödinger's) Stochastic Variational Dynamics" [5].

4) Idea of the proof: $\bar{\theta}$ solves $-h\frac{\partial\bar{\theta}}{\partial t} = H\bar{\theta}$ then $\bar{A}_h$ solves $\frac{\partial\bar{A}_h}{\partial t} - \frac{h}{2}\Delta\bar{A}_h + \frac{1}{2}(\nabla A_h)^2 - V = 0$, an equation of "Dynamic programming" whose solution is the minimal value of the Action functional $\bar{I}_{h,\hat{B}_*}$.

The probabilistic generalization of the Hamilton principle is due to K. Yasue [6]:
The Action functional is defined here by the (absolute) expectation

$$\bar{A}_h[Z] = -E\left[\int_{-T/2}^{T/2} \left\{\frac{1}{2} DZD_*Z + V(Z)\right\} dt\right] \tag{4.4}$$

which indeed reduces to the classical one at the limit $h = 0$.
Z is critical point of $\bar{A}_h$ if $\bar{A}_h[Z+\delta Z] - \bar{A}_h[Z] = O(\delta Z)$ where $\delta Z = \delta Z(z,t)$ is an arbitrary vector field with compact support in $M \times I$.
Then Z is critical point of $\bar{A}_h$ if and only if the Newton law holds

$$\frac{1}{2}(DDZ + D_*D_*Z)(\tau) = +\nabla V \qquad \tau \in I \tag{4.5}$$

in accordance also with (4.3).

Remarks:

1) The two given variational principles are quite different from each

other. The first one uses an Action defined as a conditional expectation, and then introduces a time asymmetry absent from the second one, in which the time symmetry of the Bernstein process is fully involved. This is reflected in the different form of the Lagrangians. Yasue's result was at the origin of this area of research.

2) According to the Least Action Principle, there is no non-Markovian diffusion process relevant in this dynamical theory. Because of the definition of Bernstein processes, this is not a trivial observation.

5. Schrödinger's Program and the Foundations of Quantum Mechnics

5.1 On the Heat equation and Quantum Mechanics

Among the theoretical physicists, there is an immemorial tradition of resistance to the interpretation of quantum phenomena in terms of classical diffusion processes. This resistance is perfectly justified. Let us consider the one dimensional free diffusion equation on $M = \mathbb{R}$,

$$\frac{\partial p}{\partial t} = \frac{\hbar}{2} \Delta p \qquad (5.1,1)$$

formally analogous to the free Schrödinger equation (if the parameter t is interpreted as an imaginary time $i\tau$, $\tau > 0$). An easy computation shows that

$$E[Z_t^2] = \int_M z^2 \, p(z,t) dz$$

is a linear function of time, and therefore

$$\text{Var } Z_t = z_o^2 + h\, t\,. \qquad (5.1,2)$$

This means that such an "analogue" of a quantum free evolution is as different as possible from this one: here, the "uncertainty" in the position increases continuously with time, in flat contradiction with the quantum situation where we always have contraction then spreading of the free wave packet. Notice that (5.1,2) expresses the irreversible nature of the time evolution.

Now let us consider the free evolution of a Markovian Bernstein process Z_t constructed as before.
Its equation of motion on $I = [-\frac{T}{2}, \frac{T}{2}]$ is given by Eq. (4.3) for $V = 0$,

$$\frac{1}{2}(DDZ + D_*D_*Z)(t) = 0 \;. \tag{5.1,3}$$

It is easy to verify that the "kinetic energy" term of the Yasue Lagrangian (4.4) is conserved during the free evolution,

$$E[-\frac{1}{2}DZD_*Z](t) = e = \text{constant}. \tag{5.1,4}$$

A natural extra hypothesis is that this constant e is positive and finite. (Notice that, at the Euclidean classical limit, the initial velocity will be interpreted as purely imaginary.)
A simple computation using this shows that the variance of Z_t is quadratic in time, but with a negative leading coefficient. The Dynamics of such a Markovian Bernstein process is a real classical analogue of a Quantum Dynamics.
Let us give an example of free Stochastic Variational Dynamics. Suppose that

$$\begin{aligned} p_{-T/2}(x)dx &= \left[\pi\left(\frac{a^2-T^2/4}{a}\right)\right]^{-1/2} e^{-\frac{a(x-v_0T/2)^2}{a^2+T^2/4}}\,dx \\ p_{T/2}(y)dy &= \left[\pi\left(\frac{a^2-T^2/4}{a}\right)\right]^{-1/2} e^{-\frac{a(y+v_0T/2)^2}{a^2+T^2/4}}\,dy \end{aligned} \tag{5.1,5}$$

$a > 0$ and v_0 being two constants.

Since we consider a free evolution, the kernel h in (3.12) is the Brownian kernel

$$h(s,x,t,y) = [2\pi(t-s)]^{-1/2}\, e^{-\frac{(y-x)^2}{2(t-s)}} \;. \tag{5.1,6}$$

Therefore, the Schrödinger system (3.12) to solve here is

$$\begin{cases} \bar{\theta}_{-T/2}(x) \int_M h(x,T,y)\, \theta_{T/2}(y)dy = p_{-T/2}(x) \\ \theta_{T/2}(y) \int_M \bar{\theta}_{-T/2}(x)\, h(x,T,y)dx = p_{T/2}(y) \end{cases} \quad . \tag{5.1,7}$$

The unique solution of this system (in the class we are looking for) is laboriously found:

$$\bar{\theta}_{-T/2}(x) = \left[\pi\left(\frac{a^2-T^2/4}{a}\right)\right]^{-\frac{1}{4}} e^{-\frac{a(x-v_oT/2)^2}{2(a^2-T^2/4)}}\, e^{v_o x - \frac{(x-v_oT/2)^2 T/2}{2(a^2-T^2/4)}} \tag{5.1,8}$$

$$\theta_{T/2}(y) = \left[\pi\left(\frac{a^2-T^2/4}{a}\right)\right]^{-\frac{1}{4}} e^{-\frac{a(y+v_oT/2)^2}{2(a^2-T^2/4)}}\, e^{-v_o y - \frac{(y+v_oT/2)^2 T/2}{2(a^2-T^2/4)}} \quad .$$

Using the forward and backward evolutions (3.5) and (3.9) of these solutions under the Brownian kernel h, we get $\bar{\theta}(x,t)$ and $\theta(x,t)$ on I and, from (3.11), the probability density of the Markovian Bernstein Z_t on this time interval,

$$p(x,t)dx = \left[\pi\left(\frac{a^2-t^2}{a}\right)\right]^{-1/2} e^{-\frac{a(x+v_ot)^2}{a^2-t^2}}\, dx \ . \tag{5.1,9}$$

This is a Gaussian distribution, with mean $-v_o t$ and variance $\frac{a^2-t^2}{2a}$. In particular, this variance is indeed of the expected form. Using the forward and backward drifts defined in (3.7) and (3.10), it may be seen that Z_t indeed solves the free Newton equation (5.1,3), for $t \in I$. Notice that this free solution cannot be extended outside of a time interval $[-a,a]$ since the variance of Z_t would become negative. However, in any $I = [-\frac{T}{2},\frac{T}{2}] \subset [-a,a]$ it is well-defined. Also observe that the time-evolution of the probability density of Z_t looks like a kind of "reciprocal" of a free quantum dynamics: till the time 0 it spreads, then contracts again.

This kind of stochastic processes appears sometimes in the context of stochastic control theory [7].

5.2 Stochastic Variational Dynamics and the Foundations of Quantum Mechanics

The key point of this paper has the character of a program and can easily be summarized: to investigate the Foundations of Quantum Mechanics, it is useful to understand (Schrödinger's) Stochastic Variational Dynamics.

Let us consider the following hypothetical problem for a theoretical physicist.

For a known potential V and $p_{-T/2}(dx)$, $p_{T/2}(dy)$ any pair of probabilities (without zeros, for simplicity), an experimentalist prepares N diffusing particles in such a way that $Np_{-T/2}(x)dx$ particles leave initially the region $(x,x+dx)$, and repeats the experience long enough to receive $Np_{T/2}(y)dy$ into $(y,y+dy)$. The problem of the theoretician is to construct the dynamical theory of these relatively improbable process, and the result is Schrödinger's Stochastic Variational Dynamics.
I claim that during the elaboration of his model, this physicist will confront the same qualitative difficulties as in Quantum Mechanics. For example, it is shown in [5] that the resulting theory is not separable, in a sense completely analogous to Quantum Mechanics

From the technical point of view, notice that Stochastic Variational Dynamics is quite hard to use because we need first to solve the Schrödinger system (3.12) and this is, in general, far to be easy. It is therefore not without interest to get rid of this difficulty in a particular realization of the dynamical framework.

Suppose that we know a (regular enough) solution of the Schrödinger equation on $I = [-\frac{T}{2}, \frac{T}{2}]$

$$i\hbar \frac{\partial\psi}{\partial t} = H\psi \qquad (5.2,1)$$

H being the Hamiltonian of (3.1). Let us represent this solution by

$$\psi(x,t) = e^{[R(x,t) + iS(x,t)]/\hbar} . \qquad (5.2,3)$$

By hypothesis, we know in particular the quantum analog of the classical initial velocity,

$$v_o = \nabla S(x,0) \ . \tag{5.2,4}$$

Therefore, we also can regard (5.2,3) as a function of this initial velocity, $\psi_{v_o} = \psi_{v_o}(x,t)$. Now we define

$$\begin{aligned}\bar{\theta}(x,t) &= \psi_{-iv_o}(x,-it) \\ &= e^{[R_{-iv_o}(x,-it) + iS_{-iv_o}(x,-it)]/h} \ .\end{aligned} \tag{5.2,5}$$

Since we know the equations of motion of R and S by Eq. (5.2,1), it is easy to find the dynamics of $\bar{R}$ and $\bar{S}$ defined by

$$\bar{R}(x,t) = R_{-iv_o}(x,-it) \quad \text{and} \quad \bar{S}(x,t) = -iS_{-iv_o}(x,t). \tag{5.2,6}$$

The result is

$$\begin{cases} \dfrac{\partial \bar{R}}{\partial t} = -\dfrac{h}{2}\Delta \bar{S} - \nabla\bar{R}\cdot\nabla\bar{S} \\[2ex] \dfrac{\partial \bar{S}}{\partial t} = -\dfrac{h}{2}\Delta\bar{R} - \dfrac{1}{2}(\nabla\bar{R})^2 - \dfrac{1}{2}(\nabla\bar{S})^2 + V \end{cases} \ . \tag{5.2,7}$$

Equations (5.2,7) are real, and then they have real solutions. Now, by construction, $\bar{\theta}(x,t) = \exp(\bar{R}(x,t) - \bar{S}(x,t))/h$ linearizes this non-linear partial differential system, and it is easy to check that $\bar{\theta}$ solves the Heat equation (3.1). Symmetrically, $\theta(x,t) = \exp(\bar{R}(x,t) + \bar{S}(x,t))/\hbar$ solves the time reversed equation $\hbar \frac{\partial\theta}{\partial t} = H\theta$.

The couple $\theta,\bar{\theta}$ is precisely the one required in Chapter 3 for constructing a Bernstein process Z_t . This means that they are, respectively, the backward and forward evolutions of the unique (couple) $\theta_{T/2}(x)$ and $\bar{\theta}_{-T/2}(x)$ of solutions of the Schrödinger system (3.12), for the given boundary probabilities

$$p(x,-\tfrac{T}{2})dx = e^{2\bar{R}(x,-\frac{T}{2})/\hbar}\,dx ,$$
$$p(y,\tfrac{T}{2})dy = e^{2\bar{R}(y,\frac{T}{2})/\hbar}\,dy . \qquad (5.2,8)$$

The construction of Chapter 3 and 4 assures us of the existence and uniqueness of the associated Markovian Bernstein process Z_t, $t \in I$. In particular, Z_t solves the Euclidean Newton equation (4.3) which follows from the Least Action Principle. The above mentioned explicit example was modelled in this way from a free solution $(V = 0)$ of Eq. (5.2,1). Knowing everything about Z_t, we can take the backward analytical continuation in time. It is shown in [5,8] that we get another diffusion process on I, denoted by X_t. This one solves another Newton equation, namely

$$\frac{1}{2}(DD_*X + D_*DX) = -\nabla V . \qquad (5.2,9)$$

Notice that the left hand "acceleration" is different from the one used in (4.3), and that the "force" is $-\nabla V$, as it must be in a real time theory (in order to be consistent with the classical limit). Observe also that this constructive approach is valid as well for the processes associated to any quantum excited states. The presence of the nodes in the boundary probabilities is therefore allowed [5].

To know everything about Z_t on I is equivalent to know everything about X_t. The real time process X_t has been discovered by Nelson twenty years ago [9]. The completely different construction sketched here opens new perspectives for the interpretation of the resulting theory ("Stochastic Mechanics"), and particularly for the extension to Quantum Field Theory. Stochastic Variational Dynamics also constitutes a call for experiments. The "Euclidean" process Z_t is a purely classical diffusion process associated (in a non-canonical way) to the Heat equation. Nevertheless, its characteristics are very close to the ones considered as exclusively relevant to Quantum Mechanics. And it is certainly much easier to do critical experiments with classical diffusive particles than with polarized photons, for example (see [10]). Since the Foundations of Quantum Mechanics are still at the origin of very controversial interpretations, it would be regrettable not to investigate, as a very close analogy of the typical quantum phenomena, this classical probabilistic model, initiated fifty years ago by E. Schrödinger, apparently in this goal.

Acknowledgements

The author would like to thank the organizing committee for this very stimulating meeting in Udine. Most of the ideas described here have their origin in invaluable discussions with Professor E. Nelson during the two last years, in Princeton University.

It is also a pleasure to benefit from the critical comments of Professors S. Albeverio, Ph. Blanchard and K. Yasue.

References:

1. E. Schrödinger, Sitzungsbericht der Preußischen Akademie, Phys. Math. Classe, 144 (1931); Ann. de l'Institut Henri-Poincaré 11, 300 (1932).
2. S. Bernstein, "Sur les liaisons entre les grandeurs aléatoires" in Verh. des intern. Mathematikerkongt. Zürich, Band 1 (1932).
3. R. Feynman and A. Hibbs, "Quantum Mechanics and Path Integrals", McGraw-Hill, New York (1965).
4. B. Jamison, Z. Wahrscheinlichkeitstheorie ver. Gebiete 30, 65 (1974); A. Beurling, Annals of Mathematics 72, 1, 189 (1960); R. Fortet, J. Math. Pures et Appl. IX, 83 (1940).
5. J.C. Zambrini, "Variational processes and stochastic versions of mechanics", to appear in J. of Math. Physics; "Stochastic Mechanics according to E. Schrödinger, to appear in Phys. Rev. A.
6. K. Yasue, J. Math. Phys. 22, 5, 1010 (1981); J. Funct. Anal. 41, 327 (1981).
7. A. Blaquiere, paper presented in this conference.
8. S. Albeverio, K. Yasue, J.C. Zambrini, in preparation.
9. E. Nelson, Phys. Rev. 150, 1079 (1966); "Quantum Fluctuations", Princeton U. Press (1985).
10. J.A. Wheeler and W.H. Zurek, Eds., "Quantum Theory and Measurements", Princeton U. Press, Princeton, N.J. (1983).

PART III

QUANTUM STOCHASTIC PROCESSES

AN INTRODUCTION TO NON STANDARD ANALYSIS AND APPLICATIONS TO QUANTUM THEORY

Sergio Albeverio
Ruhr Universität Bochum, Bochum, Federal Republic of Germany

ABSTRACT

We briefly sketch the basic theory of non-standard analysis. We also discuss some applications to quantum theory and related areas (stochastic processes, partial differential equations).

0. Introduction

The use of infinitesimal and infinite quantities has a long history in our cultural tradition, see e.g. [1] - [3] and references therein. On the more philosophical side the well known criticism of Anaxagoras (500-428 B.C.) and Zeno of Elea (ca. 460 B.C.) of the concept of continuum involve considerations of infinitesimal quantities; there is mentioning of infinite numbers in Plato (427-347 B.C) and Eudoxus (ca. 400-350 B.C.), and Archimedes (287-212 B.C.) made a quite extensive use of limiting procedures involving infinitesimal and infinite numbers. There is mentioning of "different infinities" in Thabit Ibn Qurra (836-901), and in work by the cardinal Nicolaus Cusanus (1401-1464) there is a revival of infinitesimal methods, showing e.g. that the area of a circle of unit radius is one half the length of the boundary. F. Cavalieri (1598-1647) developed quite systematically a "method of indivisibilia" for calculating areas and volumes, Galilei, Kepler, Pascal, Gregory, Wallis and other "natural philosophers" all used "infinitesimal methods", until, as well known, the very advent (around 1660-1670) of "integral and differential calculus", with G.W. Leibniz (1646-1716), I. Newton (1642-1707) and, in Japan, Seki Kowa (1642-1708). The first textbook on "calculus" in 1696, by the

Marquis de l'Hospital has the explicit mentioning of infinitesimals in the title "Analyse des infiniments petits ..." , as well as the very influential textbook "Introductio in Analysis Infinitorum" by L. Euler, published in 1748.
Analysis with infinitesimal and infinite numbers continued to be used in mathematics and physics of the 17./18 century (Jakob and Johannes Bernouilli, Huyghens ...), despite philosophical criticism e.g. by the bishop Berkeley (infinitesimals as "ghosts of departed quantities"). The motto was D'Alembert's one "continuez et la foi vous viendra" (his article for the Encyclopedia was very influential for the practical establishment of the use of calculus). Only in the XIX century with Bolzano (1781-1848) ("Paradoxien des Unendlichen"), Cauchy (1789-1857) and above all in the second half of the century with Weierstrass ("ε-δ-method"), Dedekind and Cantor a systematic "foundation" of infinitesimal calculus resulting in banishment of infinitesimals from the main stream of mathematics (and replacement by the concept of limit) was carried through. However it is well known that a calculus with infinitesimals continued to live parallel to the main stream (e.g. in Paul Du Bois-Reymond analysis, in Veronese's differential geometry, in T. Levi-Civita's formal power series, in physics, and even in algebra (Hahn)...).
A complete rigorous general and systematic development of "infinitesimal methods" has however only been possible quite recently by A. Robinson ((1918-1974), in work around 1960) using methods of mathematical logic which have their origins in papers by Löwenheim (1915) and Th. Skolem (1934) (but perhaps methods of model theory - E. Hewitt's ultrapowers 1948, Łos 1955 - and the Malcev-Henkin compactness theorem should also be mentioned). In a sense this foundation, the so called "non standard analysis" realized the original Leibniz dreams of a rigorous calculus with infinitesimal and infinite numbers. Beyond that it leads to far reaching new applications. Some of these applications came immediately with the new foundations (in fact, even before, in 1958 Schmieden and Laugwitz had a partial solution with applications to analysis; Luxemburg's lectures (1962/64) [5] were very influential in making known the method among mathematicians; Nelson's new approach (1977) [17] ("internal set theory"), besides using a set theoretical framework of great interest in itself also exerced a strong influence on research).
As with usual analysis, one can start axiomatically by postulating the existence of a field of numbers and rules how to operate with them or one can "construct" the field from "more elementary" numbers (e.g. real numbers from rational numbers). We shall use here the latter approach and give a construction of the basic field of numbers, the non standard real line ${}^*\mathbb{R}$, of non standard analysis, enlarging the reals $\mathbb{R}$ by adjonction of infinitesimal and infinite numbers. This construction uses the so called "ultrapower model" (for other approaches see e.g [10][17]-[20],[62]). In this paper we shall use the basic terminology of most books on non standard analysis (e.g. [1]-[16]). It should be noted that the books [19],[20] ,[62] and several research articles, e.g. [17],[42],[43],[54],[68],[69] use Nelson's terminology. However a translation between these languages is not difficult. This lectures are based on a joint book with J.E. Fenstad, R. Høegh-Krohn and T. Lindstrøm [4], to which we refer for a more thorough discussion and

further applications. For more extensive references we refer to this book [4] and to the bibliography of [70].

I. The number field $^*\mathbb{R}$ of non standard analysis

We shall see how one can solve, in a simple way, the following problem: Construct an enlargement of $\mathbb{R}$, with the same algebraic and order properties as $\mathbb{R}$, containing infinitesimal and infinite numbers.

Let m be a given 0-1-valued, finitely additive normalized measure defined on all subsets of $\mathbb{N}$ and vanishing on all finite subsets of $\mathbb{N}$, i.e.

$$0 \leqq m(A) \leqq 1 \ \forall A \subset \mathbb{N};\ m(\phi) = 0,\ m(\mathbb{N}) = 1;\ m\left(\bigcup_{i=1}^{n} A_i\right) = \sum_{i=1}^{n} m(A_i) \text{ if } A_i \cap A_j = \phi$$

$i \neq j$, $A_i \subset \mathbb{N}$; $m(A) = 0$ if A is finite.

Remark:

The existence of such m is actually equivalent with the socalled "ultrafilter theorem", which follows from Zorn's lemma (but is actually strictly weaker than Zorn's lemma or, equivalently, Zemelo's axiom of choice).

Definition: Sets $A \subset \mathbb{N}$ such that $m(A) = 0$ will be called small, sets $A \subset \mathbb{N}$ such that $m(A) = 1$ will be called big.

Proposition 1: a) Finite sets are small; b) complements of finite sets (the so-called cofinite sets) are big; c) there are infinite subsets of $\mathbb{N}$ which are small.

Proof: a) is part of the definition of m; b) follows from $m(A)=m(\mathbb{N}-C)=0$ if $A = \mathbb{N}-C$, A finite, which together with $M(\mathbb{N}) = 1$ yields $m(C) = 1$; c) e.g. either $A = \{2n-1, n \in \mathbb{N}\}$ or $B = \{2n, n \in \mathbb{N}\}$ is small, because if both were big, we would have $m(A) = m(B) = 1$ and then $m(A \cup B) = m(A) + m(B) = 2$, on the other hand, $A \cup B = \mathbb{N}$ hence $m(\mathbb{N}) = 1$, a contradiction.

Remark: Clearly whether $m(A) = 0$ or $m(B) = 0$ in c) depends on m. Call U the family of big subsets of $\mathbb{N}$, i.e. $U = \{A \subset \mathbb{N} \mid m(A) = 1\}$. Then b), c) above can be expressed by $U \supset \mathrm{Cof}\ \mathbb{N}$, where Cof $\mathbb{N}$ is the family of all cofinite subsets of $\mathbb{N}$, i.e. $\mathrm{Cof}\ \mathbb{N} = \{B \subset \mathbb{N} \mid \mathbb{N} - B \text{ is finite}\}$.

Proposition 2: U is an ultrafilter, i.e. U is a filter: F1) $\phi \notin U$; F2) $\mathbb{N} \in U$; F3) $A \in U$, $B \supset A \Rightarrow B \in U$; F4) $A, B \in U \Rightarrow A \cap B \in U$ and: UF) U has the maximality property with respect to inclusion or, equivalently, has the property that for any $A \subset \mathbb{N}$ we have either $A \in U$ or $\mathbb{N} - A \in U$ (but not both!).

Proof: The filter properties F1) - F4) follow easily from the properties of m. The specific ultrafilter property UF) follows from the fact that $A \subset \mathbb{N} \Rightarrow$ either $m(A) = 1$ or $m(A) = 0$, since m is 0-1-valued. In the first case we have $A \in U$, by definition of U, in the second case we have $m(\mathbb{N} - A) = 1$, hence $\mathbb{N} - A \in U$.

Proposition 3: There is a 1-1 correspondence between ultrafilters U extending Cof $\mathbb{N}$ and measures of type m, given by $m(A) = 1 \Leftrightarrow A \in U$.

Proof: Above Proposition 2 shows that, given m, U defined as

$\{A \mid m(A) = 1\}$ is an ultrafilter. Conversely, if V is an ultrafilter and we define m by $m(A) = 1$ iff $A \in V$, $m(A) = 0$ iff $A \notin V$, then we have $m(\phi) = 0$, $m(\mathbb{N}) = 1$ and $m(A \cup B) = 1$ if $A \in V$ since $A \cup B \supset A$ hence $A \cup B \in V$, hence $m(A \cup B) = m(A) + m(B)$, for $A \cap B = \phi$ (since $A \in V \to B \notin V$, hence $m(B) = 0$, when $A \cap B = \phi$, otherwise we would have a contradiction to F1), F4)).

Remark: Any 0-1-valued normalized additive measure m vanishing on finite subsets of $\mathbb{N}$ is not σ-additive.

Proof: We have $m(\bigcup_{n=1}^{\infty} \{1,\ldots,n\}) = m(\mathbb{N}) = 1$, but $m(\{1,\ldots,n\}) = 0 \ \forall n \in \mathbb{N}$.

Remark: It is sometimes useful to think of m as a (finitely additive) probability measure and use the corresponding terminology: m-almost surely, m-almost everywhere, with m-probability one ... Thus small sets are m-zero sets, large sets are m-a.s. sets.

As mentioned in the Introduction, we would like to distinguish sequences in $\mathbb{R}$ converging with different speeds to the same real number, like $\frac{1}{n}$ and $\frac{1}{n^2}$ as $n \to \infty$. The trick is to use equivalence classes modulo an ultrafilter U (like the one associated to the 0-1-measure m we started from).

Let $\mathbb{R}^{\mathbb{N}}$ be the set of all sequences with values in $\mathbb{R}$, i.e. $f \in \mathbb{R}^{\mathbb{N}} \Leftrightarrow f$ maps $\mathbb{N}$ into $\mathbb{R}$, i.e. $f(i) \in \mathbb{R} \ \forall i \in \mathbb{N}$.
We introduce the following equivalence $\sim$ between sequences f,g: $f \sim g \Leftrightarrow \{i \in \mathbb{N} \mid f(i) = g(i)\} \in U$, i.e. $f \sim g$ iff the set of components i s.t. $f(i) = g(i)$ is big, i.e. $f \sim g$ iff $f = g$ (m-)almost surely.

That $\sim$ is an equivalence relation is a consequence of the filter properties (and is very clear for the "almost sure interpretation")!

Let $\langle f \rangle$ be the equivalence class to the sequence f with respect to $\sim$, i.e. $\langle f \rangle = \langle g \rangle$ whenever $f \sim g$. We shall write $\langle f \rangle = \{f(i),\ i \in \mathbb{N}\}/U$ (since the equivalence $\sim$ is determined by U).

Let ${}^*\mathbb{R} \equiv \mathbb{R}^{\mathbb{N}}/U$ be the set of all equivalence classes $\langle f \rangle$. We define of course equality in ${}^*\mathbb{R}$ by $\langle f \rangle = \langle g \rangle \Leftrightarrow f \sim g$, and inequality in ${}^*\mathbb{R}$ by $\langle f \rangle \neq \langle g \rangle \Leftrightarrow f \not\sim g \Leftrightarrow f \neq g$ a.s. (here the ultrafilter property of U is used).

There exists a natural embedding * of $\mathbb{R}$ into ${}^*\mathbb{R}$:
$r \in \mathbb{R} \Rightarrow \langle f \rangle = \{f(i) = r \ \forall i \in \mathbb{N}\}/U = \{r,r,\ldots,r,\ldots\}/U$.

We shall also denote this sequence $\langle f \rangle$, i.e. $\{r,r,\ldots,r,\ldots\}/U$ modulo U, by ${}^*r$ or $\langle r \rangle$. We shall call ${}^\sigma\mathbb{R}$ the set of all ${}^*r$ with $r \in \mathbb{R}$. We have ${}^\sigma\mathbb{R} \subset {}^*\mathbb{R}$ (we shall see below ${}^\sigma\mathbb{R} \subsetneqq {}^*\mathbb{R}$).

$\mathbb{R}$ is a field with operations $+,\cdot$, and neutral elements 0 resp. 1. We want ${}^*\mathbb{R}$ to be a field, too. Therefore, we must introduce in ${}^*\mathbb{R}$ the operations addition and multiplication and corresponding neutral elements. It is practical to denote addition in ${}^*\mathbb{R}$ by + again and multiplication by $\cdot$, as in $\mathbb{R}$. We define + in ${}^*\mathbb{R}$ by

$\langle h\rangle = \langle f\rangle + \langle g\rangle \Leftrightarrow h = f + g$ a.s., and define $\cdot$ in ${}^*\mathbb{R}$ by

$\langle h\rangle = \langle f\rangle \cdot \langle g\rangle \Leftrightarrow h = f \cdot g$ a.s. (often later on $\cdot$ will be dropped, i.e. $\langle f\rangle\langle g\rangle$ stands then for $\langle f\rangle \cdot \langle g\rangle$). By the filter property, these definitions are independent of the chosen representatives. The neutral element for the sum in ${}^*\mathbb{R}$ is ^{*0}, the neutral element for the multiplication in ${}^*\mathbb{R}$ is ^{*1}, as easily verified. There are no zero-divisors, i.e. $\langle f\rangle\langle g\rangle = {}^*0 \Leftrightarrow \langle f\rangle = {}^*0$ or $\langle g\rangle = {}^*0$. In fact, if $\langle f\rangle \neq {}^*0$, $\langle g\rangle \neq {}^*0$ then $A \equiv \{i \mid f(i) \neq 0\}$ and $B \equiv \{i \mid g(i) \neq 0\}$ are big, so $C \equiv \{i \mid f(i)g(i) \neq 0\} \supset A \cap B$ is big (by the filter properties F3), F4)), hence defining $\langle h\rangle$ by $h(i) = f(i)g(i)$ for $i \in C$, $h(i) = 0$ for $i \in \mathbb{N} - C$, we have $\langle h\rangle = \langle f\rangle\langle g\rangle \neq {}^*0$. Thus $\langle f\rangle\langle g\rangle = {}^*0 \Rightarrow \langle f\rangle = {}^*0$ or $\langle g\rangle = {}^*0$. Conversely, if $\langle f\rangle = {}^*0$ then if $\langle h\rangle \equiv \langle f\rangle \cdot \langle g\rangle$ we have for a big set A of i's: $h(i) = f(i)g(i)$. But then, with $B \equiv \{i \mid f(i) = 0\}$ (a big set), we have $A \cap B$ big, and $h(i) = 0$ for i in this big set, hence $\langle f\rangle\langle g\rangle = {}^*0$. Similarly for $\langle g\rangle = {}^*0$.

Example: $\langle f\rangle = \{1,0,3,0,\ldots,0,2n+1,0,\ldots\}/U$; $\langle g\rangle = \{0,2,0,4, 0,2n,0,\ldots\}/U$. Then $\langle f\rangle \cdot \langle g\rangle = {}^*0$, and indeed either $\langle f\rangle = {}^*0$ or $\langle g\rangle = {}^*0$, which corresponds to the cases $A = \{2n-1, n \in \mathbb{N}\} \notin U$ resp. $\mathbb{N} - A = \{2n, n\ \mathbb{N}\} \notin U$, since in the first case $f(i) \neq 0$ only on A, hence $\langle f\rangle = {}^*0$, A being small, in the second case $g(i) \neq 0$ only on $\mathbb{N} - A$, hence $\langle g\rangle = {}^*0$, $\mathbb{N} - A$ being small. Incidentally, this example illustrates well the role of the basic ultrafilter in excluding zero divisors.
How does the inverse $\langle f\rangle^{-1}$ of $\langle f\rangle \neq {}^*0$ in ${}^*\mathbb{R}$ look like? We can define $\langle f\rangle^{-1}$ as $\langle g\rangle$, with $g(i) = 1/f(i)$ for those i for which $f(i) \neq 0$ (there is a big set of such i's), and $g(i) = 1$ (e.g.) for those i for which $f(i) = 0$. Then we are sure that $f(i)g(i) = 1$ on a big set of i's, hence indeed $\langle f\rangle\langle g\rangle = \langle g\rangle\langle f\rangle = {}^*1$, thus $\langle g\rangle = \langle f\rangle^{-1}$.
In this way then we see that ${}^*\mathbb{R}$ is made into a field of numbers. What about a total order in ${}^*\mathbb{R}$ (recall that $\mathbb{R}$ is totally ordered)?
We define for any $\langle f\rangle, \langle g\rangle \in {}^*\mathbb{R}$:

$\langle f\rangle < \langle g\rangle$ iff $f(i) < g(i)$ for a.e. i; $\langle f\rangle > \langle g\rangle$ iff $f(i) > g(i)$ for a.e. i.
We then have the

Proposition 4: ${}^*\mathbb{R}$ is totally ordered with respect to $<$ (i.e. for any given $\langle f\rangle \neq \langle g\rangle$ either $\langle f\rangle < \langle g\rangle$ or $\langle g\rangle < \langle f\rangle$, and $\langle f\rangle < \langle g\rangle$, $\langle g\rangle < \langle h\rangle$ implies $\langle f\rangle > \langle h\rangle$).

Proof: By the ultrafilter property of U we have that either is the set $\{i \mid f(i) < g(i)\}$ in U or its complement in $\mathbb{N}$, $\{i \mid f(i) \geq g(i)\}$, is in U. Hence either $\langle f\rangle < \langle g\rangle$ or $\langle f\rangle \geq \langle g\rangle$.

Example: Is $(-1,2,-1,2,\ldots)/U < (-3,3,-3,3,\ldots)/U$?
Answer: Yes if $\{2,4,6,\ldots\} \in U$; no if $\{1,3,5,\ldots\} \in U$ (exactly one of the two cases occurs, by the ultrafilter property of U).

Remark: The embedding $*$ of $\mathbb{R}$ in ${}^*\mathbb{R}$ (through ${}^\sigma\mathbb{R} \subset {}^*\mathbb{R}$) preserves the order, i.e. $r < s$ in $\mathbb{R}$ implies ${}^*r < {}^*s$. (Since ${}^*r = \{r,r,\ldots r\ldots\}/U$ and $r < s$ implies all components of ${}^*r$ are less than those of ${}^*s$).
What have we gained in going from $\mathbb{R}$ to ${}^*\mathbb{R}$?

Theorem 5: ${}^*\mathbb{R}$ is a proper extension of ${}^\sigma\mathbb{R}$ (hence of $\mathbb{R}$ as embedded in ${}^*\mathbb{R}$), i.e. ${}^*\mathbb{R} \supsetneqq {}^\sigma\mathbb{R}$. In fact, ${}^*\mathbb{R}$ contains numbers which are strictly larger than any number in ${}^{\sigma}\mathbb{R}$ (i.e. than any real number).

Proof: Let us consider $\langle f\rangle \equiv \{1,2,\ldots n,\ldots\}/U$.

We have by definition $\langle f\rangle \in {}^*\mathbb{R}$, moreover $\langle f\rangle > {}^*r\ \forall r \in \mathbb{R}$, since $\{i \in \mathbb{N} \mid f(i) > r\} = \{i \in \mathbb{N} \mid i > r\}$ is big (being a cofinite set, containing all numbers $r+1$, $r+2,\ldots$).

Thus $\langle f\rangle \notin {}^{\sigma}\mathbb{R} = \{{}^*r, r \in \mathbb{R}\}$: qed.

It will be convenient, from now on, to use a simpler notation for elements in ${}^*\mathbb{R}$, namely to denote elements of ${}^*\mathbb{R}$ by letters. Let ω be the element $\langle f\rangle = \{1,2,\ldots,n\ldots\}/U$ occurring in the above proof. We have seen above that $\omega > {}^*r\ \forall r \in \mathbb{R}$. Using the algebra and order in ${}^*\mathbb{R}$ we can then produce many other elements in ${}^*\mathbb{R}$ which are not in ${}^{\sigma}\mathbb{R}$, e.g. $\omega^2 > {}^*r$, in fact $\omega^2 > \omega$. Moreover, $n\omega > \omega$, $\forall n \in \mathbb{N}$, $n > 1$, $\omega + k > \omega\ \forall k \in \mathbb{R}_+$, etc. Moreover we can produce easily elements in ${}^*\mathbb{R}$ which are smaller than any ${}^*r$, $r > 0$, $r \in \mathbb{R}$: e.g. $\omega^{-1} = \{1, \frac{1}{2}, \ldots, \frac{1}{n}, \ldots\}/U$.

We shall need some simple definition to classify these numbers. For $s \in {}^*\mathbb{R}$ we set $|s| = s$ if $s \geq {}^*0$, $|s| = -s$ if $s \leq {}^*0$. $s \in {}^*\mathbb{R}$ is called <u>positive</u> (or nonnegative) if $s \geq {}^*0$, <u>negative</u> (or nonpositive) if $s \leq {}^*0$.

Definition: $s \in {}^*\mathbb{R}$ is called <u>infinite</u> iff $|s| > {}^*n\ \forall n \in \mathbb{N}$, <u>finite</u> iff $|s| < {}^*n$ for some $n \in \mathbb{N}$. $s \in {}^*\mathbb{R}$ is called <u>infinitesimal</u> iff either $s = {}^*0$ or $s = \omega^{-1}$ for some infinite ω.

Remark: We have s infinitesimal iff $|s| < {}^*r\ \forall r > 0$, $r \in \mathbb{R}$. Thus for $\omega = \{1,2,\ldots,n,\ldots\}/U$ we have ω^{-1} infinitesimal, but also e.g. $(\omega^{-2}) < \omega^{-1}$ infinitesimal, $r\omega^{-1}$ infinitesimal for all $r \in {}^{\sigma}\mathbb{R}$, $r \neq 0$.

Definition: $s \in {}^*\mathbb{R}$ is called <u>standard</u> (real) iff $s \in {}^{\sigma}\mathbb{R}$ (i.e. if s is a real number as embedded in ${}^*\mathbb{R}$; i.e. the components of s in a big set are all identical), $s \in {}^*\mathbb{R}$ is called <u>non-standard</u> iff $s \in {}^*\mathbb{R} - {}^{\sigma}\mathbb{R}$.

All the numbers in ${}^*\mathbb{R}$ (standard and non-standard) are called <u>hyperreal</u> (and ${}^*\mathbb{R}$ is the field of hyperreal numbers).

Notation: It is often convenient in the following to drop the distinction between $\mathbb{R}$ and ${}^{\sigma}\mathbb{R}$ i.e. to think $\mathbb{R}$ as embedded in ${}^*\mathbb{R}$. In the same spirit one writes then r instead of ${}^*r$ (in particular 0 instead of ^{*0}, 1 instead of ^{*1}).

We shall now exploit infinitesimals to introduce a "nearness" concept (a substitute of topology!) in ${}^*\mathbb{R}$.

Definition: For any given number $s \in {}^*\mathbb{R}$ one defines <u>the monad $\mu(s)$ of s</u> as follows: $\mu(s) \equiv \{t \in {}^*\mathbb{R} \mid \exists$ infinitesimal ε such that $t = s + \varepsilon\}$.

I.e. $\mu(s)$ consists of s and all numbers in ${}^*\mathbb{R}$ which differ from s only by a (positive or negative) infinitesimal (intuitively we can think of $\mu(s)$ as an "infinitesimal cloud" around s).

E.g. the monad $\mu({}^*0)$ of ^{*0} is the set of all infinitesimals.

Definition: For any two numbers $s_1, s_2 \in {}^*\mathbb{R}$ we say that s_1 is "(infinitely) close to s_2" or "(infinitely) near to s_2" iff $s_1 - s_2$ is an infinitesimal, i.e. the monads of s_1 and s_2 coincide. We then write $s_1 \approx s_2$.

E.g. for ε infinitesimal, we have $\varepsilon \approx {}^*0$. The monad $\mu(s)$ of s can be written as $\mu(s) = \{t \in {}^*\mathbb{R} \mid t \approx s\}$.

Intuitively, the finite numbers in ${}^*\mathbb{R}$ should be near to standard numbers, i.e. we should be able to extract from them a "standard part". This is the content of the following theorem, in which we denote by F the finite numbers of ${}^*\mathbb{R}$.

Theorem 6: a) There exists an uniquely defined map st (also denoted by ^{o}) from F into $\mathbb{R}$ such that for any $x \in F$ we have x-stx is infinitesimal (i.e. $x \approx \mathrm{st}x$ i.e. $x \in \mu(\mathrm{st}x)$). st is a surjective homomorphism (relative to the field operations), in particular, st (x+y) = st x + st y, st(xy) = = st x st y. st x is the least upper bound in $\mathbb{R}$ of $\{r \in \mathbb{R} \mid {}^*r < x\}$.

b) ${}^{\sigma}\mathbb{R}$ is isomorphic (hence can be identified) (as an ordered field) with the quotient F/I of F modulo the ideal I of infinitesimal numbers in F. (In short: real numbers are finite numbers modulo infinitesimals).

Proof: a) Uniqueness is clear, suppose $r, s \in \mathbb{R}$ s.t. $r \approx x$, $s \approx x$, then $r - s \approx 0$, but r,s being real this implies r = s. As to the existence, let $x \in F$ and set $S_x \equiv \{r \in \mathbb{R} \mid {}^*r < x\}$. Then $S_x \neq \emptyset$, since $|x| < {}^*s$ for some $s \in \mathbb{R}$, x being finite, hence $-s \in S_x$. s is an upper bound of S_x, and by the Dedekind completeness of $\mathbb{R}$, S_x has then a least upper bound t in $\mathbb{R}$. Then $x \leq t + r$ $\forall$ real $r > 0$, thus $x - {}^*t \leq r$. On the other hand, $t - r \leq x$ (otherwise t - r would be an upper bound to S_x, smaller than t!). Hence

$-r \leq x - t \leq r$ $\forall$ $r \in \mathbb{R}$, $r > 0$ hence $x - t \approx 0$.
This proves a).

b) is proven using that I is a maximal ideal in F and F/I is a totally ordered, archimedean field, thus isomorphic to a subfield of $\mathbb{R}$. Since it contains however $\mathbb{R}$ it must be isomorphic to $\mathbb{R}$ (for details see [5], [6], [14]).

Remark: Clearly, we have at disposal an "algebraic" way of expressing "near", rather than a topological one. ${}^*\mathbb{R}$ can be made into a normal Hausdorff topological space in an interval topology, but it is disconnected and the embedding of $\mathbb{R}$ into ${}^*\mathbb{R}$ is not topological (in fact, the topology induced on $\mathbb{R}$ by the interval topology in ${}^*\mathbb{R}$ is the discrete one). It is better altogether, at least at the beginning, not to think at all topologically when handling with ${}^*\mathbb{R}$ (those interested in these aspects might look at [7], [14], e.g.).

II. Basic Structures for elementary Analysis over ${}^*\mathbb{R}$

For analysis we need functions, spaces of functions, etc., not only a field. The first question we might like to answer is: Let F be a function from $\mathbb{R}$ into $\mathbb{R}$. How can we extend it to a function ${}^*F$ from ${}^*\mathbb{R}$ into ${}^*\mathbb{R}$?

The following definition is quite natural:

<u>Definition:</u> ${}^*F(\langle f\rangle) \equiv \langle g\rangle \Leftrightarrow F(f) = g$ a.s.

i.e. F is extended to ${}^*F$ "componentwise" in the sense that F evaluated at the sequence $\langle f\rangle$ is the sequence $\langle g\rangle$ (modulo the ultrafilter U, of course) iff $F(f(i)) = g(i)$ for a big set of components i.

E.g. ${}^*\sin x = y \Leftrightarrow \sin x_i = y_i$ for a big set of i, where $x = \{x_i, i \in \mathbb{N}\}/U$, $y = \{y_i, i \in \mathbb{N}\}/U$.

Of course functions ${}^*F$ of the above form are very special among the set of all functions from ${}^*\mathbb{R}$ into ${}^*\mathbb{R}$. Below we shall discuss more general ones. Let us first remark that in the same way as we extended unary functions $F(x)$, $x \in \mathbb{R}$ to ${}^*F(x)$, $x \in {}^*\mathbb{R}$ we can extend n-ary functions $F(x_1,\dots,x_n)$ (and relations).

We also can extend sets, and this will be important in the following.

<u>Definition:</u> To any subset E of $\mathbb{R}$ we associate a subset ${}^*E$ of ${}^*\mathbb{R}$ defined as ${}^*E = \{\langle f\rangle \in {}^*\mathbb{R} \mid f(i) \in E$ for a big set of $i\}$. Thus ${}^*E$ consists of numbers for which a big set of components are in E.

<u>Remark:</u> We immediately realize that ${}^*\mathbb{R}$ as defined in Section I is indeed the * of the set $\mathbb{R}$, since all numbers in ${}^*\mathbb{R}$ have components in $\mathbb{R}$. For any $E \subset \mathbb{R}$ let us set ${}^\sigma E = \{{}^*r \mid r \in E\}$. Then ${}^\sigma E$ coincides, for $E = \mathbb{R}$, with ${}^\sigma\mathbb{R}$ as defined in Section I. Similarly as we proved that ${}^\sigma\mathbb{R} \subsetneqq {}^*\mathbb{R}$ we can show ${}^\sigma E \subset {}^*E$ and, see e.g. [4], [3], [11], [13], that ${}^\sigma E = {}^*E$ iff E is finite. E.g. for $E = [0,1]$ we have ${}^\sigma E$ = standard numbers in $[0,1]$, whereas ${}^*E = \{x \in {}^*\mathbb{R} \mid 0 \leq x \leq 1\}$ contains all positive infinitesimals, the monads of all real in $(0,1)$ and the part of the monad of ^{*1} consisting of numbers in ${}^*\mathbb{R}$ less or equal ^{*1}. In fact, by the definition of ${}^*E$ we have $x \in {}^*E \Leftrightarrow x(i) \in E$ for a big set of i, and this by the definition of E means $0 \leq x(i) \leq 1$ for a big set of i. ε positive infinitesimal means that its i-th component, for a.e. i, $\varepsilon(i)$, is $< \frac{1}{n}$, $n \in \mathbb{N}$, hence $\varepsilon \in {}^*E$, etc. It is easy to verify that * is a homomorphism of Boolean algebras (i.e. ${}^*(A \cap B) = {}^*A \cap {}^*B$, etc.) and is injective (i.e. ${}^*A = {}^*B$ iff $A = B$). It might help the intuition to realize that the following are true and easily verified (cfr. [4]):

$E \subset \mathbb{R}$ open $\Leftrightarrow$ all monads of numbers in E are in ${}^*E$

$E \subset \mathbb{R}$ closed $\Leftrightarrow$ all finite x in ${}^*E$ have standard parts in E

$E \subset \mathbb{R}$ compact $\Leftrightarrow$ all x in ${}^*E$ are finite and their standard parts are in E.

Particular sets of the form ${}^*E$ are the "<u>hyperfinite integers</u>" ${}^*\mathbb{N}$. The numbers in ${}^*\mathbb{N} - \mathbb{N}$ are infinite and are called <u>infinite (hyperfinite) integers</u>. E.g. the number $\omega = \{1,2,\dots,m\dots\}/U$ introduced in Section 1 is an infinite hyperfinite integer. Also $\omega - 1 \in {}^*\mathbb{N} - \mathbb{N}$, in fact $\omega - k \in {}^*\mathbb{N} - \mathbb{N}\ \forall k \in \mathbb{N}$. We summarize what we have achieved up to now, in going from $\mathbb{R}$ to ${}^*\mathbb{R}$ and extending sets and functions by * in the following "transfer principle", stated informally:

"Any statement φ involving $+, \cdot, 0, 1, =, <$, real functions, real subsets and real numbers, and the logical and set theoretical operations "=", "$\in$", "and", "or", "not", "$\Rightarrow$", $\forall x$, $\exists y$ for x,y real variables is true in $\mathbb{R}$ iff φ taken with * everywhere is true in ${}^*\mathbb{R}$."

We have given this statement informally, it is however easy to provide a precise formulation using a bit of terminology from elementary logic, see e.g. [4], [6], [7].

Remark: Since we have been using a model for $^*\mathbb{R}$ the transfer principle is here an (easy) theorem. In an axiomatic approach it is a postulate, see e.g. [9].

Examples: $\sin^2 x + \cos^2 x = 1$ is true for all $x \in \mathbb{R}$, hence is true "by transfer" for all $x \in {}^*\mathbb{R}$ (with $\sin^2$ replaced by $^*\sin^2$, $\cos^2$ by $^*\cos^2$, 1 by ^{*1}, and = by the corresponding operations in $^*\mathbb{R}$).

As an application of the principle we might discuss continuity properties of functions.

Proposition 7: Let f be a real function on $\mathbb{R}$. f is continuous at $c \in \mathbb{R} \Leftrightarrow {}^*f(x) \approx {}^*f({}^*c)$ for all $x \in {}^*\mathbb{R}$ s.t. $x \approx {}^*c$.

Proof: $\Rightarrow$: From the assumption we have that for each $n \in \mathbb{N}$ there exists $\delta_n > 0$ such that for all $x \in \mathbb{R}$, $|x-c| < \delta_n \Rightarrow |f(x)-f(c)| < 1/n$. Then by transfer for each $n \in {}^*\mathbb{N}$ $\exists \delta_n > {}^*0$ such that for all $x \in {}^*\mathbb{R}$, $|x-{}^*c| < {}^*\delta_n \Rightarrow |{}^*f(x) - {}^*f({}^*c)| < {}^*1/n$. If now $x \approx {}^*c$ for $x \in {}^*\mathbb{R}$ then $|x - {}^*c| < {}^*\delta_n$, hence $|{}^*f(x) - {}^*f({}^*c)| < {}^*1/n$, thus ${}^*f(x) \approx {}^*f({}^*c)$.

$\Leftarrow$: By assumption there exist $\varepsilon, \delta > 0$ infinitesimal such that whenever $|x - {}^*c| < \delta$ then $|{}^*f(x) - {}^*f({}^*c)| < \varepsilon$. By transfer for given $\varepsilon > 0$, $\varepsilon \in \mathbb{R}$ there exists $\delta > 0$, $\delta \in \mathbb{R}$ such that $|x-c| < \delta \Rightarrow |f(x)-f(c)| < \varepsilon$, hence f is continuous. □

What is then the characterization of uniform continuity? (a quite important problem within the old "infinitesimal methods": recall the historical difficulties, the "interactions" between Abel and Cauchy, see e.g. [72]).

Proposition 8: f is uniformly continuous on an interval I of $\mathbb{R} \Leftrightarrow {}^*f(x) \approx {}^*f(y)$ for all $x,y \in {}^*I$ such that $x \approx y$.

Proof: An immediate consequence of transfer.

Application: $f(x) = x^{-1}$ in (0,1) is not uniformly continuous, since for $x = \omega^{-1} \in {}^*\mathbb{N} - \mathbb{N}$ we have $x^2 \approx x \approx 0$ but ${}^*f(x) = \omega \not\approx {}^*f(x^2) = \omega^2$!

As another exercise in transfer we might mention the following non-standard expression of convergence for sequences:

s_n a real sequence, r a real number:

$s_n \to r$ as $n \to \infty \Leftrightarrow \mathrm{st}\, {}^*s_\omega = {}^*r \quad \forall \omega \in {}^*\mathbb{N} - \mathbb{N}$,

where ${}^*s_\omega$ is the extension of the function (sequence) $s: \mathbb{N} \to \mathbb{R}$ to ${}^*\mathbb{N} \to {}^*\mathbb{R}$.

Here as a further illustration of elementary uses of non standard analysis, let us mention the translation of two other elementary notions in non standard analysis terms:

a) Derivative of function at a point:

The derivative $\frac{df}{dx}(x)$ of the real function f at x exists and is finite iff $\frac{{}^*f(x+\Delta x)-{}^*f(x)}{\Delta x}$ is finite for all infinitesimals $\Delta x \neq 0$, and its standard part is independent of Δx. Then $st\left(\frac{{}^*f(x+\Delta x)-{}^*f(x)}{\Delta x}\right) = \frac{df}{dx}(x)$.

b) Riemann integration of a continuous function f on [0,1]:

$$\int_a^b f(x)dx = st \, {}^*\!\!\sum_{k=1}^{1/\Delta x} {}^*f(k\Delta x)\, \Delta x \quad \forall \Delta x \neq 0,\ \Delta x \approx 0$$

and st independent of Δx.

The proofs are not difficult and quite intuitive, see e.g. [4] [8][16]. Having this all of standard calculus can be easily rewritten and reinterpreted in non standard analysis terms, regaining in particular the original intuition of the "method of infinitesimals. Besides this gain in intuition, one might mention nice non standard proofs of Peano existence theorem for first order differential equations $\frac{d}{dt} u = f(t,u)$ for continuous f(by polygonal approximation and transfer, without Ascoli-Arzelà) and resolution of non uniqueness in differential equations (in equation like e.g. $\frac{d}{dt} u = 3u^{2/3}$) by "allowing infinitesimal data", see [4], [19], [20], [32]-[34], [38], [42], [43], [54], [65], [69].

Remark: One often uses the convention to drop * on functions ${}^*F$ which are extensions of functions F on $\mathbb{R}$, e.g. to use the symbol exp for the real-valued exponential function on $\mathbb{R}$ as well as its extension from ${}^*\mathbb{R}$ to ${}^*\mathbb{R}$.

Remark: Although the formalism given above suffices for "elementary classical analysis", for more advanced applications, like in the theory of stochastic processes and applications to quantum (field) theory, we should learn to apply the *-extension not only to $\mathbb{R}$, subsets of $\mathbb{R}$, numbers in $\mathbb{R}$, functions on $\mathbb{R}$ as we did until now but also to such objects as spaces of functions (like $C_0^\infty(\mathbb{R})$), spaces of measures, distributions etc. This will be discussed shortly in the next section.

III. Tools for more advanced analysis: Internal and external quantities

Until now we have only considered numbers of $\mathbb{R}$, subsets of $\mathbb{R}$, functional relations on $\mathbb{R}$, and we have only (as seen easily at closer look) used $\exists,\forall$ on real variables. For more advanced analysis, measure, probability and integration theory, functional analysis etc. one needs also such concepts as spaces of functions, families of sets etc. To be able to incorporate such structures into the non standard framework some care has to be taken. In fact whereas for sentences on numbers in ${}^*\mathbb{R}$ "by transfer" all laws as for those in $\mathbb{R}$ hold, some care has to be taken for subsets of ${}^*\mathbb{R}$: e.g. whereas any subset of $\mathbb{R}$ which is upper bounded has a least upper bound, this in not true for arbitrary subsets of ${}^*\mathbb{R}$, e.g. the infinitesimals are upper bounded by ${}^*r$ for all real $r > 0$, but there is no least upper bound. "Transfer" here is only possible to certain subsets of ${}^*\mathbb{R}$, namely the so called "internal" ones. These are the subsets of ${}^*\mathbb{R}$ which are elements of sets of the form ${}^*A$, for $A \subset \mathbb{R}$. Sets of the form ${}^*A$ for $A \subset \mathbb{R}$ are called (*-)standard, so that internal sets are elements of (*-)standard sets.

What happens to sets of sets, sets of functions e.g.? We imagine we will have to introduce also here a concept of "*-standard" and "internal", but how to procede systematically? It is at this point that some elementary set theoretical and logical terminology is required, and it is here that most not lògically inclined readers of non standard texts usually are put off. However the formalism can be reduced to a minimum, see [4] and, for a short introduction [70]. The basic idea is to realize that all structures for analysis are contained in the so called "superstructure" $V(\mathbb{R})$ over $\mathbb{R}$. This contains all sets formed inductively from $\mathbb{R}$ in a finite number of steps, by successively taking subsets of the preceeding ones. E.g. a function f can be realized as a graph $<x,f(x)>$, hence as an element of $\mathbb{R} \times \mathbb{R}$, which in turn can be realized as an element in the union of the point subsets and two point subsets of $\mathbb{R}$ (by $<x,y> \to \{x\} \cup \{x,y\} \in P(\mathbb{R})$, with $P(\cdot)$ the power set). So if we define inductively $V_o(\mathbb{R}) = \mathbb{R}$, $V_{n+1}(\mathbb{R}) = V_n(\mathbb{R}) \cup P(V_n(\mathbb{R}))$, and set

$V(\mathbb{R}) = \bigcup_{n=0}^{\infty} V_n(\mathbb{R})$, then it is not difficult to realize that all structures encountered in analysis can be found in $V(\mathbb{R})$ (e.g. spaces of functions, spaces of distributions as spaces of continuous linear functionals of functions etc.).

The elements of the superstructure $V(\mathbb{R})$ over $\mathbb{R}$ are called <u>entities</u>. Entities in $V_2(\mathbb{R})$ are e.g. $r \in \mathbb{R}$, the subset of $\mathbb{R}$ consisting of all even numbers, the family of all finite subsets of $\mathbb{R}$, of all finite closed intervals in $\mathbb{R}$.

In the previous sections we saw the utility of embedding $\mathbb{R}$ in $^*\mathbb{R}$ and extending subsets of $\mathbb{R}$ and functions (and relations) on $\mathbb{R}$ to such on $^*\mathbb{R}$. How can we extend this to the whole superstructure $V(\mathbb{R})$ in a similar useful way? How to define $^*A$ for A arbitrary in $V(\mathbb{R})$? One thing is sure, we can expect $^*A$ to belong to $V(^*\mathbb{R})$, the superstructure over $^*\mathbb{R}$ (e.g. this is so for $^*A$ with $A \in \mathbb{R}$ or $A \subset \mathbb{R}$).

It turns out that one can define * as a map from $V(\mathbb{R})$ <u>into</u> $V(^*\mathbb{R})$, with range the so called <u>*-standard subsets</u> of $V(^*\mathbb{R})$, all contained in the largest one, $^*V(\mathbb{R})$ (thus $^*V(\mathbb{R}) \subset V(^*\mathbb{R})$, but $^*V(\mathbb{R}) \neq V(^*\mathbb{R})$). * reduces to the operation we had defined before on elements $r \in \mathbb{R}$, on subsets $A \subset \mathbb{R}$ and on functions from $\mathbb{R}$ into $\mathbb{R}$. Moreover $^*A \in V_{n+1}(^*\mathbb{R}) - V_n(^*\mathbb{R})$ if $A \in V_{n+1}(\mathbb{R}) - V_n(\mathbb{R})$ and $B \in A \in {}^*V_n(\mathbb{R})$, $n \geq 1 \Rightarrow B \in {}^*V_{n-1}(\mathbb{R})$. The entities B s.t. $B \in {}^*A$ for some $A \in V_n(\mathbb{R})$ (i.e. the elements of <u>*-standard entities</u>) are called <u>internal entities</u>. The latter implication says in particular that elements of internal entities are internal. As anticipated above the internal entities are precisely those to which we can transfer properties which hold for ordinary entities, and roughly are formalizable using logical symbols, arithmetic symbols $+, \cdot, <$, set theoretical symbols $=$, $\in$ and $\forall$, $\exists$ applied on entities in $V(\mathbb{R})$, e.g. for subsets of $\mathbb{R}$ the sentence "if it has a lower bound then it has also a biggest lower bound" is true, and this hold also for internal subsets of $^*\mathbb{R}$ (but as we mentioned above not all subsets of $^*\mathbb{R}$, e.g. not for the subset of all infinitesimals;

in fact this already implies that not all subsets of $^*\mathbb{R}$ are internal). The entities in $V(^*\mathbb{R})$ which are not internal are called <u>external</u>. We just saw that the set of infinitesimals must be external. Other examples of external entities are $\mathbb{R}$, $\mathbb{N}$, $^*\mathbb{R} - \mathbb{R}$, $^*\mathbb{N} - \mathbb{N}$; $\mathbb{Z}$,..., the set of all finite numbers. It is important to gain some intuition about internal entities (for a precise construction of $^*V(\mathbb{R})$ and of internal entities see [4], [6], [7] [11]). * standard implies internal, the internal subsets of $^*\mathbb{R}$ are the elements of $^*P(\mathbb{R})$. st is an external mapping, χ_A for A external is an external function. If A and B are internal then $A \cap B$, $A \cup B$, A-B and $A \times B$ are internal.

As recognizable by a more explicit construction of internal entities, see e.g. [4], internal subsets of $^*\mathbb{R}$ can be characterized by equivalence classes (modulo U), of sequences of the form $\{A_1, A_2, \ldots, A_n, \ldots\}$, $A_i \subset \mathbb{R}$, *-standard subsets of $^*\mathbb{R}$ are equivalence classes of sequences of the form $\{A, A, \ldots, A, \ldots\}$, $A \subset \mathbb{R}$). In particular $[^*a, ^*b]$, $a,b \in \mathbb{R}$ is *-standard, $[a,b]$, $a,b \in {^*\mathbb{R}}$ is internal and not *-standard if a or b are not in $\mathbb{R}$ i.e. are nonstandard. sin ωx, ω infinite is an internal $^*\mathbb{R}$-valued function on $^*\mathbb{R}$ (not a *-standard one), (a function is internal iff its graph is internal). Internal subsets of $^*\mathbb{N}$ have "underflow" and "overflow" properties, e.g. if they contain $\mathbb{N}$ they must contain also some infinite number and if they contain all infinite numbers they must contain also some finite numbers ... This might all sound a little mysterious, at first reading, but some practice in recognizing where "transfer" can be used will help understanding the meaning of internal sets and their applications. We shall now go over to the discussion of some applications of non standard analysis to subjects related to quantum mechanics.

IV Some applications to quantum mechanics and related areas

IV.1 How to treat a singular interaction by non standard analysis

We shall begin by looking at some examples from the theory of Schrödinger operators with singular coefficients, of the type arising e.g. in nuclear physics, solid state physics, electromagnetism, optics, polymer physics, see e.g. [4], [36], [44]-[46] and references therein. Let us consider first a heuristic operator of the type $-\Delta + \lambda\delta(x)$, $x \in \mathbb{R}^d$ in $L^2(\mathbb{R}^d, dx)$. To define it properly one has, in non-standard analysis, basically two procedures, either by "discretization" (taking a hyperfinite lattice $\varepsilon\mathbb{Z}^d$ "realization", ε infinitesimal, of $\mathbb{R}^d$ and replacing consequently Δ and $\delta(x)$ by their hyperdiscrete analogues) or by "a smooth non-standard realization" of the singular coefficient, in this case of $\delta(x)$, realized e.g. as $\delta_\varepsilon(x) \equiv (\frac{4}{3}\pi\varepsilon^3)^{-1}\ \chi(|x|/\varepsilon)$, with χ the characteristic function of the unit ball $\{|x| \leq 1\}$ in $\mathbb{R}^3$. Let us give a few details of both procedures.
First, we describe shortly the latter method, for d = 3. The operator we want to realize acts thus in $L^2(\mathbb{R}^3)$. We want to use δ_ε as a "perturbation of the Laplacian". Let $^*L^2(\mathbb{R}^3)$ be the *-standard Hilbert space in which operators are going to be defined, let Δ be the self-adjoint operator

obtained by transfer to ${}^*L^2(\mathbb{R}^3)$ from the self-adjoint Laplacian in $L^2(\mathbb{R}^3)$. For any finite $\lambda_\varepsilon \in {}^*\mathbb{R}$, $H_\varepsilon = -\Delta + \lambda_\varepsilon \delta_\varepsilon(x)$ is a well-defined self-adjoint operator in ${}^*L^2(\mathbb{R}^3)$ (by transfer, $\lambda_\varepsilon \delta_\varepsilon$ being *-bounded). Again by transfer, we can split, in the same way as with $L^2(\mathbb{R}^3)$, ${}^*L^2(\mathbb{R}^3)$ into a rotationally symmetric part ${}^*\mathcal{H}_s$ and its orthogonal complement ${}^*\mathcal{H}_s^\perp$. H_ε acts as $-\Delta$ in ${}^*\mathcal{H}_s$ and in fact its standard part (defined e.g. through its resolvent) is $-\Delta$ in ${}^*\mathcal{H}_s^\perp$. The restriction of H_ε to ${}^*\mathcal{H}_s$ is unitary equivalent (in the sense of ${}^*L^2(\mathbb{R}^3)$) with $A_\varepsilon = -\frac{d^2}{dr^2} + \lambda_\varepsilon \delta_\varepsilon(r)(|x| = r)$, acting in ${}^*L^2(\mathbb{R}_+, dr)$, with Dirichlet boundary conditions at $r = 0$.

For $\operatorname{Im} k^2 \neq 0$, $\operatorname{Re}\sqrt{k^2} > 0$ we have for any finite $x, y \in {}^*\mathbb{R}_+$, once more by transfer:

$$2\sqrt{k^2}\,(A_\varepsilon - k^2)^{-1}(x,y) = \sin(\sqrt{k^2-\lambda}\,x)\,a_-^{-1}(c_-\exp(\sqrt{\lambda-k^2}\,y) + c_+\exp(-\sqrt{\lambda-k^2}\,y))$$

for $x \leq y \leq \varepsilon$; $= a_-^{-1}\sin(\sqrt{k^2-\lambda}\,x)\exp(-\sqrt{k^2}\,y)$

for $x \leq \varepsilon \leq y$; $= [\exp(\sqrt{k^2}\,x) + a_-^{-1} a_+ \exp(-\sqrt{k^2}\,x)]\exp(-\sqrt{k^2}\,y)$,

for $\varepsilon \leq x \leq y$.

This Green's function is found by transfer and Sturm-Liouville theory for the situation with ε standard, the coefficients of the trigonometric functions being fixed by continuity conditions at $r = \varepsilon$. From these formulae we see easily that for $\lambda_\varepsilon / (\frac{4}{3}\pi\varepsilon^3)$ finite the standard part of $(A_\varepsilon - k^2)^{-1}(x,y)$, $x \neq y$ is simply $(-\frac{d}{dr^2} - k^2)^{-1}(\mathrm{st}\,y, \mathrm{st}\,y)$ with $-\frac{d^2}{dr^2}$ acting in $L^2(\mathbb{R}_+, dr)$ with Dirichlet boundary conditions at 0. Thus in this case we do not get any interaction. For $\lambda_\varepsilon / (\frac{4}{3}\pi\varepsilon^3)$ infinite then the standard part of $(A_\varepsilon - k^2)^{-1}(x,y)$ exists only when $a_\pm$ are finite and a_- is non-infinitesimal, which then requires $\sin(\sqrt{k^2-\lambda}\,\varepsilon)$ and $\sqrt{k^2-\lambda}\cos(\sqrt{k^2-\lambda}\,\varepsilon)$ both finite. But then $\cos(\sqrt{k^2-\lambda}\,\varepsilon)$ must be infinitesimal, which yields $\sqrt{k^2-\lambda}\,\varepsilon = (\gamma + \frac{1}{2})\pi + \eta$ for some infinitesimal η and any $\gamma \in \mathbb{R}$ such that $\sqrt{k^2-\lambda}\cos(\sqrt{k^2-\lambda}\,\varepsilon)(-1)^\gamma$ is near standard. An elementary calculation then shows that this is only possible when

$$\lambda_\varepsilon = \frac{4}{3}\pi[-(\gamma + \frac{1}{2})\pi^2\varepsilon + 4\pi\alpha\varepsilon^2 + \beta\varepsilon^3]$$

for arbitrary $\beta \in \mathbb{R}$. In this case the standard part of $(A_\varepsilon - k^2)^{-1}(x,y)$ is independent of β, γ and H_ε defines, by taking standard parts in the resolvent, a self-adjoint, lower bounded operator H^{α}, depending on the parameter $a \in \mathbb{R}$. H^{α} is thus a one-parameter family of realizations of "$-\Delta + \lambda\delta(x)$". It can be shown that H^{α} coincides with the limit, in the norm resolvent sense, of $-\Delta + \lambda_{\varepsilon}\delta_{\varepsilon}(x)$ for $\varepsilon > 0$ real, as $\varepsilon \downarrow 0$, when λ_{ε} is chosen as above. In a sense, however, to think of H^{α} as $-\Delta$ perturbed by a potential with infinitesimal support is closer to the physical intuition.

For more details about this model and various extensions using non-standard analysis see [17], [58] (see also [4],[44],[46] for further information about the study of models of this type).

Below we shall look at the approach to singular perturbations using "hyperfinite discretization". Before doing so, however, we would like to introduce a little bit of non-standard measure and integration theory, which is also useful for our later discussion of probabilistic applications.

IV.2 A basic instrument from non-standard measure and integration theory: the Loeb measure

The natural measures in non-standard analysis are additive internal, not σ-additive, measures. Loeb realized in '75 that to any such measure one can associate a standard σ-additive measure (on a non-standard space): this was a breakthrough, which made it possible to apply non-standard, even hyperfinite, tools to measure theory and related domains, like the theory of stochastic processes (but also, the study of certain non-linear partial differential equations through the concept of "Loeb solution" [47], [4] as well as to statistical mechanics [4], [35], [41], [48], [56], [60]).

Let X be some internal set, $\mathcal{A}$ an algebra of internal subsets of X, ν a finitely additive measure on $(X,\mathcal{A})$ with values in $^*[0,1]$, with $\nu(X) = 1$. $(X,\mathcal{A},\nu)$ is called an internal measure space. Can one associate to ν a σ-additive $[0,1]$-valued measure? The answer is yes, for a simple reason. It is namely possible to show in our model of non-standard analysis discussed in Sect. I-III that the following "saturation property" holds:

Theorem: If $A_n \neq \phi$ are internal for all $n \in \mathbb{N}$ then $\bigcap_{n\in\mathbb{N}} A_n \neq \phi$.

The proof is not difficult, it needs of course a precise definition of internal entities, which we have not given, see e.g. [4], [11], [70]

The theorem says intuitively that internal sets are rich. E.g. $\bigcap_{n\in\mathbb{N}} (0,n) = \phi$ ((0,n) are external, since $(0,n) = \{x \in \mathbb{R} \mid 0 < x < n\}$, but $\mathbb{R}$ is external), whereas $\bigcap_{n\in\mathbb{N}} {}^*(0,n)$ = (positive infinitesimals different from zero) $\neq \phi$ ($^*(0,n)$ are internal, even *-standard, being of the form $\{x \in {}^*\mathbb{R} \mid 0 < x < n\}$ with $^*\mathbb{R}$ $*$-standard). In fact, as an immediate corollary of the above we have:

Corollary 1: Internal elements of $^*V(\mathbb{R})$ are either finite or uncountable.

Proof: Assume A is internal and A is countable, i.e. $A = \{a_n, n \in \mathbb{N}\}$, $a_i \in V(\mathbb{R})$. Define A_o A, A_n $A - \{a_1, \ldots, a_n\}$. Then the A_i are internal and $A_n = \emptyset$, on the other hand, by saturation $\cap A_n \neq \emptyset$ if $A_n \neq \emptyset$. Hence to avoid contradiction we must have $A_{n_o} = \emptyset$ for some n_o (and then for all $n \geq n_o$), but then A is finite. Or we have to reject the "ad absurdum" assumption, i.e. A is uncountable.

Now let $(X, \mathcal{A}, \nu)$ be an "internal measure space". We shall see, using the saturation theorem, that ν can be used to obtain a standard, σ-additive measure, "near ν". Let $^o\nu$ be the standard additive, $[0,1]$-valued normalized measure on the algebra $\mathcal{A}$ defined by $o\nu(A) = st(\nu(A)), \forall\ A \in \mathcal{A}$.

Theorem (Loeb): $^\circ\nu$ has a Caratheodory extension to a σ-additive probability measure $L(\nu)$ on the σ-algebra generated by . $(X, \sigma(\mathcal{A}), L(\nu))$ is a probability space. $L(\nu)$ is "near ν" in the sense that, to any $A \in \mathcal{A}, \exists\ B \in \sigma(\mathcal{A})$ such that $L(\nu)(A \Delta B) = 0$, where Δ means symmetric difference of sets.

Proof: Having the saturation theorem, the proof is immediate from Caratheodory's criterium. In fact, from $A_n \in \mathcal{A}$, $A_n \downarrow \emptyset$ we have by saturation that there exists an $n_o \in \mathbb{N}$ such that $A_n = \emptyset$ for all $n \geq n_o$ (otherwise $\bigcap_{n \in \mathbb{N}} A_n \neq \emptyset$, the A_n being internal!). But then of course $^\circ\nu(A_n) = 0 \forall n \geq n_o$, hence $^\circ\nu(A_n) \to 0$ as $n \to \infty$, thus Caratheodory's theorem condition is satisfied and $^\circ\nu$ has a σ-additive extension to a probability measure on $\sigma(\mathcal{A})$. ■

The measure $L(\nu)$ (or often its completion) is called Loeb measure associated with ν. Correspondingly $(X, \sigma(\mathcal{A}), L(\nu))$ is called Loeb measure space. The following is easily shown, see [4]:

Corollary: For any $L(\nu)$-integrable real-valued function f on X there exists an internal function F from X into $^*\mathbb{R}$ s.t. st $F(x) = f(x)$ $L(\nu)$-a.e. $x \in X$, and viceversa. $\int |F|\, d\nu$ is finite and $\int_A F d\nu \approx 0$ if $\nu(A) \approx 0$. One has $\int F d\nu \approx \int f dL(\nu)$. F is called "lifting" of f.

Remark: As mentioned above, Loeb's construction is also possible when ν is not $^*[0,1]$-valued, see [4], [7], [11]

The usefulness of Loeb measures is that they are in the above sense entirely approximable by internal additive ones, which are easy to operate with, as we shall now see in the particularly important case of the one yielding a realization of Lebesgue measure (as a "uniform measure") on [0,1] (say). Such (special) Loeb measures are associated with an important class of internal measure spaces $(X, \mathcal{A}, \nu)$, the so called hyperfinite ones. An internal set X is called hyperfinite if there exists an internal bijective map of a "hyperfinite segment" of $^*\mathbb{N}$, $\{n \in {^*\mathbb{N}} \mid n \leq \omega$ for some $\omega \in {^*\mathbb{N}}\}$, onto X ($\omega$ is then called the "internal cardinality" of X). Hyperfinite sets behave in a sense as finite sets, although they can be used to "modellize a continuum". E.g. a hyperfinite model of the interval [0,1] is the hyperfinite set X describes as follows: Let ε^{-1} be some hyperfinite integer, i.e. $\varepsilon^{-1} \in {^*\mathbb{N}} - \mathbb{N}$ and let $X = \{k\varepsilon \mid k \in {^*\mathbb{N}} \cup 0, k \leq \varepsilon^{-1}\}$.

Then st X is a surjection onto [0,1]. Let, for any internal subset A of X, $\nu(A) \equiv |A|/|X|$, with $|\cdot|$ denoting internal cardinality (so that $|X| = (\varepsilon)^{-1} + 1$). $(X, \mathcal{A}, \nu)$ is a hyperfinite internal probability space (with $\mathcal{A}$ the algebra of internal subsets of X). Thus ν is "counting measure" on X.

The corresponding Loeb space $(X, \sigma(\mathcal{A}), L(\nu))$ is nothing but a hyperfinite model of the Lebesgue measure space associated with Lebesgue measure on [0,1], in the sense that, for any measurable $A \subset [0,1]$, $st^{-1}(A) \equiv \{x \in X \mid st\ x \in A\}$ is Loeb measurable, and viceversa, and $\int_A dx = L(\nu)(st^{-1}(A))$. This gives substance to the intuition that Lebesgue measure is uniform measure (in fact it is "near" the counting measure ν on X).

Similarly for any Lebesgue integrable f_o there is a lifting F s.t. $f \equiv st\ F = f_o \circ st$ is $L(\nu)$ integrable, and F is S-ν-integrable and viceversa, and we have

$$\int F d\nu = \sum_{x \in X} |X|^{-1} F(x) \approx \int_X f(x) dL(\nu)(x) = \int_o^1 f_o(y) dy.$$

We can thus approximate Lebesgue's integral by hyperfinite sums.

Remark: A similar construction holds for general Radon spaces, cf. [4]. We shall use below Loeb measures in the discussion of hyperfinite Dirichlet forms as well as in the discussion of stochastic processes.

IV.3 Return to Schrödinger operators. Dirichlet forms.

We shall now take up again the theme of IV.1, looking at the study of singular differential operators, of the form $-\Delta + \lambda\delta(x)$ e.g. In this study it is often useful to look at the basic mappings to be constructed as quadratic forms rather than operators (e.g. as operator in $L^2(\mathbb{R}, dx)$, $\lambda\delta(x)$, $x \in \mathbb{R}$ does not make sense, it is however a small form perturbation of $-\Delta$, and hence $-\Delta + \lambda\delta(x)$ is well-defined as lower bounded quadratic form in this 1-dimensional case). Thus it is quite natural to try to develop a theory of hyperfinite quadratic forms. The fact that even in 3 dimensions $-\Delta + \lambda\delta(x)$, realized as H^{α}, (cf. Sect. IV.1), is, modulo a finite, α-dependent constant, unitary equivalent with a Dirichlet form (in the sense of [51], [50], see also e.g. [4], [53]), and the approach to elliptic operators via Dirichlet forms is, at least in the symmetric case, the most powerful one, motivates the idea of trying to construct a theory of hyperfinite Dirichlet forms. This has been done [4], [49] and we shall here give some sketch of this approach.

Let $S = \{s_1, \ldots, s_N\}$ be an internal subset of ${}^*\mathbb{R}^d$, $N \in {}^*\mathbb{N} - \mathbb{N}$ (think of S as a hyperfinite lattice "realization" of $\mathbb{R}^d$). Let $m: S \to {}^*\mathbb{R}_+$ be an internal function, which we shall think of as a measure on S. We then write for any internal $A \subset S$: $m(A) = \sum_{s_i \in A} m(s_i)$.

Let $L^2(s,m)$ be the internal space of all internal functions $f: S \to {}^*\mathbb{R}$ with the inner product $\langle f,g\rangle \equiv \sum_{i=1}^{N} f(i)g(i)m_i$.

Let Q be an $N \times N$ stochastic matrix ("transition matrix") s.t. $Q = ((q_{ij})), i,j=1,\dots,N, q_{ij} \geq 0$, $\sum_{j=1}^{N} q_{ij} = 1$ and such that Q is symmetric with respect to m in the sense that $m_i q_{ij} = m_j q_{ji}$ for all i,j. Let Δt be a positive infinitesimal. $Q^{\Delta t}$ is, by transfer, a well-defined operator on $L^2(S,m)$ given by $(Q^{\Delta t}f)(i) = \sum_{j=1}^{N} q_{ij} m_j f(j)$.

Let $A \equiv [I-Q^{\Delta t}]/\Delta t$. Then A is a symmetric positive operator which generates the "hyperfinite Dirichlet form" $E(f,g) = \langle A^{1/2}f, A^{1/2}g\rangle$ (this is in analogy with the standard theory of (discrete) Dirichlet forms developed originally by Beurling and Deny. $Q^{k\Delta t}$, $k \in \mathbb{N}$ is a hyperfinite version of the semigroup generated by A, with time replaced by the discretized version $k\Delta t$, $k \in \mathbb{N}$). The domain $D(E)$ of E is defined as the set of all $f \in L^2(S,m)$ s.t. $E_1(f,f) \equiv E(f,f) + \langle f,f\rangle$ is finite and minimal among all $f' \approx f$. It turns out that E with domain $D(E)$ is always closed (in the norm given by $\|f\|_1 \equiv E_1(f,f)^{1/2}$). (This is based on the fact that for all $f \in D(E)$, $t = k\Delta t \approx 0$, $k \in \mathbb{N}$, $\langle([I-Q^t]/t)f,f\rangle \approx E(f,f)$. In fact, if $\{f_n, n \in \mathbb{N}\}$ is an internal extension of an E_1-Cauchy sequence of elements f_n of $D(E)$, let $N \in {}^*\mathbb{N} - \mathbb{N}$ so that $st|f_n - f_N| \to 0$ as $n \to \infty$ in $\mathbb{N}$. Let $f = f_N$, we show $f \in D(E)$. In fact, if f were at absurdum not in $D(E)$, then by the above there would exists an infinitesimal t and an $\varepsilon > 0$, $\varepsilon \in \mathbb{R}$ s.t. $\|f-Q^t f\|_1 < \varepsilon/3$. Choose n so large that $\|f-f_n\|_1 < \varepsilon/3$, then, as easily shown, $\|Q^t f - Q^t f_n\|_1 < \varepsilon/3$. An $\varepsilon/3$-argument yields then, using $\|f_n - Q^t f_n\| \approx 0$, a contradiction). We shall now see how one can associate to hyperfinite Dirichlet forms standard Dirichlet forms. Let $F(L^2(s,m))$ be the set of all elements in $L^2(S,m)$ with finite norm. The set of equivalence classes modulo infinitesimals with representatives in $F(L^2(S,m))$ is a Hilbert space with scalar product given by the standard part of the one in $L^2(S,m)$. We define the standard form E_{st} induced by E to be the symmetric bilinear form on $F(L^2(S,m))/\approx$ given by $E_{st}(v,v) = \inf_{f \in v} \{st\, E(f,f)\}$, with domain $D(E_{st}) = \{v \in F(L^2(S,m)/\approx | v \cap D(E) \neq \emptyset\}$.
The argument giving the closedness of E also yields that E_{st} is closed. If $m\{s \in S \mid \|s\| < r\}$ is finite for all finite r, we can define a Radon measure $\tilde{m}$ on all measurable sets B of $\mathbb{R}^d$ by $\tilde{m}(B) = L(m)(st^{-1}(B))$, with $L(m)$ the Loeb measure to m. The standard part of E is the Dirichlet form on $L^2(\mathbb{R}^d,\tilde{m})$ given by $E(f,f) = E_{st}(\tilde{f},\tilde{f})$, with $\tilde{f} = {}^*f \restriction S$, for any bounded continuous real-valued f on $\mathbb{R}^d$. E is a standard Dirichlet form (in the sense of [S1], i.e. E is a closed densely defined positive bilinear form with contraction property $E(f^\#,f^\#) \leq E(f,f)$, $f^\# = (f \vee 0 \wedge 1)$. Equivalently the associated semigroup p_t in $L^2(\mathbb{R}^d,\tilde{m})$ is sub-Markov in the sense $0 \leq f \leq 1 \Rightarrow 0 \leq p_t f \leq 1$).

In this way from hyperfinite Dirichlet forms one produces standard Dirichlet forms, and viceversa one can associate to every standard Dirichlet form a hyperfinite Dirichlet form. One can then exploit the hyperfinite Dirichlet forms (which behave "by transfer" very similarly to finite Dirichlet forms) to give a new (and in a sense "more intuitive") construction of a strong Markov, Hunt process associated with the standard Dirichlet form (for the standard construction see [51]) and the development of a stochastic calculus for hyperfinite Dirichlet forms. For details on this see [4], [49]. A particular case is $S = \{z \in \delta\, {}^*\mathbb{Z}^d,\ |z_i| < \delta^{-2},\ i = 1,\dots,d\}$, δ a positive infinitesimal.

Let U be the set of unit vectors in ${}^*\mathbb{R}^d$ of the form $(0,\dots,\pm 1,0,\dots 0)$ and define, for $s \in S$, $e \in U$ and any internal function $f: S \to {}^*\mathbb{R}$

$$D_e f(s) = \frac{f(s+\delta e)-f(s)}{\text{sign}(e)\,\delta},$$

with sign e the sign of the non-zero component of e.

Hyperfinite forms of the type $F(f,g) = \frac{1}{4} \sum_{s\in S} \sum_{e\in U} D_e g(s) D_e f(s) \nu(s)$,

with ν an internal measure on S are called <u>hyperfinite energy forms</u>. It is not difficult to show that, for some constant $\eta \in {}^*\mathbb{R}$, with $\Delta t \equiv \eta \frac{\delta^2}{d}$, $F(f,g) = \eta^{-1} E(f,g)$, with E the hyperfinite Dirichlet form defined as above, with $m(s) = \nu(s) + (4d)^{-1} \sum_{e \in U} \nu(s+\delta e)$ (so that $L(m)\circ st^{-1} = L(\nu)\circ st^{-1}$!), $Q = ((q_{s,s'}))$, $s,s' \in S$, $q_{s,s'} = 0$ if $|s-s'| \neq \delta$ or $m(s) = 0$, $q_{s,s'} = [\nu(s) + \nu(s')]\,/[4dm(s)]$ if $|s-s'| = \delta$ and $m(s) \neq 0$.

The process associated with E has standard part which is a Hunt process with stationary measure $L(\nu)\circ st^{-1}$. In the case where $\eta = 1$ with a suitable choice of ν one gets the diffusion process associated with a standard energy form $\frac{1}{2}\int_{\mathbb{R}^d} \langle df,dg\rangle \varphi^2 dx$, $\varphi > 0$ a.e., $\varphi \in L^2_{loc}(\mathbb{R}^d)$, in $L^2(\varphi^2 dx)$.

<u>Remark:</u> A similar construction can be applied to the case where $\mathbb{R}^d$ is replaced by a differentiable manifold.

Let us shortly describe, following [49], [75], a non standard procedure for discussing the case where M is a fractal set, i.e. the case of diffusions on fractal sets. We shall briefly discuss the case of a Koch curve M. In this case the state space S consists of the 4N + 1 vertices of the "N-th approximation" of M, with N an infinitely large integer. We take $q_{ij} = 0$ if i,j are not near neighbors, $q_{ij} = \frac{1}{2}$ if i,j are near neighbors and i is not an endpoint, $q_{ij} = 1$ if i,j are near neighbors and i is an endpoint. Moreover $m_i = 4^{-N}$ if i is not an endpoint, $m_i = \frac{1}{2} 4^{-2N}$ if i is an endpoint, $\Delta t = 4^{-2N}$.

Take $N \in {}^*\mathbb{N} - \mathbb{N}$. With this choice then the associated Markov chain $X(k\Delta t)$, $k = 1,\dots,1/\Delta t$ with state space S is near standard (so that st $X(k\Delta t)$ exists) and nontrivial.

It is natural to call the process $\tilde{X}$ defined by $\tilde{X}(\mathrm{st}\, k\Delta t) \equiv \mathrm{st}\, X(k\Delta t)$ a Brownian motion on the fractal set M. For further discussion see [49], [73].

Let us now come to the description of how the above hyperfinite approach copes with the problem of giving a meaning to "operators" of the form $-\Delta + \lambda\delta(x)$ in $L^2(\mathbb{R}^d, dx)$. More generally, we try to find operators associated with formal quadratic forms $E(f,g) \equiv E_o(f,g) + \int_C \lambda fg d\rho$, with E_o a known quadratic form (e.g. the one associated with $H_o = -\Delta$), C a set of $\mathbb{R}^d$ of Lebesgue measure zero and ρ a measure concentrated on C. The point is to find hyperfinite quadratic forms $\tilde{E}$, $\tilde{E}_o$ on $L^2(S,m)$, S hyperfinite, m internal positive function, s.t. $\tilde{E}(\tilde{f},\tilde{g}) = \tilde{E}_o(\tilde{f},\tilde{g}) + \int_B \tilde{\lambda}\tilde{f}\tilde{g}\, d\tilde{\rho}$, with $B \subset S$ internal s.t. st B = C, $\tilde{\lambda}$ internal function resp. measure on B s.t. $\rho = L(\tilde{\rho}) \circ \mathrm{st}^{-1}$, f,g "versions" of $\tilde{f},\tilde{g}$ in $L^2(S,m)$. The generator $\tilde{L}$ of $\tilde{E}$ is given by $(\tilde{L}\tilde{f})(i) = \tilde{L}_o\tilde{f}(i) + \tilde{\lambda}(i)\tilde{f}(i)\frac{\tilde{\rho}(i)}{m(i)}$, with $\tilde{L}_o$ the generator associated with E_o. With our case C = {0}, λ infinitesimal, in mind (corresponding to the interaction $\lambda\delta(x)$) we allow λ to be non-standard.

Here is a sketch of how to handle such singular hyperfinite perturbation problems. A computation using transfer applied to Neumann series yields, for Im $z \neq 0$, $(\tilde{L} - z)^{-1} = G_z + \hat{G}^*_z(\tilde{\lambda}^{-1} - G'_z)^{-1}\hat{G}_z$, with $G_z \equiv (\tilde{L}_o - z)^{-1}$, $\hat{G}_z, G'_z, \hat{G}^*_z$ denoting G_z as operator $L^2(S) \to L^2(B)$, resp. $L^2(B) \to L^2(B)$ resp. $L^2(B) \to L^2(S)$.

The whole trick is now to find $\tilde{\lambda}$ s.t. the standard part of the operator $(\tilde{\lambda}^{-1} - G')^{-1}$ exists (in the sense that the kernel has a standard part on finite elements outside the diagonal) and the standard part $(H - z)^{-1}$ of $(\tilde{L} - z)^{-1}$ exists and is different from the one of $(\tilde{L}_o - z)^{-1}$ (which is in the above case $(H_o - z)^{-1}$). A sufficient condition for this is the existence of a $z_o \in {}^*\mathbb{R}$ such that $\mathrm{st}\,[\sum_i G_{z_o} G_z(\cdot - i)\,\tilde{\rho}(i)] \in L^2(\mathbb{R}^d, \rho)$.

In such cases one gets a self-adjoint family of operators $H_{\tilde{\lambda}}$ in $L^2(\mathbb{R}^d, m)$ parametrized by internal functions $\tilde{\lambda}$ s.t. $\tilde{\lambda}(x)^{-1} - \sum_i G_{z_o}(x - i)\,\tilde{\rho}(i)$ has a standard part. In the case d = 3 C = {0}, ρ = Lebesgue measure, in which case $H_{\tilde{\lambda}}$ is formally $-\Delta + \lambda\delta(x)$, we get that $\tilde{\lambda}$ is infinitesimal negative, in fact of the form $\tilde{\lambda}^{-1} = -G_o(0) + \tilde{\alpha}$, with $\tilde{\alpha} \in \mathbb{R}$ ($\tilde{\alpha}$ can be chosen constant).

We thus recover by the hyperfinite approach the result we obtained above by smooth non-standard perturbations concerning the definition of "$-\Delta + \lambda\delta(x)$".

<u>Remark:</u> The "hyperfinite" as well as the "smooth" non-standard approaches to $-\Delta + \lambda\delta(x)$ can be extended to the study of operators of the form $-\Delta + \sum_{y\in Y} \lambda_y \delta(x-y)$, with Y some finite subset of $\mathbb{R}^d$, see [58], [4].

The case of some infinite Y can also be handled, below we shall treat one example where Y is the path of a Brownian motion in $\mathbb{R}^d$. It would be very worthwhile to have a systematic treatment of the case Y infinite, this has not yet been done (see however [4]).

Remark: The treatment of above singular Schrödinger operators $-\Delta + \sum_{y\in Y} \lambda_y \delta(\cdot - y)$ is just an example of the kind of problems which can, and partly have been, handled by non-standard analysis. In fact, many other problems can be treated, like e.g. in detailed studies of classical dynamical systems (exploiting "infinitesimal coefficients")([43] and references therein) or in the treatment of singular Sturm-Liouville problems with "measure coefficients" given in [72], [34], [58], which has also been stimulus to new developments on the standard side, see [74], [75], [76] and references therein (and [46]). Moreover, we would like to mention that also in the case of non-linear partial differential equations some of the methods of non-standard analysis have been useful, e.g. in classical boundary layers problems [1], in the asymptotic study of differential equations [77], [54] and recently in a breakthrough for the study of Boltzmann's equation of kinetic gas theory, through very important work by Arkeryd, see e.g. [47], [4] and references therein. Here the decisive element is the construction of a suitable "Loeb" solution (a non-standard tool based on Loeb measures, playing a role somewhat similar to "generalized solutions" in the linear case, this joined with as much as possible control on standard parts to extract from such a solution a classical solution). It seems to us that more of these tools should prove to be useful in handling problems of non-linear functional analysis (turbulence problem in hydrodynamics, e.g.) (non-standard versions of [78], e.g.?).

We shall now leave the domain of differential equations to take up the discussion of some examples of applications in stochastic analysis and quantum field theory.

IV.4 Some applications in stochastic analysis and quantum field theory

Here is Anderson's hyperfinite model for Brownian motion (see [4]). Let T be the above hyperfinite model of [0,1], i.e. $T = \{k\Delta t \mid k \in {}^*\mathbb{N}, k \leq (\Delta t)^{-1}\}$ for a fixed $(\Delta t)^{-1} \in {}^*\mathbb{N} - \mathbb{N}$.

We look at T as "time set". Let $\Omega = \{-1,+1\}^T$. Let $\mathcal{A}$ be the internal subsets of the internal set Ω, let P be the product $\prod_{t\in T} P_t$ of symmetric Bernoulli measures (P_t the Bernoulli distribution of the component $\omega(t)$ of $\omega \in \Omega$).

Let $B(\omega,t) \equiv \sum_{s=0}^{t-\Delta t} \omega(s)\sqrt{\Delta t}$; $\omega \in \Omega$. This is a hyperfinite random walk, moving by $\pm\sqrt{\Delta t}$ right or left at each infinitesimal time interval Δt. B is an internal stochastic process. It is a theorem then that st $B(\omega,t)$ exists (L(P)-a.s.) and calling it $b(\omega,\text{st } t)$ we have that b is a process indexed by [0,1], with values in $\mathbb{R}$ (and underlying probability space $(\Omega,\sigma(\mathcal{A}), L(P))$. b is a realization of Brownian motion on $\mathbb{R}$ and $(\Omega,\sigma(\mathcal{A}), L(P))$ is a realization of Wiener measure.

The proof is by showing that the independent increments property of B gives by transfer the same property for b (using the definition of L(P)) and proving that b(u)-b(v), $u \geq v$ has mean zero and variance u-v, by realizing that with u = st s, v = st s, $\alpha \in \mathbb{R}$:

$$\int_\Omega e^{i[b(u)-b(v)]\alpha}\, dL(P) = \text{st} \int_\Omega e^{i[B(s)-B(t)]\alpha}\, dP$$

and computing the internal integral using the definition of B and P. Of course a similar result holds for Brownian motion on $\mathbb{R}^d$. Moreover, there is no difficulty extending to the infinite dimensional case, cf. [4]. The above hyperfinite realization of Brownian motion (and other diffusion processes) has found many applications to stochastic analysis (see e.g. [..], [7], [11], [12], [22], [29], [30], [38], [41], [48], [49], [62], [53], [55], [57], [59], [60], [61], [70])

As particularly striking recent applications of non-standard methods in stochastic analysis, let us mention Cutland's work on stochastic control, Hoover's, Keisler's, Lindstrøm's and Perkin's work on stochastic differential equations, and Lindstrøm's diffusion on fractals, Perkins' proof of Levy's formula for local times in terms of excursions, Stoll's characterization of (higher dimensional time) Lévy Brownian motion.

In the domain of random fields, important results have been obtained, particularly by Kessler (on the problem of the global Markov property of random fields and on a non-standard characterization of certain generalized random fields).

In [30], [50], [52], [57], [59] and more extensively in [4] we discussed an application of non-standard analysis to the study of certain stochastic Schrödinger operators which arise in the study of polymers. The problem is to give a meaning to the operator $-\Delta + \lambda \int_0^t \delta(x-b)(s))ds$, with b Brownian motion in $\mathbb{R}^d$, in $L^2(\mathbb{R}^d, dx)$ (formally this is of the type discussed in IV.3, with Y a path of Brownian motion). This is the Schrödinger operator for a quantum mechanical particle moving in the stochastic potential created by a "polymer", modellized by a path of a Brownian motion, the interaction being of the "$\lambda\delta$-type". By using the hyperfinite model for b and the above hyperfinite Dirichlet form construction of "$-\Delta + \lambda\delta(x)$", a meaning to the above operator can be found, for $d \leq 5$, λ infinitesimal negative for d = 4,5.

In [4] we have given a hyperfinite version of Symanzik's polymer representation for quantum fields φ, with classical interaction depending only on φ^2 (like the $\lambda\varphi^4_d$ model). Using our result on above "polymer operator" for $d \leq 5$, we have given a (partial) (non-trivial)costruction of the $\lambda\varphi^4_d$ and $\lambda(\varphi_1^2\varphi_2^2)_d$ models, $d \leq 5$, with λ infinitesimal negative.

Hyperfinite lattice quantum field models are discussed in [4], with applications to exponential interactions and gauge fields. For a non-standard approach to perturbation theory in quantum electrodynamics see [23], [24]. We like to close by mentioning Stoll's work on hyperfinite random walks and polymer measures (obtaining in particular a new construction, with better control on approximations, of Symanzik's-Varadhan's result on Edwards 2-dimensional polymer measure) ([41]). Another object where non-standard tools can come to play a role is the Feynman path integral (and anyhow this is a question one is often asked when speaking on non-standard analysis). Here is a short comment on this subject. It is possible to express the solution of Schrödinger equation for a particle moving in $\mathbb{R}^s$, $i\hbar \frac{\partial}{\partial t}\psi = H\psi$, $H = -\frac{\hbar^2}{2m}\Delta + V$, with initial condition $\psi(0) = \varphi$, by

$$\psi \approx \int_{^*\mathbb{R}^{sn}} \exp\left(\frac{i}{\hbar}\, {}^*S_t(\gamma_n+x)\right) {}^*\varphi(\gamma_n(0)+x)\; {}^*d\gamma_n \quad \text{for some } n \in {}^*\mathbb{N} - \mathbb{N},$$

with $S_t(\gamma_n) = \frac{m}{2}\int_0^t \dot{\gamma}_n^2(\tau)d\tau - \int_0^t {}^*V(\gamma_n(\tau))d\gamma$, $\gamma_n(\tau)$ the piecewise linear continuous path $[0,t] \to \mathbb{R}^s$ equal to x_j for $\tau = j\frac{t}{n}$, $j=0,\dots,n$,

$$d\gamma_n = \prod_{j=0}^{n-1} \left(\frac{2\pi i\hbar t}{mn}\right)^{-s/2} d\gamma_n\left(j\frac{t}{n}\right).$$

This holds at least for V s.t. H is the strong resolvent limit for $k \to \infty$ of Hamiltonians $-\frac{\hbar^2}{2m}\Delta + V_k$ for which Lie-Trotter formula holds, cfr. [31], [4] The classical limit is discussed by taking $\hbar \approx 0$. We note that the above non-standard realization of Feynman path integral is an internal one, however the "standard part" of the complex internal measure $\exp(\frac{i}{\hbar}\, {}^*S_t(\gamma_n + x))\, {}^*d\gamma_n$ does not make sense.
Finally we like to mention Harthong's use of standard analysis to discuss the problems of the classical limit of quantum mechanics as well as([68]) problems of wave propagation. Other uses of non-standard analysis within the domain of quantum theory have been in spectral resolutions of operators [ ], [ ] [28] (the continuum spectrum on the same footing as the discrete one, in a hyperfinite realization of Hilbert space) and in problems connected with singular functions arising in quantum field theory [4], in particular in the theory of renormalization [4], [24]-[25], [39], [40], [41], [67]
However it is clear that many more uses are possible.
We hope we have encuraged the reader to learn the language of non-standard analysis and add this tool to the techniques at her/his disposal when facing mathematical problems like the ones posed e.g. by quantum theory. Besides the fact that working with infinitesimals is great fun, it is a rewarding experience, and will certainly be even more so as many more of the potential uses of it, some of which we mentioned above, will be developed.

Acknowledgements

I am very grateful to the organizers, and in particular to Simon Diener and George Lochak, for inviting me to speak over this theme at a most stimulating conference, covering so many exciting subjects.
It is a special pleasure to thank heartily Jens-Erik Fenstad, Raphael Høegh-Krohn and Tom Lindstrøm for the joy of collaboration over many years on our joint book on non-standard analysis, on which these lectures are

based. Skilful typing by Mrs. Mischke and Richter is also gratefully acknowledged.

References

[1] A. Robinson, Non-Standard Analysis, North-Holland, Amsterdam (1966)
[2] D. Laugwitz, Infinitesimalkalkül, Bibliographisches Institut, Mannheim (1978)
[3] a) D. Laugwitz, The Theory of Infinitesimals, Accademia Nazionale Lincei Roma (1980)
b) D. Laugwitz, Grundbegriffe der Infinitesimalmathematik bei Leonhard Euler, pp. 459-483 in Folkerts und Lindgren, Edts., Mathemata, Franz Steiner Verlag, Stuttgart (1985);
[4] S. Albeverio, J.E. Fenstad, R. Høegh-Krohn, T. Lindstrøm, Non standard methods in stochastic analysis and mathematical physics, Acad. Press New York (1986)
[5] W.A.J. Luxemburg, Non-standard Analysis, Lectures on A. Robinson's Theory of Infinitesimals and Infinitely Large Numbers, Caltech Bookstore, Pasadena, rev. 1964
[6] M. Davis, Applied Nonstandard Analysis, J. Wiley, New York (1977)
[7] A. Hurd, P.A. Loeb, An introduction to nonstandard real analysis, Academic Press, New York (1985)
[8] J.M. Henle, E.M. Kleinberg, Infinitesimal Calculus, MIT-Press, Cambridge (1979)
[9] H.J. Keisler, Foundations of infinitesimal calculus, Prindle, Weber and Schmidt, Boston (1976)
[10] H.J. Keisler, An infinitesimal approach to stochastic analysis, Mem. Am. Math. Soc. 297 (1984)
[11] K.D. Stroyan, J.M. Bayod, Foundations of infinitesimal stochastic analysis, North-Holland, Amsterdam (1986)
[12] A.E. Hurd, Ed., Nonstandard Analysis - Recent Developments, Lect. Notes in Maths. 983, Springer, Berlin (1983)
[13] K.D. Stroyan, W.A.J. Luxemburg, Introduction to the theory of infinitesimals, Academic Press (1976)
[14] E. Zakon, Remarks on the nonstandard real axis, pp. 195-227 in Ref. [71]
[15] W.A.J. Luxemburg, What is nonstandard analysis? Amer. Math. Monthly 80, 38-67 (1973)
[16] J. Keisler, Elementary Calculus, Prindle, Weber & Schmidt, Boston (1976)
[17] E. Nelson, Internal set theory, Bull. Am. Math. Soc. 83, 1165-1198 (1977)
[18] M. Richter, Ideale Punkte, Monaden und Nichtstandard-Methoden, Vieweg (1982)
[19] F. Diener, Cours d'Analyse Non-standard, Université d'Oran, Office Publ. Univ. (1983)
[20] R. Lutz, M. Goze, Non standard analysis, Lect. Notes Math. 881, Springer (1981)
[21] S. Nagamachi, T. Mishimura, Linear canonical transformations on Fermion Fock space with indefinite metric, Tokushima Osaka Preprint (1984)

[22] N.J. Cutland, Nonstandard measure theory and its applications, Bull. London Math. Soc. 15, 529-589 (1983)
[23] R. Fittler, Some nonstandard quantum electrodynamics, Helv. Phys. Acta 57, 579-609 (1984)
[24] R. Fittler, More nonstandard quantum electrodynamics, FU Berlin, Preprint (1985)
[25] Li Bang-He, Nonstandard analysis and multiplication of distributions, Scientia Sinica 21, 561- (1978)
[26] Ph. Blanchard, J. Tarski, Renormalizable interactions in two-dimensions and sharp-time fields, Acta Phys. Austr. 49, 129-152 (1978)
[27] A. Voros, Introduction to nonstandard analysis, J. Math. Phys. 14, 292-296 (1973)
[28] M.O. Farrukh, Applications of nonstandard analysis to quantum mechanics, J. Math. Phys. 16, 177- (1975)
[29] N.J. Cutland, Infinitesimal methods in control theory, deterministic and stochastic, Hull Preprint (1985), Acta Appl. Math. (1986)
[30] J. Oikkonen, Harmonic analysis and nonstandard Brownian motion in the plane, Math. Scand. (1986)
[31] A. Sloan, The strong convergence of Schrödinger progagators, Trans. Am. Math. Soc. 264, 557-570 (1981)
[32] A.L. MacDonald, Sturm-Liouville theory via nonstandard analysis, Ind. Uni. Math. J. 25, (1976)
[33] J. Harthong, L'analyse non-standard, La Recherche 148, 1194-1201 (1983)
[34] B. Birkeland, A singular Sturm-Liouville problem treated by non-standard analysis, Math. Scand. 47 (1980)
[35] J.E. Fenstad, Non-standard methods in stochastic analysis and mathematical physics, Jber. d. Dt. Math. Vereins 82, 167, 180 (1980)
[36] S. Albeverio, Non-standard analysis: Polymer models, quantum fields, Acta Phys. Austr. Supp. 26, 233-254 (1984)
[37] R.F. Hoskins, Standard and Nonstandard Mathematical Analysis, Ellis Horwood, Chichester (1986)
[38] J.E. Fenstad, The discrete and the continuum in the mathematics and natural sciences, Oslo Preprint, Nov. 1985
[39] S. Moore, Non-standard applications of non-standard analysis, Bogotà Preprint
[40] L.M. Pecora, A nonstandard infinite dimensional vector space approach to Gaussian functional measures, J. Math. Phys. 23, 969-982 (1982)
[41] A. Stoll, Self-repellent random walks and polymer measures in two dimensions, Diss., Bochum (1985); and to appear in Proc. BiBoS Symp. II, Lect. Notes Math., Springer (Eds. S. Albeverio, Ph. Blanchard, L. Streit)
[42] A.K. Zvonkin, M.A. Shubin, Nonstandard analysis and singular perturbations of ordinary differential equations, Russ. Math. Surv. 39, 69-131 (1984)
[43] III Rencontre de géometrie du Schnepfenried, Vol. 2, Astérisque, 109-110 (1983)
[44] S. Albeverio, R. Høegh-Krohn, Schrödinger operator with point interactions and short range expansions, Physica 124A, 11-28 (1984)

[45] S. Albeverio, F. Gesztesy, R. Høegh-Krohn, H. Holden, Some exactly solvable models in quantum mechanics and the low energy expansions, pp. 12-28 in "Proceedings of the Second International Conference on Operator Algebras, Ideals and Their Applications in Theoretical Physics", Ed. H. Baumgärtel et al., Teubner, Leipzig (1984)
[46] S. Albeverio, F. Gesztesy, R. Høegh-Krohn, H. Holden, Solvable models in quantum mechanics, book in preparation
[47] L. Arkeryd, Loeb solutions of the Boltzmann equation, Arch. Rat. Mech. Anal. 86, 85-98 (1984)
[48] C. Keßler, Nonstandard methods in random fields, Diss., Bochum (1984)
[49] T. Lindstrøm, Non-standard energy forms and diffusions on manifolds and fractals, to appear in Proc. Ascona Conf. "Stochastic Processes in Classical and Quantum Systems", Eds. S. Albeverio, G. Casati, D. Merlini, Lect. Notes Phys. Springer (1986)
[50] S. Albeverio, R. Høegh-Krohn, Diffusion fields, quantum fields and fields with values in Lie groups, pp. 1-98 in M. Pinsky (Ed.) Stochastic Analysis and Applications, M. Dekker, New York (1985)
[51] M. Fukushima, Energy forms and diffusion processes, pp. 65-97 in "Mathematics + Physics, Lectures on Recent Results", Vol. I, Ed. L. Streit, World Scientific Publ. (1985)
[52] S. Albeverio, J.E. Fenstad, R. Høegh-Krohn, W. Karwowski, T. Lindstrøm, Perturbations of the Laplacian supported by null sets, with applications to polymer measures and quantum fields, Phys. Letts. 104 (1984)
[53] S. Albeverio, Some points of interaction between stochastic analysis and quantum theory, BiBoS-Preprint, to appear in Proc. Conf. "Stochastic Systems and Applications", Ed. K. Helmes (1986)
[54] M. Diener, C. Lobry, Eds., Analyse non standard et représentation du réel, Actes de L'Ecole d'Eté, OPU (Alger - CNRS (Paris)(1984)
[55] J.E. Fenstad, Lectures on stochastic analysis with applications to mathematical physics, Proc. Simposio Chileno Log. Mat., Santiago (1986)
[56] a) C. Keßler, The global Markov property for lattice spin systems in the case of uniqueness (in preparation)
b) C. Keßler, The global Markov property of a convex combination of GMP-states (in preparation)
[57] T. Lindstrøm, Nonstandard analysis and perturbation of the Laplacian along Brownian paths, pp. 180-200 in "Stochastic Processes - Mathematics and Physics", Proc. BiBoS I, Eds. S. Albeverio, Ph. Blanchard, L. Streit, Lect. Notes Maths. 1158, Springer (1985)
[58] S. Albeverio, J.E. Fenstad, R. Høegh-Krohn, Singular perturbations and non-standard analysis, Trans. Am. Math. Soc. 252, 275-295 (1979)
[59] S. Albeverio, Ph. Blanchard, R. Høegh-Krohn, Newtonian diffusions and planets, with a remark on non-standard Dirichlet forms and polymers, pp. 1-24 in "Stochastic Analysis and Applications", Eds. A. Truman, D. Williams, Lect. Notes Maths. 1095, Springer (1984)
[60] C. Keßler, Hyperfinite representation of generalized random fields, Bochum, Preprint (1984)
[61] G.F. Lawler, A self-avoiding random walk, Duke Math. J. 47, 655-692 (1980)

[62] A. Robert, Analyse non standard, Presses Polyt. Romandes XVII, Lausanne (1985)
[63] B. Birkeland, D. Normann, A non-standard treatment of the equation y' = f(y,t), Mat. Sem. Oslo (1980)
[64] E. Perkins, Stochastic processes and nonstandard analysis, in Ref. [12]
[65] W.A.J. Luxemburg, A. Robinson, Contributions to Non-Standard Analysis, North-Holland, Amsterdam (1972)
[66] C.E. Francis, Applications of non-standard analysis to relativistic quantum mechanics, J. Phys. A 14, 2539-2551 (1981)
[67] J. Tarski, Short introduction to nonstandard analysis and its physical applications, pp. 225-229 in "Many Degrees of Freedom in Field Theory", Ed. L. Streit,Plenum, New York (1978)
[68] J. Harthong, Etudes sur la mécanique quantique, Astérisque 111 (1984)
[69] P. Cartier, Perturbations singularières des équations différentielles ordinaires et analyse non-standard, Astérisque 92-93 (1982)
[70] S. Albeverio, Non-standard analysis: applications to probability theory and mathematical physics, to appear in Mathematics and Physics, Ed. L. Streit, World Scient. Publ., Singapore (1986)
[71] W.A.J. Luxemburg, Ed., Applications of Model Theory to Algebra, Analysis and Probability, Holt, New York (1969)
[72] D. Laugwitz, Cauchy and infinitesimals, Darmstadt Preprint (1985)
[73] L. Nottale, J. Schneider, Fractals and non-standard analysis, J. Math. Phys. 25, 1296-1300 (1984)
[74] J. Brasche, Perturbation of Schrödinger Hamiltonians by measures self-adjointness and lower semiboundedness, J. Math. Phys. 26, 621-626 (1985); and paper in preparation.
[75] J. Persson, Fundamental theorems for linear measure differential equations, Lund Preprint (1985) (to appear in Math. Scand.)
[76] A.N. Kochubei, Elliptic operators with boundary conditions on a subset of measure zero, Funct. Anal. Appl. 16, 137-139 (1982)
[77] V. Komkov, C. Waid, Asymptotic behavior of non-linear inhomogeneous equations via non-standard analysis, Ann. Pol. Math. 28, 67-87 (1973)
[78] S. Albeverio, D. Merlini, R. Høegh-Krohn, Euler flows, associated generalized random fields and Coulomb systems, pp. 197-215 in "Infinite dimensional analysis and stochastic processes", Ed. S. Albeverio, Pitman (1985)

QUANTUM THEORY OF CONTINUOUS OBSERVATIONS
SOME SIGNIFICANT EXAMPLES

Giovanni Maria Prosperi
University of Milan, Milan, Italy

1. Introduction

Ordinary Quantum Mechanics has to do with an idealized situation in which a system is prepared in a given state at an initial time t_0, it is left to evolve freely for some time and it is submitted to some kind of measurement at a single final time t_1 or at certain well separated subsequent times $t_1, t_2, \ldots$. In any case the single process is considered as pratically istantaneous.

Such idealization is well suited for treating certain kind of experimental situations, like the scattering experiments, but it is not sufficiently general to handle appropriately other very interesting ones.

Among the cases which do not fit in the above mentioned scheme we may typically mention the counting rate experiments and the macroscopic description of the evolution of a large body.

In both cases the system may be considered as continuously taken under observation for the entire duration of the experiment. In the first one the counters are supposed to be always alight and the final outcome can be naturally expressed in terms of the time distribution of the hittings. In the second one light or some other physical signal may be thought as continuously impinging on the large system and providing continuous informations on the values of a certain number of collective variables which are said to specify its *macroscopic state.*

The consideration of this second example is particular interesting in principle. In fact according to Bohr the experimental set up and the outcoming of an experiment have to be described in classical terms, i.e. in terms of a number of quantities having well defined values at any time. This circumstance compels us towards a kind of dualistic picture which could be composed if certain quantities were treated at least formally as continuously observed.

A formalism specifically suited for handling counting rate experiments has been proposed by Davies and Srinivas [1,2]. A more general one for treating any kind of continuous observation has been developed by our group in Milan [3–8]. In our formalism the outcome of the experiment is expressed in terms of the complete time development of a set of quantities. Davies result can be recovered as a particular case. Related to our approach it is the somewhat different one by Ghirardi, Rimini and Weber [9] in which the system is considered as observed at very close discrete times distribuited at chance [10].

The present paper is devoted to a breef introduction to our general formalism and to a sketch of some of the most recent results ([6–7]). Three examples shall be treated in particular: 1) the continuous observation of a kind of coarse grained position for a particle; 2) the case of the counting rate experiments; 3) the continuous observation of the molecular distribution function for a dilute gas.

2. General formalism

In order to introduce the formalism for continuous observations and to establish the notations I need to recall briefly the more general formulation of Quantum Mechanics based on the idea of effect and operation which has been developed by various authors and in particular by Haag and Kastler, Ludwig, Davies and Holevo ([1,11]).

In such more general formulation a set of p compatible observables, abstractly denoted by $A = (A_1, A_2, \dots A_p)$, is associated to an *effect valued measure* (e.v.m.) $\hat{F}_A(T)$, defined on the class $\mathcal{B}(\mathbf{R}^p)$ of the Borel subsets of $\mathbf{R}^p$, and an apparatus S_A for observing A to a similar *operation valued measure* (o.v.m.) $\mathcal{F}_{S_A}(T)$

I recall that by the terms *effect* and *operation* we mean a bounded selfadjoint operator with the property

$$0 \leq \hat{F} \leq \hat{I} \tag{2.1}$$

and a linear mapping of the space of the trace class operators $\mathbf{T}(\mathbf{H})$ into itself which is *completely positive* and *trace decreasing*,

$$\mathrm{Tr}(\mathcal{F}\hat{X}) \leq \mathrm{Tr}\hat{X}\,, \tag{2.2}$$

respectively (*). Similarly by an e.v.m. and an o.v.m. we mean a mapping from $\mathcal{B}(\mathbf{R}^p)$ into the family of the effects or of the operations respectively, such that

$$\hat{F}(\cup_{j=1}^{\infty} T_j) = \sum_{j=1}^{\infty} \hat{F}(T_j) \quad \text{and} \quad \mathcal{F}(\cup_{j=1}^{\infty} T_j) = \sum_{j=1}^{\infty} \mathcal{F}(T_j) \tag{2.3}$$

for $T_i \cap T_j \neq 0$ if $i \neq j$.

The specific e.v.m. $\hat{F}_A(T)$ and o.v.m. $\mathcal{F}_{S_A}(T)$ associated to A and S_A must be supposed normalized,

$$\hat{F}_A(\mathbf{R}^p) = \hat{I}\,, \quad \mathrm{Tr}\ \mathcal{F}_{S_A}(\mathbf{R}^p)\hat{X} = \mathrm{Tr}\hat{X}\,, \tag{2.4}$$

and related by the equation

$$\hat{F}_A(T) = \mathcal{F}'_{S_A}(T)\hat{I}\,, \tag{2.5}$$

(*) In the above, by $\mathbf{H}$ I have denoted the Hilbert space associated to the system and by $\hat{I}$ the identity in $\mathbf{H}$; I recall also that by a positive mapping we intend a mapping which transforms positive operator in positive operator i.e. $\hat{X} \geq 0 \Rightarrow \mathcal{F}\hat{X} \geq 0$, and by *completely* positive mapping a mapping which remains positive when extended to the tensorial product of $\mathbf{H}$ with any finite dimensional complex space.

where $\mathcal{F}'$ denotes the adjoint of $\mathcal{F}$ (remember that the dual space of $\mathbf{T}(\mathbf{H})$ is the space $\mathbf{B}(\mathbf{H})$, of the bounded operators in $\mathbf{H}$), i.e.

$$\mathrm{Tr}\ (\hat{F}_A(T)\hat{X}) = \mathrm{Tr}\ (\mathcal{F}_{S_A}(T)\hat{X})\ . \qquad [2.5']$$

Obviously there are *many* o.v.m.'s corresponding to *one* e.v.m. by [2.5] or [2.5′]. This agrees with the fact that we may conceive many different kinds of apparatus S_A for observing the same set of quantites A. A particular simple o.v.m. associated with a given $\hat{F}_A(T)$ is provided by the equation

$$\mathcal{F}_{S_A}(T)\hat{X} = \int_T (d\hat{F})^{\frac{1}{2}}\hat{X}(d\hat{F})^{\frac{1}{2}}\ , \qquad [2.6]$$

if the integral exists. However a priori this choice must not be considered priviledged in respect of any other one.

In the Heisenberg picture we set

$$\hat{F}_A(T,t) = e^{i\hat{H}t}\hat{F}_A(T)e^{-i\hat{H}t}\ , \qquad [2.7a]$$

$$\mathcal{F}_{S_A}(T,t)\hat{X} = e^{i\hat{H}t}[\mathcal{F}_{S_A}(T)(e^{-i\hat{H}t}\hat{X}e^{i\hat{H}t})]e^{-i\hat{H}t}\ . \qquad [2.7b]$$

Then the probability of observing a set of values $A \in T$ by the apparatus S_A at a time t_1 is assumed to be given by

$$P(A \in T, t_1|W) = T_r(\hat{F}_A(T,t_1)\hat{W}) = \mathrm{Tr}\ [\mathcal{F}_{S_A}(T,t_1)\hat{W}]\ , \qquad [2.8]$$

$\hat{W}$ being the ***statistical*** (or ***density***) ***operator*** representing the state of the system. Furthermore, if result $A \in T$ has been actually found, as a result the state of the system is modified in the following way

$$\hat{W} \to \mathcal{F}_{S_A}(T,t_1)\hat{W}/\mathrm{Tr}\ [\mathcal{F}_{S_A}(T,t_1)\hat{W}]\ . \qquad [2.9]$$

Obviously ordinary text book Quantum Mechanics is recovered by requiring the e.v.m. $\hat{F}_A(T)$ be a ***projection valued measure*** and by assuming [2.6] as a general rule under the hypothesis of a pure discrete spectrum.

Note also that from eq. [2.8] we obtain for the expectation values of the various A_s

$$< A_s >= \mathrm{Tr}\{\hat{O}_s(t)\hat{W}\}\ , \qquad [2.10]$$

having set

$$\hat{O}_s(t) = e^{i\hat{H}t}\hat{O}_s e^{-i\hat{H}t}\quad ,\quad \text{and}\quad \hat{O}_s = \int_{\mathbf{R}^p} d\hat{F}(x)x_s\ . \qquad [2.11]$$

So even in the present formulation a set of symmetric (even if not necessarily selfadjoint) operators is associated to a set of compatible observables. There are however two important differences with the ordinary formulation: a) in general the operators $\hat{O}_1, \ldots, \hat{O}_p$ do not commute each other (in general $\hat{F}_A(T)\hat{F}_A(S) \neq \hat{F}_A(S)\hat{F}_A(T)$). b) since now the decomposition [2.11] is no longer unique, there are many e.v.m.'s, and so many different sets of compatible observables $A \equiv$

$(A_1, \dots A_p)$, associated to the same set of operators. Obviously all such sets correspond to a single set of classical quantities.

In some sense we could think of the above $A_1, \dots, A_p$ as corresponding to a kind of simultaneuous coarse grain observation of the p commuting or non commuting quantities associated to $\hat{O}_1, \dots, \hat{O}_p$ in the ordinary formulation.

As a concequence of eq's [2.8] and [2.9], the *joint probability* of observing for A a *sequence of results* at certain subsequent times $t_0 < t_1 \dots t_n$ can be written as

$$P(A \in T_N, t_N; \dots; A \in T_1, t_1; A \in T_0, t_0 | W) = \\ \mathrm{Tr}\, [\mathcal{F}_{S_A}(T_N, t_N) \dots \mathcal{F}_{S_A}(T_1, t_1) \mathcal{F}_{S_A}(T_0, t_0) \hat{W}] \qquad [2.12]$$

which generalizes a well known formula by Wigner. Notice that, setting

$$\mathcal{F}(T_N, t_N; \dots; T_0, t_0) = \mathcal{F}_{S_A}(T_N, t_N) \dots \mathcal{F}_{S_A}(T_0, t_0) \qquad [2.13a]$$

and

$$\hat{F}(T_N, t_N; \dots; T_0, t_0) = \mathcal{F}'(T_0, t_0) \dots \mathcal{F}'_{S_A}(T_N, t_N) \hat{I} \;, \qquad [2.13b]$$

eq. [2.12] takes the form

$$P(A \in T_N, t_N; \dots; A \in T_0, t_0 | W) = \mathrm{Tr}\, [\hat{F}(T_N, t_N; \dots; T_0, t_0) \hat{W}] = \\ \mathrm{Tr}\, [\mathcal{F}(T_N, t_N; \dots; T_0, t_0) \hat{W}] \;. \qquad [2.14]$$

Since eq's [2.14] obviously define an o.v.m. and e.v.m. on $\mathcal{B}(\mathbf{R}^{p(N+1)})$, a sequence of observations at subsequent times is treated on the same foot as a single observation at one time.

This last circumstance is particularly intersting for us, since it suggests the possibility of treating in a significant way the somehow limit situation of a system kept continuously under observtion for a certain time interval. It is well know that such a limit situation would bring to unavoidable paradoxes in the framework of ordinary text book Quantum Mechanics.

In order to achieve this aim we have to generalize the concepts of e.v.m. and o.v.m. replacing the space $\mathbf{R}^p$ of the possible values for a set of quantities at a definite time by the functional space $\mathbf{Y}$ of the *possible complete hystories* $x(t) \equiv (x^1(t), \dots, x^n(t))$ for a similar set of quantities in an entire time interval (t_i, t_f). Correspondingly we have also to replace the class $\mathcal{B}(\mathbf{R}^p)$ of subsets of $\mathbf{R}^p$ by an appropriate σ-algebra of subsets of $\mathbf{Y}$. For this purpose we find convenient to set $t_i = -\infty$, $t_f = +\infty$ and to identify $\mathbf{Y}$ with the cartesian product $\mathbf{E}' = \mathcal{D}' \times \dots \times \mathcal{D}'$ of n identical factor $\mathcal{D}'$, $\mathcal{D}'$ being the space of the Schwarz ordinary distributions. Note that $\mathbf{E}'$ is the dual space of $\mathbf{E} = \mathcal{D} \times \dots \times \mathcal{D}$, $\mathcal{D}$ being the space of the infinitely differentiable functions with compact support in $\mathbf{R}$.

For any given element $h(t) \equiv (h_1(t), \dots, h_2(t))$ of $\mathbf{E}$ ad any trajectory $x(t) \in \mathbf{E}'$ we may define the *time average*

$$x_h = \int dt\; h_s(t) x^s(t) \qquad [2.15]$$

This quantity can also be assumed as a *coordinate* which partially specifies the trajectory. If we choose l different linearly independent elements of $\mathbf{E}, h^{(1)}, \ldots, h^{(l)}$, we may introduce l different coordinates, $x_{h^{(1)}}, \ldots, x_{h^{(l)}}$, for $x(t)$ and correspondingly consider the subset of $\mathbf{E}'$

$$C(h^{(1)}, \ldots, h^{(l)}; B_l) = \{x(t) \in \mathbf{E}' : (x_{h^{(1)}}, \ldots, x_{h^{(l)}}) \in B_l\} , \qquad [2.16]$$

B_l being a Borel set in $\mathbf{R}^l$. The subsets of the form [2.16] for any choice of l, B_l and $h^{(1)}, \ldots, h^{(l)}$ are called *Cylinder sets.* They generate a σ-algebra which we shall denote by $\sum$; furthermore we shall denote by $\sum_{t_0}^{t_1}$ the σ-algebra generated by the cylinder sets for which $h^{(1)}, \ldots, h^{(l)}$ have support in the interval (t_0, t_1).

Then to the continuous observation of a set of quantities we associate a mathematical structure which we call *Operation Valued Stochastic Process* (O.V.S.P.) and denote by

$$\{\mathbf{E}', \ \Sigma_{t_0}^{t_1}, \ \mathcal{F}(t_1, t_0; \ \cdot \ )\} .$$

Such structure is defined in the following way

1) For any time interval (t_0, t_1) an o.v.m. $\mathcal{F}(t_1, t_0; M)$ and a related e.v.m. $F(t_1, t_0; M) = \mathcal{F}'(t_1, t_0; M)\hat{I}$ are given on $\sum_{t_0}^{t_1}$ and the probability of observing a result $x(t) \in M$ is expressed by

$$P(M|W, t_0) = \text{Tr } [F(t_1, t_0; M)\hat{W}] = \text{Tr } [\mathcal{F}(t_1, t_0; M)\hat{W}] , \qquad [2.17]$$

if the system is prepared in the state W before the time t_0.

2) The composition law (cf. eq. [2.13a])

$$\mathcal{F}(t_2, t_0; N \cap M) = \mathcal{F}(t_2, t_1; N)\mathcal{F}(t_1, t_0; M) \qquad [2.18]$$

holds for $M \in \sum_{t_0}^{t_1}$ and $N \in \sum_{t_1}^{t_2}$ (note that $N \cap M \subset \sum_{t_0}^{t_2}$).

3) The conditional probability of finding $x(t) \in N \in \sum_{t_1}^{t_2}$, if $x(t) \in M \in \sum_{t_0}^{t_1}$ has been observed, is given by (cf. eq. [2.9])

$$P(N|M; W, t_0) = \text{Tr } [\hat{F}(t_2, t_1; N)\mathcal{F}(t_1, t_0; M)\hat{W}]/\text{Tr } [\mathcal{F}(t_1, t_0; M)\hat{W}] \qquad [2.19]$$

4) The time translatiom equation (cf. eq. [2.7b])

$$\mathcal{F}(t_1 + \tau, t_0 + \tau; M_\tau)\hat{X} = e^{i\hat{H}\tau}[\mathcal{F}(t_1, t_0; M)(e^{-i\hat{H}\tau}\hat{X}e^{i\hat{H}\tau})]e^{-i\hat{H}\tau} \qquad [2.20]$$

holds, where $M_\tau = \{x(t); x(t) = x'(t - \tau), x'(t) \in M\}$.

5) $\mathcal{F}(t_1, t_0; M)$ is normalized; i.e., if we set $\mathcal{G}(t_1, t_0) = \mathcal{F}(t_1, t_0; \mathbf{E}')$ the equation

$$\text{Tr } [\mathcal{G}(t_1, t_0)\hat{X}] = \text{Tr } \hat{X} \qquad [2.21]$$

holds.

Note that if we put $M = \mathbf{E}'$ in eq. [2.19], by eq. [2.21] we have

$$P(N|\mathbf{E}'; W, t_0) = \text{Tr } [\hat{F}(t_2, t_1; N)\mathcal{G}(t_1, t_0)\hat{W}] . \qquad [2.22]$$

So the mapping $\mathcal{G}(t_1, t_0)$ describes the modification produced on the state of the system by the action of the apparatus when no notice is taken of the result; briefly it describes the *disturbance* by the apparatus.

3. Poissonian and Gaussian O.V.S.P.

In the preceding section we have introduced a formalism for treating continuous observations on an axiomatic basis. There remains to show that an object $\mathcal{F}(t_1, t_0; M)$ satisfying all the requirements we have introduced actually exists and to produce significant examples.

For this purpose we find convenient to introduce the *characteristic functional* related to the probability distribution defined by eq. [2.17]. We set

$$L(t_1, t_0; [\xi(t)]|W) = \int dP([x(t)]|W, t_0) \exp \{i \int_{t_0}^{t_1} dt\ \xi_s(t) x^s(t)\} \qquad [3.1]$$

for any $\xi(t) \in \mathbf{E}$. Such quantity has the following important properties:
1) Positivity (it follows by the positivity of $P(M|Wt_0)$),

$$\sum_{ij} c_i^* L(t_1, t_0; [\xi^{(i)}(t) - \xi^{(j)}(t)]|W) c_j \geq 0\ , \qquad [3.2]$$

for any choice of the test functions $\xi^{(1)}(t), \xi^{(2)}(t), \dots$ and of the complex number $c_1, c_2, \dots$.
2) Normalization (it follows by [2.21])

$$L(t_1, t_0; 0|W) = 1 \qquad [3.3]$$

A general theorem (Milnos Theorem) in the theory of the so called *generalized stochastic process* ([12]) states that conversely if a functional $L(t_1, t_0; [\xi(t)]|W)$ satisfies [3.2], [3.3] and certain regularity conditions, it is the characteristic functional of a probability distribution $P(M|W; t_0)$. In practice, first we construct the probability density for the quanties $x_{h^{(1)}}, \dots, x_{h^{(l)}}$ as

$$p(x_1, h^{(1)}; \dots; x_l, h^{(l)}|Wt_0) =$$

$$= \frac{1}{(2\pi)^l} \int dk_1 \dots dk_l\ exp(-i \sum_{j=1}^{l} k_j x_j) L(t_1, t_0; [\sum_{j=1}^{l} k_j h^{(j)}(t)]) \equiv \qquad [3.4]$$

$$\equiv \int dP([x(t)]|W, t_0) \delta(x_1 - x_{h^{(1)}}) \dots \delta(x_l - x_{h^{(l)}})$$

then we obtain the probability for a cylinder set $P(C(h^{(1)} \dots h^{(l)}; B_l)|Wt_0)$ and finally we extend it to the entire $\sum_{t_0}^{t_1}$.

Mimicing the above procedure we may define a *characteric functional operator* (C.F.O.)

$$\mathcal{G}(t_1, t_0; [\xi(t)]) = \int d\mathcal{F}(t_1, t_0; [x(t)] \exp \{i \int_{t_0}^{t_1} dt\ \xi_s(t) x^s(t)\} \qquad [3.5]$$

which is related to L by the equation

$$L(t_1, t_0; [\xi(t)]|W) = \mathrm{Tr}\ \{\mathcal{G}(t_1, t_0; [\xi(t)])\hat{W}\} \qquad [3.6]$$

and which has the following properties:

$$\sum_{ij} c_i^* \mathcal{G}(t_1, t_0; [\xi^{(i)}(t) - \xi^{(j)}(t)]) c_j \ : \text{ completely positive} \qquad [3.7]$$

$$\mathcal{G}(t_1, t_0; 0) = \mathcal{F}(t_1, t_0; \mathbf{E}') = \mathcal{G}(t_1, t_0) : \text{ trace preserving} \qquad [3.8]$$

which correspond to [3.2] and [3.3].

Furthermore if $\xi_1(t)$ and $\xi_2(t)$ are two elements of $\mathbf{E}$ with support in (t_0, t_1) and (t_1, t_2) respectively, from eq. [2.18] we have

$$\mathcal{G}(t_2, t_0; [\xi_1(t) + \xi_2(t)]) = \mathcal{G}(t_2, t_1; [\xi_2(t)]) \mathcal{G}(t_1, t_0; [\xi_1(t)]) \ , \qquad [3.9]$$

whilst a time translation equation similar to [2.20] can be written also for $\mathcal{G}$.

Then, if we have a mapping $\mathcal{G}(t_1, t_0; [\xi(t)])$ in $\mathbf{T}(\mathbf{H})$ satisfing [3.7]-[3.9] and the time traslation equation, an associated $\mathcal{F}(t_1, t_0; M)$ with all the required properties can be constructed starting form the operatorial equation corresponding to eq. [3.4]. The problem of constructing on O.V.S.P. is so reduced to the simpler problem of costructing a *characeristic function operator.*

In order to solve the last problem let us introduce two additional hypothesis (which obviously amount to restrict the class of C.F.O. we are able to take into consideration). First we assume that $\mathcal{G}(t_1, t_0; [\xi])$ can be extended to functions not vanishing at t_0 and t_1, then for any $\xi(t) \in \mathbf{E}$ eq. [3.9] can be written

$$\mathcal{G}(t_2, t_0; [\xi(t)]) = \mathcal{G}(t_2, t_1; [\xi(t)]) \mathcal{G}(t_1, t_0; [\xi(t)]) \ . \qquad [3.10]$$

Furthermore we assume that [3.10] can be put in the differential form

$$\frac{\partial}{\partial t} \mathcal{G}(t, t_0; [\xi(\tau)]) = \mathcal{K}(t; \xi(t)) \mathcal{G}(t, t_0; [\xi(\tau)]) \qquad [3.11]$$

from which $\mathcal{G}(t_1, t_0; [\xi(t)])$ can be reobtained as

$$\mathcal{G}(t_1, t_0; [\xi(t)]) = T \exp \int_{t_0}^{t_1} dt \mathcal{K}(t; \xi(t)) \ , \qquad [3.12]$$

T denoting the time ordering prescription.

The problem is now to characterize the class of the operators $\mathcal{K}(t; \xi(t))$ for which $\mathcal{G}(t_1, t_0; [\xi(t)])$ as given by [3.12] satisfies [3.7] and [3.8]. Presently we are not able to solve this problem in full generality but we can produce an already interesting subclass for which the above conditions are met.

First let us set $\xi(t) = 0$ in [3.11]. Then we obtain

$$\frac{\partial}{\partial t} \mathcal{G}(t_1, t_0) = \mathcal{L}(t) \mathcal{G}(t_1, t_0) \qquad [3.13]$$

with $\mathcal{L}(t) = \mathcal{K}(t; 0)$ and we must find under what assumptions on $\mathcal{L}(t)$ the mapping $\mathcal{G}(t_1, t_0)$ defined by [3.13] turns out to be completely positive and trace preserving (cf. [3.8]). This last problem has been already studied in an different context in the litteracture ([13,14]). It is found that, if $\mathcal{L}(t)$ is bounded, it must be of the form

$$\mathcal{L}(t)\hat{X} = -i[\hat{K}(t),\hat{X}] - \frac{1}{2}\sum_{j=1}^{Q}\{\hat{R}_j(t),\hat{X}\} + \sum_{j=1}^{Q}\hat{R}_j(t)\hat{X}_j(t)\hat{R}_j^+(t) \ , \qquad [3.14]$$

with $\hat{K}(t) = \hat{K}^+(t)$ and $\hat{R}_1(t), \hat{R}_2(t), \ldots$ bounded operators. If $\mathcal{L}(t)$ is not bounded eq. [3.14] (with $\hat{K}$, $\hat{R}_1$, ... not bounded) turns out to be still a sufficient condition in order that $\mathcal{G}(t,t_0)$ has the two required properties (a part some pathological cases). Once that [3.14] has been assumed it can be shown that even [3.7] is satisfied if in turn $\mathcal{K}(t,\xi(t))$ is assumed to be of the form

$$\begin{aligned}\mathcal{K}(t,\xi(t))\hat{X} = \mathcal{L}(t)\hat{X} + \sum_{j=i}^{P}(e^{i\alpha^s\xi_s(t)} - 1)\hat{R}_j(t)\hat{X}\hat{R}_j^+(t) + \\ + \sum_{j=P+1}^{Q}[i\alpha_j^s\xi_s(t)(\hat{R}_j(t)\hat{X} + \hat{X}\hat{R}_j^+(t)) - \frac{1}{2}(\alpha_j^s\xi_s(t))^2\hat{X}] + i\beta^s\xi_s(t)\hat{X}\end{aligned} \qquad [3.15]$$

(where $\alpha_1, \ldots \alpha_\mu, \beta$ are arbitrary vectors in $\mathbf{R}^n$). This last result has been obtained in the above generality using techniques developed in the so called Quantum Stochastic Calculus ([6,15–17]).

For sake of analogy with the numerical stochastic processes and for reasons which shall be apparent in a moment, the second term in eq. [3.15] is said the *Poissonians term*, whilst the third one is said the *Gaussian term*.

Notice finally that $\hat{K}(t)$, $\hat{R}_1(t)$, ... must be simply Heisemberg operators $\hat{K}(t) = \exp\,(i\hat{H}t)\hat{K}\exp(-i\hat{H}t)$ etc., in order the time traslation prescription be satisfied. Furthermore $\hat{K}$ itself can be often reabsorbed in a redifinition of $\hat{H}$ and without loss of generality can be assumed to vanish.

Let us now try to understand the meaning of the result we have obtained. Notice that's from eq. [3.1] it follows.

$$< x^{s_1}(t^{(1)})x^{s_2}(t^{(2)})\ldots x^{s_l}(t^{(l)}) >= (-i)^l \frac{\delta^l L(t_1,t_0;[\xi(t)]|W)}{\delta\xi_{s_1}(t^{(1)})\ldots\delta\xi_{s_l}(t^{(l)})}\bigg|_{\xi=0} \ . \qquad [3.16]$$

Then combining such equation with [3.6] and [3.12] we obtain in particular

$$< x^s(t) >= -i\mathrm{Tr}\left\{\frac{\delta\mathcal{K}(t,\xi)}{\delta\xi_s}\mathcal{G}(t,t_0)\hat{W}\right\}\bigg|_{\xi=0} = \mathrm{Tr}\ \{\hat{O}^s(t)\mathcal{G}(t,t_0)\hat{W}\} \qquad [3.17]$$

with

$$\hat{O}^s = \sum_{j=1}^{P}\alpha_j^s\hat{R}_j^+\hat{R}_j + \sum_{j=P+1}^{Q}\alpha_j^s(\hat{R}_j + \hat{R}_j^+) + \beta^s \qquad [3.18]$$

Eq. [3.18] is analogous to eq. [2.10] and qualifies our O.V.S.P. as corresponding to a (coarse grained) continuous observation of the set of quantities related to the operators $\hat{O}^1, \hat{O}^2, \ldots \hat{O}^l$. Notice however that we can write similarly

$$
\begin{aligned}
< x^s(t)x^{s'}(t') >= \delta(t-t')\mathrm{Tr}\left\{ \frac{\delta^2 \mathcal{K}(t,\xi)}{\delta\xi_s\delta\xi_{s'}} \mathcal{G}(t,t_0)\hat{W} \right\}\Bigg|_{\xi=0} + \\
+\theta(t-t')\mathrm{Tr}\left\{ \frac{\delta \mathcal{K}(t,\xi)}{\delta\xi^s} \mathcal{G}(t,t') \frac{\delta \mathcal{K}(t',\xi)}{\delta\xi^{s'}} \right\}\Bigg|_{\xi=0} + \\
+\theta(t'-t)\mathrm{Tr}\left\{ \frac{\delta \mathcal{K}(t',\xi)}{\delta\xi^{s'}} \mathcal{G}(t',t) \frac{\delta \mathcal{K}(t,\xi)}{\delta\xi^{s}} \mathcal{G}(t,t_0)\hat{W} \right\}\Bigg|_{\xi=0}
\end{aligned}
\qquad [3.19]
$$

and the occurence of the $\delta(t-t')$ in the first term of such equation shows that only time averages of the kind defined by [2.15] have an actual meaning and that for not sufficiently smooth weight functions the expected fluctuations are very large.

Notice also that in the pure poissonian case $P = Q$ the operators $\hat{O}^s$, as defined by [3.18], are positive definite, whilst in the gaussian case they can have a priory eigevalues with both signes.

4. Specific examples

a) *Continuous observation of the position of a particle.* We can identify the operators $\hat{O}^s$ in [3.18] with the ordinary operator $\hat{q}_j$ corresponding to the coordinate of a particle, if we set $n = P = 3$, $Q = 0$ (pure gaussian case), $\hat{K} = 0$ and

$$
\hat{R}_j = \hat{R}_j^+ = \sqrt{\frac{\gamma}{2}}\hat{q}_j \quad , \quad \alpha_j^s = \frac{1}{\sqrt{2\gamma}}\delta_j^s . \qquad [4.1]
$$

We obtain then

$$
\mathcal{K}(t,\xi(t)) = -\frac{\gamma}{4}[\hat{q}_j(t),[\hat{q}_j(t), \cdot\ ]] + i\xi_j(t)\{\hat{q}_j(t), \cdot\ \} - \frac{1}{4\gamma}\xi_j(t)\xi_j(t) . \qquad [4.2]
$$

In particular setting

$$
q_j^h = \int dt\ h(t) q_j(t) \qquad [4.3]
$$

and identifying $h(t)$ with $\frac{1}{\Delta t}\chi_{(t,t+\Delta t)}(t')$ ($\chi_{(t,t+\Delta t)}(t')$ denoting the characteristic function of the interval $(t, t+\Delta t)$), in the limit of small Δt we have from [3.19]

$$
< (q_j^h - < q_j^h >)^2 >= \frac{\gamma}{2\Delta t} + Tr\{(\hat{q}_j(t) - < \hat{q}_j(t) >)^2 \mathcal{G}(t,t_0)\hat{W}\} . \qquad [4.4]
$$

Here the second term is essentially the ordinary quantum mechanical fluttuation $(\Delta q_{QM})^2$, whilst the first one is typical of our theory. This first term comes from the δ-term in [3.19] and diverges for $\Delta t \to 0$ (being negligible on the contrary for $\Delta t >> \gamma/(\Delta q_{QM})^2$).

The constant γ occuring in [4.21] gives a measure of the disturbance produced by the apparatus in the evolution of the particles. If γ is related in an appropriate way with the cross section of the molecules and the density of the gass (the liquid) filling the apparatus, the present example can be considered as describing the formation of a track in a Wilson chamber (in a bubble chamber). To this examples it is strictly related even the treatment of Ghirardi et al. ([9]).

b) *Counting experiment.* Let us consider a system of identical particles, let be $\{u_j(\vec{x})\}$ a set of ortogonal one-particle eigenstates, spanning a subspace **S** of the corresponding Hilbert space, and let be C a counter which reacts to particles, the state of which is in **S**. In practice **S** may be thought as corresponding to the energy range of sensibility and to the spatial disposition of C (more realistically an effect should be associated to C rather than a projection operator, but this is immaterial for the present purpose). In the non relativistic second quantization formalism the number of particles N occuring in **S** is represented by

$$\hat{N}(t) = \sum_j \hat{a}_j^+(t)\hat{a}_j(t) \ , \qquad [4.5]$$

where

$$\hat{a}_j(t) = \int d^3\vec{x}\ u_j^*(\vec{x})\hat{\psi}(\vec{x},t) \ , \qquad [4.6]$$

$\hat{\psi}(\vec{x},t)$ being the field operator. If we denote by τ the characteristic life time of a state in **S** (i.e., in practice, the time spent by a particle inside C) and assume the counter to have effinciency 1 in its range of sensibility, the number of hits in a time interval (t_1, t_2) would be given by the quantity

$$N_c(t_2,t_1) = \frac{1}{\tau}\int_{t_1}^{t_2} dt\ N(t) \ . \qquad [4.7]$$

The appropriate O.V.P.S. can be obtained setting in [3.18], $n = 1$, $P = Q$ (*purely poissonian* case), $\alpha = \tau$, $\hat{R}_j = \frac{1}{\sqrt{\tau}}\hat{a}_j$. In fact we have then

$$\mathcal{K} = -\frac{1}{2\tau}\sum_j\{\hat{a}_j^+\hat{a}_j, \cdot\} + \frac{1}{\tau}e^{i\tau\xi(t)}\sum_j \hat{a}_j \cdot \hat{a}_j^+ \ , \qquad [4.8]$$

so $\hat{O}(t) \equiv \hat{N}(t)$ and the probability for the occurrence of a number N_c of counts belonging to the interval B would be given by [4.6] taking $l = 1$ and $h(t) \equiv \frac{1}{\tau}\chi_{(t_1,t_2)}(t)$. We find

$$\begin{aligned} P(N_c \in B|Wt_0) &= \frac{1}{2\pi}\int_B dx\ e^{-ikx}\ \mathrm{Tr}\ \{\mathcal{G}(t_2,t_1;[k\frac{1}{\alpha}\chi_{(t_1,t_2)}])\mathcal{G}(t_1,t_0)\hat{W}\} = \\ &= \sum_{n\in B} \mathrm{Tr}\ \{\mathcal{N}_n(t_2,t_1)\mathcal{G}(t_1,t_0)\hat{W}\} \ , \end{aligned} \qquad [4.9]$$

where we have set

$$\begin{cases} \mathcal{N}_n(t_2,t_1) = \int_{t_1}^{t_2} dt^{(n)} \int_{t_1}^{t^{(n)}} dt^{(n-1)} \dots \\ \qquad \int_{t_2}^{t''} dt' \, \mathcal{U}(t_2,t^{(n)}) \mathcal{I}(t^{(n)},t^{(n-1)}) \dots \mathcal{I}(t')\mathcal{U}(t',t_0) \\ \mathcal{U}(t'',t') = T \exp(-\frac{1}{2\tau} \sum_j \int_{t'}^{t''} dt \{\hat{a}_j^+(t)\hat{a}_j(t), \cdot \}) \\ \mathcal{I}(t) = \frac{1}{\tau} \sum_j \hat{a}_j(t) \cdot \hat{a}_j^+(t) \end{cases} \qquad [4.10]$$

and we have used the expansion

$$\mathcal{G}(t_2,t_1;[k\frac{1}{\tau}\chi_{(t_1,t_2)}]) = \sum_{n=0}^{\infty} \frac{1}{n!} T[(\int_{t_1}^{t_2} dt \; e^{i\tau k \frac{1}{\tau}\chi_{(t_1,t_2)}(t)} \mathcal{I}(t))^n \; \mathcal{U}(t_2,t_1)] =$$
$$= \sum_{n=0}^{\infty} e^{ikn} \int_{t_1}^{t_2} dt^{(n)} \int_{t_1}^{t^{(n)}} dt^{(n-1)} \dots \qquad [4.11]$$
$$\int_{t_1}^{t''} dt' \, \mathcal{U}(t_2,t^{(n)}) \mathcal{I}(t^{(n)}) \mathcal{U}(t^{(n)},t^{(n-1)}) \dots \mathcal{I}(t')\mathcal{U}(t',t_0) \, .$$

Notice that the quantity N_c can take only integer values according to [4.9] as required; for this result the identification $\alpha = \tau$ was crucial. The last term in [4.9] coincides with the corresponding expression as obtained by Davies and Srinivas in the context of their own formalism ([1,2]). A generalization of the treatment to the second quantization of a relativistic wave equation is trivial.

c) *Phase space molecular distribution function.* In the more general formulation of Quantum Mechanics we have recalled in section 2, the *density of effects* for a coarse grained simultaneous measurement of the position and the momentum of a single particle may be taken as

$$\hat{f}_1(\vec{x},\vec{p}) = \hat{U}_1(\vec{x},\vec{p})\hat{\rho}_1\hat{U}_1^+(\vec{x},\vec{p}) \, , \qquad [4.12]$$

where

$$\hat{U}_1(\vec{x},\vec{p}) = \exp i(\vec{p}\cdot\hat{\vec{q}}_1 - \vec{x}\cdot\hat{\vec{p}}_1) \qquad [4.13]$$

($\hat{\vec{q}}_1$ and $\hat{\vec{p}}_1$ being the ordinary one-particle position and momentum operators) and $\hat{\rho}_1$ denotes any fixed positive trace-one operator which we shall assume spectralized as

$$\hat{\rho}_1 = \sum_j |j > \omega_j < j| \quad (\omega_j \geq 0, \sum_j \omega_j = 1) \, . \qquad [4.14]$$

The above statement means that the quantity

$$\mathrm{Tr}\{\hat{f}_1(\vec{x},\vec{p};t)\hat{W}_1\} \qquad [4.15]$$

has to be interpreted as the phase space density of probability for the considered particle prepared in the state W_1. According to general rules for going from first to second quantization quantities the *actual distribution function* for a system of identical particles can be represented by the operator

$$\begin{aligned}\hat{f}(\vec{x},\vec{p};t) &= \int d^3\vec{y}d^3\vec{y}'\hat{\psi}^+(\vec{y},t) < \vec{y}|\hat{f}_1(\vec{x},\vec{p})|\vec{y}' > \hat{\psi}(\vec{y}',t) = \\ &= \sum_j \hat{R}_j^+(\vec{x},\vec{p};t)\hat{R}_j(\vec{x},\vec{p};t) ,\end{aligned} \qquad [4.16]$$

where we have set

$$\hat{R}_j(\vec{x},\vec{p};t) = \sqrt{\omega_j}\int d^3\vec{y} < j|\hat{U}_1^+(\vec{x},\vec{p})|\vec{y} > \hat{\psi}(\vec{y},t) . \qquad [4.17]$$

The simplest O.V.S.P. corresponding to a continuous observation of $f(\vec{x},\vec{p};t)$ can then be defined by

$$\begin{aligned}K &= -\frac{1}{2}\gamma\sum_j \int d^3\vec{x}\ d^3\vec{p}\ \{\hat{R}_j^+(\vec{n},\vec{p};t)\hat{R}_j(\vec{x},\vec{p};t),\ \cdot\ \}+ \\ &+\gamma\sum_j \int d^3\vec{x}\ d^3\vec{p}\ e^{\frac{i}{\gamma}\xi(\vec{x},\vec{p};t)}\hat{R}_j(\vec{x},\vec{p};t)\ \cdot\ \hat{R}_j^+(\vec{x},\vec{p};t) ,\end{aligned} \qquad [4.18]$$

wich is again purely Poissonian in agreement with the positive character of the quantity defined by [4.16]. In eq. [4.18] we have obviously identified the two indices s and j occuring in eq.'s [2.15] and [3.15] with the set of continuous variables $\vec{x}$, $\vec{p}$ and the mixed set j , $\vec{x}$, $\vec{p}$ respectively and we have replaced α_j^s by $\alpha_{j\vec{x}\ \vec{p}}^{\vec{x}'\ \vec{p}'} = \frac{1}{\gamma}\delta^3(\vec{x}-\vec{x}')\ \delta^3(\vec{p}-\vec{p}')$.

References

1) E.B. Davies, *Quantum theory of open systems.* (Academic London, 1976).
2) E.B. Davies, IEEE Trans. Inf. Theory **23** (1977), 530; M.D. Srinivas, J. Math. Phys. **18** (1977), 2138; M.D. Srinivas and E.B. Davies, Opt. Acta **28** (1981), 981.
3) A. Barchielli, L.Lanz and G.M. Prosperi, Nuovo Cimento, **72 B** (1982), 79; Found. of Physics, **13** (1983), 779.
4) G. Lupieri, J. Math. Phys. **24** (1983), 2329.
5) A. Barchielli, Nuovo Cimento, **74 B** (1983), 113; Phys. Rev. **D 32** (1985), 347; preprint IFUM 311/FT (1985).
6) A. Barchielli and G. Lupieri, J. Math. Phys. **26** (1985), 2222; Lect. Notes in Math. 1136 (Springer, Berlin, 1986), 57.
7) L. Lanz, O. Melsheimer and S. Penati, preprint Univ. di Milano 1985.
8) Revue papers: G.M. Prosperi, Lect. Notes in Math. **1055** (Springer, Berlin, 1984), 301; A. Barchielli, L. Lanz and G.M. Prosperi, Proc. I.S.Q.M. Tokyo 1984, 165; Proc. Chaotic Behaviour in Q.S. Como 1984 (Plenum, New York, 1985), 321.
9) A. Rimini, Proc. Theoretical Physics Meeting - Amalfi, (ESI, Napoli, 1984), 275.
G.C. Ghirardi, A. Rimini and T. Weber, I.C.T.P. preprint, IC/85/292.
10) Prof. V.P. Belavkin has kindly comunicated me that continuous observation can be treated even in the context of his formalism on control theory. (Cf. this volume).
11) K. Kraus, *States, Effects, and Operations*, Lect. Notes in Phys., **190** (Springer, Berlin, 1983);
G. Ludwig, *Foundations of Quantum Mechanics* (Springer, Berlin, 1982);
A.S. Holevo, *Probabilistic and Statistical Aspects of Quantum Theory* (North Holland, Amsterdam, 1982);
E. Prugovecki, *Stochastic Quantum Mechanics and Quantum Spacetime* (Reidel, Pordrecht and Boston, 1983).
12) I.M. Gel'fand and N. Ya Vilenkin, Generalized Functions, Application of Harmonic Analysis, vol. 4 (Academic New York and London, 1964);
M.C. Reed, Lect. Notes in Phys. **25** (Springer, Berlin, 1973).
13) V. Gorini, A. Kossakorowki and E.C.G. Sudarshan, J. Math., Phys. **17** (1976), 821.
14) G. Lidblad, Comm. Math. Phys. **48** (1976), 119.
15) R.L. Hudson and K.R. Parthasarathy, Comm. Math. Phys. **93** (1984), 301; Acta Appl. Math. **2** (1984), 353.
16) K.R. Parthasarathy, preprint Indian Statistical Inst., New Delhi 1985.
17) A.S. Holevo, preprint Steklov Math. Inst. Moskow, 1986.

CONDITIONAL EXPECTATION AND STOCHASTIC PROCESSES IN QUANTUM PROBABILITY

Werner Stulpe
Technische Universität Berlin, Federal Republic of Germany

ABSTRACT

Proceeding in close analogy to classical probability theory, quantum concepts of conditional expectations, conditional distributions, stochastic processes, and a Markov property of these are introduced by means of instruments. Examples concerning successive measurements and the random walk of a particle are discussed.

1. INTRODUCTION

Whereas most of the basic notions of classical probability theory have a unique counterpart in the probabilistic framework of quantum mechanics, a uniform meaning of less elementary concepts like conditional expectations and stochastic processes does not exist in the quantum case. Many different formulations of quantum conditional expectations or quantum stochastic processes have been given. It is the aim of this contribution to add a further formulation where it is crucial that the concepts introduced here are as close as possible to the classical ones.

As basic elements of a statistical physical theory I will comprehend the statistical ensembles of physical systems and the classes of statistically equivalent realistic measurements with the only two outcomes 0 and 1 which are called effects. The set of ensembles and the set of effects are connected by a probability functional assigning to every ensemble and to every effect the probability for the outcome 1. There are some other notions, namely observables, operations and instruments, which are also important, but less fundamental than ensembles and effects. Observables are related to measuring apparatuses with more than two outcomes, operations and instruments to preparative measurements.

All these concepts were introduced systematically in the axiomatics of Ludwig and coworkers [1,2] in the late sixties as well as in the so called operational approach of Davies, Edwards, Lewis, and others [3,4] about 1970. These formulations, worked out in the framework of base-norm and order-unit-norm spaces, concern the general structure of a statistical theory. In this paper we ristrict our considerations to Hilbert space quantum mechanics. In chapter 2 the notions mentioned above are repeated. In chapter 3 we introduce quantum conditional expectations and distributions and interpret them in terms of successive measurements. Quantum stochastic processes and their Markovicity are defined in Chapter 4 where also three examples are discussed. Finally, we shall give some concluding remarks.

2. BASIC NOTIONS OF QUANTUM PROBABILITY

In quantum mechanics, the statistical ensembles are described by (normalized) density operators acting in some separable complex Hilbert space X. The density operators W form a convex set K in the space $C_s^1(X)$ of all self-adjoint trace class operators. By the trace functional, the space $B_s(X)$ of all bounded self-adjoint operators in X is dual to the ordered Banach space $C_s^1(X)$. The effects correspond to the operators $A \in \mathcal{L}_s(X)$ with $0 \leq A \leq 1$ (1 being the unit operator in X). These operators A form a convex set L. The probability for the outcome 1 of an effect $A \in L$ in an ensemble $W \in K$ is given by tr WA.

A measurement is preparative if the microsystems interacting with the measuring apparatus are not absorbed. Preparative 0-1-measurements are formally described by operations. An operation is defined as a positive linear map $T: C_s^1(X) \to C_s^1(X)$ such that $0 \leq \mathrm{tr}\, TW \leq 1$ for all $W \in K$ (equivalently, T is positive and linear with $\|T\| \leq 1$). Let an ensemble $W \in K$ be given. Then we interpret as follows:

(i) $\mathrm{tr}\, TW = \mathrm{tr}\,(TW)1 = \mathrm{tr}\, WT'1$ is the probability for the outcome 1 (T' denotes the adjoint map of T with respect to the trace functional)

(ii) for $TW \neq 0$ $\frac{TW}{\mathrm{tr}\, TW}$ is the ensemble obtained by selection of systems according to the outcome 1.

As a consequence of (i), $A := T'1 \in L$ is the effect "measured by T".

Observables describe statistically equivalent measuring apparatuses with arbitrarily many outcomes and are defined as effect-valued measures on an arbitrary measurable space (M, Σ), i.e. an observable is a map $F: \Sigma \to L$, $B \mapsto F(B)$, with the properties

(i) $F(\Phi) = 0$, $F(M) = 1$

(ii) $F\left(\bigcup_{i=1}^{\infty} B_i\right) = \sum_{i=1}^{\infty} F(B_i)$ for $B_i \cap B_j = \Phi$ $(i \neq j)$

where the sum converges in the strong operator topology. M is the "value space" of F, and $B \to P_W^F(B) := \mathrm{tr}\, WF(B)$ is the probability distribution of F in an ensemble $W \in K$.

Instruments are associated with preparative measurements of observables. An instrument J is an operation-valued measure on (M, Σ), i.e. J is a map $B \mapsto J(B)$ assigning to every measurable set an operation such that the conditions

(i) $J(\Phi) = 0$

(ii) $J(M)W \in K$ for $W \in K$

(iii) $J\left(\bigcup_{i=1}^{\infty} B_i\right)V = \sum_{i=1}^{\infty} J(B_i)V$ for $B_i \cap B_j = \Phi$ $(i \neq j)$ and $V \in C_s^1(X)$

are fulfilled, the sum converging in the trace norm. Note that

(i) $B \to F(B) := J'(B)1$ is the observable determined by J

(ii) given an ensemble W K, $\frac{J(B)W}{\mathrm{tr}\, J(B)W}$ is the ensemble obtained by selection of systems according to outcomes in the set $B \in \Sigma$.

The σ-additivity of J, for instance, means physically that one can form mixtures of ensembles corresponding to selection procedures according to outcomes in disjoint sets B_i.

3. QUANTUM CONDITIONAL EXPECTATIONS AND DISTRIBUTIONS

Quantum conditional expectations are usually defined in the context of von Neumann algebras. Also the duals of instruments are sometimes understood as quantum conditional expectations. By means of instruments we will introduce a notion of quantum conditional expectation which differs from these concepts, but which is as close as possible to classical probability theory.

Let a Hilbert space X, a measurable space (M,Σ), an instrument J on (M,Σ) determining the observable $F:=J'(.)1$, and an effect A L be given. To define quantum conditional expectations, consider the successive measurement of F and A in a fixed ensemble $W\in K$. This means that A is measured in the ensemble $J(M)W$ obtained by non-selective measurement of F. The probability for occurrence of values of the observable F in the set $B\in\Sigma$ is given by

$$P_W^F(B) = \mathrm{tr}\ WF(B) = \mathrm{tr}\ WJ'(B)1 = \mathrm{tr}\ J(B)W \qquad (3.1)$$

and the joint probability for occurrence of values of F in B and for the outcome 1 of the effect A by

$$P_W^F(B)\ \mathrm{tr}\left(\frac{J(B)W}{\mathrm{tr}\ J(B)W}A\right) = \mathrm{tr}\ (J(B)W)A \quad .$$

Now let A be an arbitrary element of $B_s(X)$. $B\rightarrow \mathrm{tr}\ (J(B)W)A$ defines a bounded signed measure on Σ which is continuous with respect to the probability measure P_W^F. Indeed, because of (3.1) and $J(B)W\geq 0$ it follows from $P_W^F(B)=0$ that $\mathrm{tr}\ (J(B)W)A=0$. Hence, we can apply Radon-Nykodym's theorem and obtain the following statement.

Proposition: For every $A\in B_s(X)$ there exists a Σ-measurable function $E_W^J(A):M\rightarrow\mathbb{R}$ such that

$$\mathrm{tr}\ (J(B)W)A = \int_B E_W^J(A)\ dP_W^F \qquad (3.2)$$

holds true for all $B\in\Sigma$. This function is determined uniquely P_W^F-a.e.

Definition: We call $E_W^J(A)$ a version of the quantum conditional expectation of A under the hypothesis J for the given ensemble W, the equivalence class of all $E_W^J(A)$ is the quantum conditional expectation of A. For an effect $A\in L$ $E_W^J(A)$ is also called a version of the quantum conditional probability of A.

Let us remark that this definition is completely analogous to the definition of the corresponding notions in classical probability [5,6]. The type of quantum conditional expectations established here has been introduced implicitly in the paper of Cycon and Hellwig [7] and explicitly in contributions given by Hellwig, the author, and Ozawa [8-11].

To give an interpretation of quantum conditional expectations assume that there is a set $B\in\Sigma$ and a version $E_W^J(A)$ such that $E_W^J(A)$ being constant on B and $P^F_W(B)\neq 0$. Then Eqs. (3.2) and (3.1) yield for all $x\in B$

$$E^J_W(A)(x) = \text{tr}\left(\frac{J(B)W}{\text{tr}\, J(B)W}A\right) \quad .$$

This shows that for an effect A L the value of the version $E^J_W(A)$ of the quantum conditional probability at any point $x\in B$ coincides with the probability for the outcome 1 of A in the ensemble obtained by the selection procedure according to outcomes of F in the set B. Especially, B may consist of one point $x\in M$ only. These remarks motivate the denotation "quantum conditional probability" respectively "quantum conditional expectation".

Let us consider the successive measurement of two observables F and G in the ensemble W, the first one associated with the instrument J defined on (M,Σ) and the second one defined on a second measurable space (M',Σ'). In this situation the joint probabilities

$$\text{tr}\,(J(B)W)G(B') = \int_B E^J_W(G(B'))\, dP^F_W$$

$(B\in\Sigma,\ B'\in\Sigma')$ are of interest. For versions $E^J_W(G(B'))$ of the conditional probabilities of $G(B')$ the following equalities and unequalities hold true P^F_W-a.e.:

(i) $0\leq E^J_W(G(B'))\leq 1$

(ii) $E^J_W(G(\Phi))=0,\ E^J_W(G(M'))=1$

(iii) $E^J_W(G(B'_1))\leq E^J_W(G(B'_2))$ for $B'_1\subseteq B'_2$

(iv) $E^J_W\left(G\left(\bigcup_{i=1}^{\infty} B'_i\right)\right) = \sum_{i=1}^{\infty} E^J_W(G(B'_i))$ for $B'_i\cap B'_j=\Phi$ $(i\neq j)$.

The proof is an easy exercise. These properties do not imply that $B'\to E^J_W(G(B'))(x)$ is a probability measure on Σ' for P^F_W-almost every $x\in M$ because the equation (iv), for instance, only holds true up to a null set depending on the sequence of disjoint sets B'_i and, in general, there are more than countably many such sequences. However, under a technical assumption it is always possible to choose the versions $E^J_W(G(B'))$ in such a manner that they define a probability measure for each $x\in M$. In this case the map $P: M\times\Sigma'\to[0,1]$ given by

$$P(x,B') := E^J_W(G(B'))(x)$$

is a Markov kernel, i.e. P is a measurable function in the first argument and a probability measure in the second one.

Definition: A Markov kernel P on $M\times\Sigma'$ is called a quantum conditional distribution of the observable G under hypothesis of the instrument J for the given ensemble W if for every $B'\in\Sigma'$ $P(.,B')$ is a version of the quantum conditional expectation of $G(B')$, i.e.

$$P(.,B') = E_W^J(G(B'))$$

P_W^F-a.e.

In view of the following theorem we remark that a polish space is a topological space the topology of which can be derived from a complete and separable metric.

Theorem: Let M' be a polish space and Σ' the σ-algebra of its Borel sets. Then there exists a quantum conditional distribution P of G under the hypothesis J for the given W. Moreover, P is unique in the sense that two quantum conditional distributions differ only on a set $N\times\Sigma'$ where N is a P_W^F-null set.

The theorem is the generalization of a corresponding classical theorem [5,6]. Its proof is not trivial, but it is mainly a transfer of the classical proof. This transfer as well as the definition of quantum conditional distributions is due to the author [11]. As we shall see in the next chapter, there are important examples of observables and instruments for which the quantum conditional expectations and distributions can even be calculated.

4. QUANTUM STOCHASTIC PROCESSES AND A MARKOV PROPERTY

4.1. Definition

A uniform meaning of quantum stochastic processes does not exist. Quantum stochastic processes are mostly introduced in the framework of C^*- and von Neumann Algebras. Quantum dynamics itself is sometimes understood to be a quantum stochastic process. The notion "quantum stochastic process" also arises in the so called stochastic interpretation of quantum mechanics. Here we will define quantum stochastic processes as one-parameter families of instruments. Moreover, we give a Markov property using quantum conditional distributions.

Definition: Let X be a Hilbert space and T a non-empty subset of $\mathbb{R}_0^+$. If $W\in K$ is a fixed ensemble and for every $t\in T$ an instrument J_t is defined on the same measurable space (M,Σ), then we call $(W,\{J_t\}_{t\in T})$ a quantum stochastic process on the "state space" (M,Σ) with Hilbert space X, given the ensemble W. We call this process Markovian if

(i) for any $s,t\in T$, $s\leq t$, there exists a quantum conditional distribution P_{st} of the observable $F_t:=J_t'(.)1$ under the hypothesis of J_s, given the ensemble W

(ii) by a suitable choice of the conditional distributions P_{st}, for any

$r,s,t\in T$, $r\leq s\leq t$, the Chapman-Kolmogorov equations

$$P_{rt}(x,B) = \int P_{st}(y,B)\, P_{rs}(x,dy) \tag{4.1}$$

($x\in M$, $B\in\Sigma$) are satisfied.

Again, this definition of quantum stochastic processes and Markovicity is very close to classical probability theory [5,6]. Conditional distributions of a classical Markov process, sometimes called transition probabilities of the process, fulfil the Chapman-Kolmogorov equations. However, a classical stochastic process with conditional distributions fulfilling the Chapman-Kolmogorov equations may not necessarily be Markovian. Thus we see that, in a certain sense, Markovicity in quantum mechanics is more general than in classical probability.

Quantum stochastic processes are comprehended as one-parameter families of instruments by Davies [4], Srinivas [12], Gudder [13], and Hellwig [7-9]. The definition of Markovian quantum stochastic processes bases on an idea of Hellwig [8,9].

Quantum stochastic processes may be employed in connection with

(i) the description of successive measurements of several observables or of one observable developing in time

(ii) the quantum mechanical description of the random walk of a particle

(iii) the microscopic explanation of statistical thermodynamic processes of macroscopic systems.

We will look at some examples for quantum stochastic processes now.

4.2. Successive Measurements of Observables by von Neumann-Lüders Instruments

Let $M:=\mathbb{N}$ if the Hilbert space X is infinite-dimensional and $M:=\{1,2,\dots,N\}$ if the dimension is N, let Σ be the power set of M, and let for each element t of the non-empty set $T\subseteq\mathbb{R}_0^+$ a complete orthonormal system $\{\psi_{it}\}_{i\in M}$ be given. Define a quantum stochastic process $(W,\{J_t\}_{t\in M})$ by an arbitrary fixed density operator $W\in K$ and the atomic von Neumann-Lüders instruments J_t on (M,Σ) corresponding to $\{\psi_{it}\}_{i\in M}$ i.e.

$$J_t(B)V := \sum_{i\in B} P_{\psi_{it}} V P_{\psi_{it}} = \sum_{i\in B} \langle\psi_{it}|V\psi_{it}\rangle P_{\psi_{it}} \tag{4.2}$$

($V\in C_s^1(X)$, $B\in\Sigma$, $P_{\psi_{it}} := |\psi_{it}\rangle\langle\psi_{it}|$). The observables $F_t:=J_t'(.)1$ are the very simple discrete projection-valued measures given by

$$F_t(B) = \sum_{i\in B} P_{\psi_{it}} \quad . \tag{4.3}$$

For this example a quantum conditional distribution P_{st} of F_t under the hypothesis J_s, given W, can easily be calculated. Inserting (4.2) and

(4.3) in Eq. (3.2) we obtain for $A \in B_s(X)$ and all $B \in \Sigma$

$$\sum_{i \in B} \langle \psi_{is} | W \psi_{is} \rangle \operatorname{tr} P_{\psi_{is}} A = \sum_{i \in B} E_W^J(A)(i) \langle \psi_{is} | W \psi_{is} \rangle \quad .$$

From this it follows that

$$E_W^{J_s}(A)(i) := \langle \psi_{is} | A \psi_{is} \rangle \tag{4.4}$$

($i \in M$) defines a version of the quantum conditional expectation of A. A quantum conditional distribution of F_t is given by

$$P_{st}(i,B) := \langle \psi_{is} | F_t(B) \psi_{is} \rangle \tag{4.5}$$

($i \in M$, $B \in \Sigma$, $s,t \in T$, $s \leq t$) respectively by

$$P_{st}(i,\{j\}) = |\langle \psi_{is} | \psi_{it} \rangle|^2 \tag{4.6}$$

($i,j \in M$). If W is effective with respect to the observable F_s (i.e. $\operatorname{tr} WF_s(B) = 0$ if and only if $F_s(B)=0$), then $E_W^{J_s}(A)$ and P_{st} are uniquely determined and necessarily of the form (4.4) and (4.5). $E_W^{J_s}(A)$ and P_{st} do not depend on W which is due to the fact that the projection operators associated with J_s are one-dimensional. The numbers $P_{st}(i,\{j\})$ given by (4.6) are the elements of a stochastic matrix and are often be interpreted as transition probabilities.

Let us consider the joint probabilities

$$\mu_{rt}(B,B') := \operatorname{tr} (J_r(B)W)F_t(B') = \operatorname{tr} J_t(B')J_r(B)W$$

$$\mu_{rst}(B,B',B'') := \operatorname{tr} (J_s(B')J_r(B)W)F_t(B'') = \operatorname{tr} J_t(B'')J_s(B')J_r(B)W$$

related to the successive measurement of the observables F_r and F_t respectively F_r, F_s, and F_t in the ensemble W ($B,B',B'' \in \Sigma$, $r,s,t \in T$, $r \leq s \leq t$). More explicitly, we have

$$\begin{aligned} \mu_{rt}(B,B') &= \sum_{i \in B} \sum_{j \in B'} \langle \psi_{ir} | W \psi_{ir} \rangle P_{rt}(i,\{j\}) \\ \mu_{rst}(B,B',B'') &= \sum_{i \in B} \sum_{j \in B'} \sum_{k \in B''} \langle \psi_{ir} | W \psi_{ir} \rangle P_{rs}(i,\{j\}) P_{st}(j,\{k\}) \end{aligned} \tag{4.7}$$

where $P_{rt}(i,\{j\})$ etc. is given by (4.6). These formulae look like the representation of the finite-dimensional distributions of a classical Markov process by its transition probabilities and an initial distribution [5,6]. However, until now the considered quantum stochastic process may not be Markovian. The representation (4.7) of μ_{rt} and μ_{rst} is again due to the special choice of the instruments J_t.

Finally, we have to investigate the physical meaning of Markovicity of the process $(W,\{J_t\}_{t\in T})$. For simplicity, let us assume now W to be effective with respect to all observables F_t. Then the conditional distributions P_{st} are uniquely determined. By an easy calculation it turns out that the Chapman-Kolmogorov equations

$$P_{rt}(i,B) = \sum_{k\in M} P_{rs}(i,\{k\})P_{st}(k,B)$$

are equivalent to

$$\mu_{rst}(B,M,B') = \mu_{rt}(B,B') \quad . \tag{4.8}$$

Hence, $(W,\{J_t\}_{t\in T})$ is a Markovian quantum stochastic process if and only if the joint probabilities for successive measurement of any two observables F_r and F_t are not changed by an additional intermediate non-selective measurement of a third observable F_s. Remember that projective properties as (4.8) hold true for the finite-dimensional distributions of any classical stochastic process.

4.3. Successive Spin-$\frac{1}{2}$ Measurements

To give a more concrete but simple example which is a special case of the preceding one take a two-dimensional Hilbert space X and three unit vectors $n_1,n_2,n_3\in\mathbb{R}^3$ and consider the quantum stochastic process $(W,\{J_t\}_{t=1,2,3})$ where W is an arbitrary density operator in X and J_t the von Neumann-Lüders instrument corresponding to the spin-$\frac{1}{2}$ measurement in the direction n_t. Then it can be shown that this process is Markovian if and only if

$$(n_1\times n_2)\cdot(n_2\times n_3) = 0$$

holds true |9|. Especially, this condition is fulfilled if n_1,n_2,n_3 are mutually orthogonal.

4.4. A Problem of Quantum Statistical Thermodynamics

Let us consider the discrete random walk of a macroscopic particle on a lattice. Thus we have the thermodynamical state space $M:=\mathbb{Z}^N$ with its power set as σ-algebra Σ and transition probabilities $P_{st}(x,B)$ from $x\in\mathbb{Z}^N$ at time s into some $y\in B$ at time t. From the microscopic point of view, the following assumptions may be obvious:

(i) there is a Hilbert space X and a Hamiltonian H associated with the particle

(ii) there is a distinguished not necessarily projection-valued observable F on Σ describing position measurements and for every macroscopic state $x\in\mathbb{Z}^N$ a distinguished density operator $\Phi(x)$ describing the localization of the particle such that

$$\operatorname{tr}\Phi(x)F(B) = \varepsilon_x(B) := \begin{cases} 1 & \text{for } x\in B \\ 0 & \text{for } x\notin B \end{cases} \tag{4.9}$$

holds true

(iii) the transition probabilities are given by

$$\mathbb{P}_{st}(x,B) = P_{t-s}(x,B) \tag{4.10}$$

$(s,t\in\mathbb{R}_0^+,\ s\leq t)$ where

$$P_t(x,B) := \operatorname{tr}\left(\Phi(x)e^{iHt}F(B)e^{-iHt}\right) \tag{4.11}$$

$(t\in\mathbb{R}_0^+)$.

Since the transition probabilities are homogeneous in time, we only have to deal with the Markov kernels P_t on $(\mathbb{Z}^N,\Sigma)$, P_0 being the unit kernel i.e. $P_0(x,B)=\varepsilon_x(B)$. Macroscopic experience leads to the further assumption

(iv) $\{P_t\}_{t\in\mathbb{R}_0^+}$ is a Markov semigroup, i.e. for all $s,t\in\mathbb{R}_0^+$ the Chapman-Kolmogorov equations

$$P_{s+t}(x,B) = \sum_{y\in\mathbb{Z}^N} P_s(x,\{y\})P_t(y,B) \tag{4.12}$$

are satisfied.

By (4.10), Eqs. (4.12) are equivalent to the Chapman-Kolmogorov equations (4.1) for the kernels $\mathbb{P}_{st}$.

The whole story can also be described by a quantum stochastic process. Such a process $(W_\mu,\{J_t\}_{t\in\mathbb{R}_0^+})$ is defined by the density operator $W := \sum_{x\in\mathbb{Z}^N}\mu(\{x\})\Phi(x)$ according to an initial probability distribution μ of the particle and by the nuclear instruments J_t according to

$$J_t(B)V := \sum_{x\in B} \operatorname{tr}(VF_t(B))\ \Phi_t(x)$$

where $B\in\Sigma$, $V\in C_s^1(X)$, $t\in\mathbb{R}_0^+$, and

$$\Phi_t(x) := e^{iHt}\Phi(x)e^{-iHt}$$
$$F_t(B) := e^{iHt}F(B)e^{-iHt} \quad .$$

As an easy calculation shows, for any $s,t\in\mathbb{R}_0^+$, $s\leq t$, the Markov kernel

$$\mathbb{P}_{st} = P_{t-s} = \operatorname{tr}\Phi_s(.)F_t(.)$$

is just a conditional distribution of the observable F_t under the hypothesis J_s. Hence, Markovicity of the random walk process implies Markovicity of the associated quantum stochastic process.

It remains the question whether $\{P_t\}_{t\in\mathbb{R}_0^+}$ can be a Markov semigroup, i.e. whether the assumption (iv) can be fulfilled if the kernels P_t are given by (ii) and (iii).

Theorem: Let the Hamiltonian H be a bounded operator. If the family $\{P_t\}_{t\in\mathbb{R}_0^+}$ defined by $P_t(x,B):=\mathrm{tr}\ \Phi(x)F_t(B)$ is a Markov semigroup on $(\mathbb{Z}^N,\Sigma)$ with P_0 being the unit kernel, then it must be trivial, i.e.

$$P_t = P_0$$

for all $t\in\mathbb{R}_0^+$.

Proof: By differentiation we obtain

$$\frac{dP_t(x,B)}{dt}\Big|_{t=0} = i\ \mathrm{tr}\ (\Phi(x)HF(B)-\Phi(x)F(B)H) \quad .$$

The condition $P_0(x,B)=\mathrm{tr}\ \Phi(x)F(B) =\varepsilon_x(B)$ implies $\Phi(x)F(B)=F(B)\Phi(x)$, hence

$$\frac{dP_t(x,B)}{dt}\Big|_{t=0} = 0 \quad . \tag{4.13}$$

Assume now $\{P_t\}_{t\in\mathbb{R}_0^+}$ to be a Markov semigroup. From (4.12) and (4.13) we find

$$\frac{dP_s(x,B)}{ds} = \frac{dP_{s+t}(x,B)}{dt}\Big|_{t=0} = \sum_{y\in\mathbb{Z}^N} P_s(x,\{y\})\frac{dP_t(y,B)}{dt}\Big|_{t=0} = 0$$

for all $s\in\mathbb{R}_0^+$. This is equivalent to

$$P_s(x,B) = P_0(x,B) \quad .$$

Q.e.d.

If the Hamiltonian is not bounded, then the proof requires additional technical assumptions. There are quantum stochastic processes of considered type which do not fulfil these additional assumptions and which are non-trivial and even Markovian as the author has shown by examples. However, these examples are far from every physics.

5. CONCLUSION

We see by the examples discussed above that the introduced concepts of quantum conditional expectations and distributions seem to be interesting. Also the given notions of a quantum stochastic process and its

Markovicity have interesting, though quite different physical meanings. However, it is rather difficult to construct quantum Markov processes of physical significance.

The general framework for statistical theories [1-4], founded on the duality of base-norm and order-unit-norm spaces and sometimes called "generalized probability theory", contains classical probability theory and quantum mechanics as special cases. It is suitable for the attempt of transferring notions of classical probability to quantum probability as well as for the comparison of both theories. All probabilistic notions introduced in this paper can also be defined in generalized probability [7,11]. So one may hope that a formulation of quantum mechanics which is not the usual Hilbert space formulation admits more interesting examples of quantum stochastic Markov processes.

Acknowledgements

The author would like to thank Prof. K.-E. Hellwig, M. Singer as well as U. Grimmer and V. Perlick for interesting discussions, stimulating comments, encouragement and support.

REFERENCES

1. Ludwig, G.: Foundations of Quantum Mechanics I, Springer-Verlag, New York 1983.
2. Ludwig, G.: An Axiomatic Basis for Quantum Mechanics, Vol. 1: Derivation of Hilbert Space Structure, Springer-Verlag, Berlin 1985.
3. Davies, E. B. and J. T. Lewis: An Operational Approach to Quantum Probability, Commun. Math. Phys., 17 (1970), 239-260.
4. Davies, E. B.: Quantum Theory of Open Systems, Academic Press, London 1976.
5. Bauer, H.: Wahrscheinlichkeitstheorie und Grundzüge der Maßtheorie, de Gruyter, Berlin 1977.
6. Breiman, L.: Probability, Addison-Wesley, Reading, Massachusetts 1968.
7. Cycon, H. and K.-E. Hellwig: Conditional expectations in generalized probability theory, J. Math. Phys., 18 (1977), 1154-1161.
8. Hellwig, K.-E.: Conditional Expectations and Duals of Instruments, in: Grundlagen der Exakten Naturwissenschaften, Bd. 5 (Ed. H. Neumann), Bibliographisches Institut, Mannheim 1981, 113-124.
9. Hellwig, K.-E. and W. Stulpe: A Formulation of Quantum Stochastic Processes and Some of its Properties, Found. Phys., 13 (1983), 673-699.
10. Ozawa, M.: Concepts of conditional expectations in quantum theory, J. Math. Phys., 26 (1985), 1948-1955.
11. Stulpe, W.: Bedingte Erwartungen und stochastische Prozesse in der generalisierten Wahrscheinlichkeitstheorie, unpublished, Berlin 1985.
12. Srinivas, M. D.: Foundations of a quantum probability theory, J. Math. Phys., 16 (1975), 1672-1685.
13. Gudder, S.: Stochastic Methods in Quantum Mechanics, North Holland, New York 1979.

CONTINUOUS MONITORING OF QUANTUM SYSTEMS

Asher Peres
Israel Institute of Technology, Haifa, Israel

ABSTRACT

The "quantum Zeno paradox" is explained and is illustrated by some examples. It may occur in measurements of finite duration. However, not every continuous monitoring of a quantum system is a "measurement" (as defined by von Neumann). A continuous interaction with a measuring apparatus does not necessarily stop the evolution of a quantum system.

1. THE QUANTUM ZENO PARADOX

The evolution of a closed (isolated) quantum system is unitary:

$$\psi(t) = e^{-iHt/\hbar}\,\psi(0), \tag{1}$$

where H is the Hamiltonian of the system (assumed time independent) and $\psi(t)$ is its state vector at time t. It can readily be shown [1] that, for small enough t,

$$|(\psi(0),\psi(t))|^2 \geqslant cos^2(Wt/\hbar), \tag{2}$$

where

$$W^2 = (H\psi(0),H\psi(0))-(\psi(0),H\psi(0))^2 \tag{3}$$

is the square of the energy uncertainty corresponding to the preparation $\psi(0)$. Eq. (1) is valid as long as $t<\pi\hbar/2W$.

Let us now imagine an ideal apparatus which tests whether, after a time t, the physical system still is in its initial state $\psi(0)$. (Note that there is no way to test *what is* the state vector [2,3]. One can only test *whether or not* the state vecor is *within a prescribed subspace* of Hilbert space.) It then follows from (2) that, for very short t, the apparatus will yield a positive answer with a probability

$$P(t) > 1-(Wt/\hbar)^2. \tag{4}$$

Thus, if many such measurements are continually performed with arbitrarily short time separations between them - in the limit, if the quantum system is continuously monitored - $P(t)$ can be kept arbitrarily close to 1 for arbitrarily long times.

This remarkable result [4-7] can be considered as a kind of quantum version of the ancient Zeno paradox [8]. It has caused considerable confusion and many authors [9] have even attempted to modify the formalism of quantum theory in order to avoid the "paradox." For good measure, other authors [10] have invoked this "quantum Zeno effect" to predict the inobservability of the proton decay, notwithstanding the fact that proton decays are routinely observed [11,12].

In this paper, I show that the Zeno effect is real and can easily be displayed experimentally. I then show how to describe the continuous interaction of a measuring apparatus with a quantum system. Finally, I show that *not* every continuous monitoring is a "measurement" in the technical sense of this word [13,14]. For example, a radiation detector placed near a radioactive atom does *not* continuously "measure" whether the atom decays.

2. SIMPLE EXAMPLES OF THE ZENO EFFECT

Consider a beam of slow neutrons passing through a sequence of Stern-Gerlach [15] magnets, as in Fig. 1. All the magnets are parallel and, after each one of them, the neutrons found with their spins down are dumped. The first magnet can be considered as the apparatus which

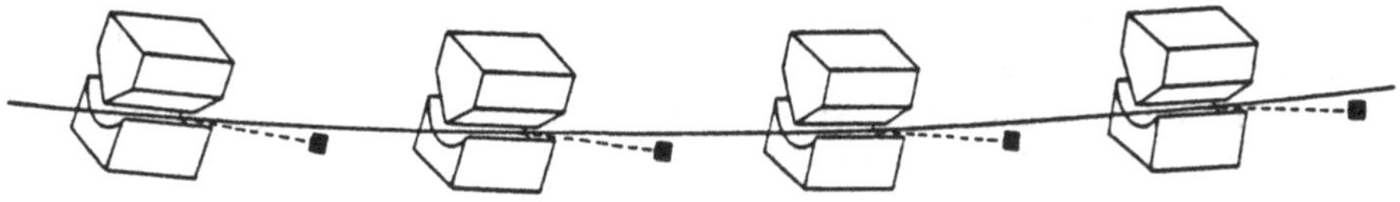

Fig. 1. Experimental verification of the Zeno effect: a beam of particles passing through a sequence of Stern-Gerlach magnets have their polarization "frozen" even in the presence of an external magnetic field.

prepares $\psi(0)$. The Hamiltonian acting on the neutrons as they travel from one magnet to the next is

$$H = (p^2/2m)-\vec{\mu}\cdot\vec{B}, \tag{5}$$

where $\vec{\mu}$ is the magnetic moment of the neutrons and $\vec{B}$ the magnetic field of the Earth. Typically, $\mu B/\hbar$ is of the order of 10^4 sec^{-1}. The velocity of the neutrons (e. g., thermal neutrons from a nuclear reactor) is about 2200 m/sec. Therefore, if the magnets are separated by much less than 22 cm, the magnetic moment of the neutrons will rotate only by a small angle $\alpha<\mu Bt/\hbar$ as they travel from one magnet to the next. Thus, the probability of a neutron being dumped after the next magnet is only α^2. Nearly all the neutrons survive the multiple Stern-Gerlach selector.

An example which is even simpler than that (that is, *experimentally* simpler - its complete theory is very complicated) is a beam of polarized photons passing through an optically active liquid [6]. If polarization selectors are placed close enough to each other, *all* the photons pass, with their polarization unaffected by the liquid.

3. MEASUREMENTS OF FINITE DURATION

Until now, we have considered each "measurement" as an instantaneous event. In other words, it was assumed that the interaction of the measuring instrument with the quantum system was so strong that it could be made arbitrarily brief and one could therefore neglect $Ht/\hbar$ *during* the measurement. This drastic simplification may not always be justified. Coupling constants occuring in nature are finite, and sometimes very small [16]. It may therefore be necessary to couple the measured system and the apparatus during a *finite*, possibly long time. For example, the historic Stern-Gerlach experiment [15] used a beam of silver atoms. The gedankenexperiment discussed in the preceding Section involves neutrons, having a much smaller magnetic moment. A magnet a thousand times more powerful would be needed to obtain a similar deflection angle.

This problem is not specific to quantum theory. It may arise in everyday life, e. g., when a photographer takes a snapshot of a moving object. However, quantum theory introduces some novel features, because a measurement is not only a passive observation, but also the *preparation* of a new state. The detailed dynamical theory of such quantum measurements of finite duration can be worked out explicitly [17]. It turns out that the result of the measurement is *not*, in general, the time average of the observed quantity. It may not be one of the eigenvalues of the operator representing the dynamical variable being measured. And at least *some* outcomes, those which do correspond to the eigenvalues, appear "frozen" by the Zeno effect. I shall not repeat here the details of the calculations, since they can be found in the literature [17].

4. EXAMPLE OF A CONTINUOUS MONITORING WHICH IS NOT A MEASUREMENT

The following paradigm is often proposed [9] as a proof of nonexistence of the Zeno effect. Consider a radioactive atom together with a perfect detector. The atom is continuously monitored by the detector and yet, this does not prevent it from decaying, eventually.

This argument is misleading because the mere *presence* of a detector capable of registering the decay is *not* a measurement (in the technical sense of this term [13,14]). A complete quantum mechanical treatment [18] of the atom+detector system shows that the final state of the detector - i. e., after many mean lifetimes of the atom - is a *superposition* of states corresponding to different decay times, distributed according to the familiar exponential decay law. Obviously, the final wavefunction only gives statistical information on what may happen: quantum theory is unable to predict when an *individual* atom will decay.

The point is that the *continuous interaction* between the detector and the decay products is *not a measurement* (let alone the equivalent of a large number of consecutive measurements). A measurement of the type capable of causing the Zeno effect is a very brief and intense interaction between the observed system (the radioactive atom) and a macroscopic apparatus. This interaction causes different states of the atom to be *correlated to macroscopically distinguishable states* of the apparatus. It is essential for the consistency of the von Neumann formalism [13,14] that these final states be macroscopically distinguishable, i. e., *incoherent* [12,19]. Otherwise, there may be no Zeno effect [7].

To conclude, let me present a very simple one-dimensional model for a radioactive atom and its detector. The "atom" is represented by a particle of mass m and position x_1, in a piecewise constant potential

$$\begin{aligned} V_1(x_1) &= V_0 \quad \text{if} \quad a<x_1<2a, \\ &= \infty \quad \text{if} \quad x_1<0, \\ &= 0 \quad \text{otherwise.} \end{aligned} \tag{6}$$

Here, $V_0 \gg \hbar^2/ma^2$ so that there are long lived metastable states in the potential well.

Likewise, the "detector" is represented by a particle of mass M and position x_2, in a piecewise constant potential

$$V_2(x_2) = V_0 \quad \text{if} \quad 0<x_2<R \quad \text{or} \quad R+b<x_2<R+c,$$
$$= \infty \quad \text{if} \quad x_2<0, \tag{7}$$
$$= 0 \quad \text{if} \quad R<x_2<R+b \quad \text{or} \quad x_2>R+c.$$

Here, $R \gg a$ is the distance between the atom and the detector, and I have chosen the same V_0 as before, for simplicity. Moreover, $M(c-b)^2 \gg ma^2$, so that the detector would have a much longer lifetime than the atom, if the two did not interact.

Finally, there is an interaction term

$$V_{\text{int}}(x_1-x_2) = -V_0 \quad \text{if} \quad |x_1-x_2|<d,$$
$$= 0 \quad \text{otherwise}. \tag{8}$$

Fig. 2 shows a topographic map of the potential

$$V(x_1,x_2) = V_1(x_1)+V_2(x_2)+V_{\text{int}}(x_1-x_2). \tag{9}$$

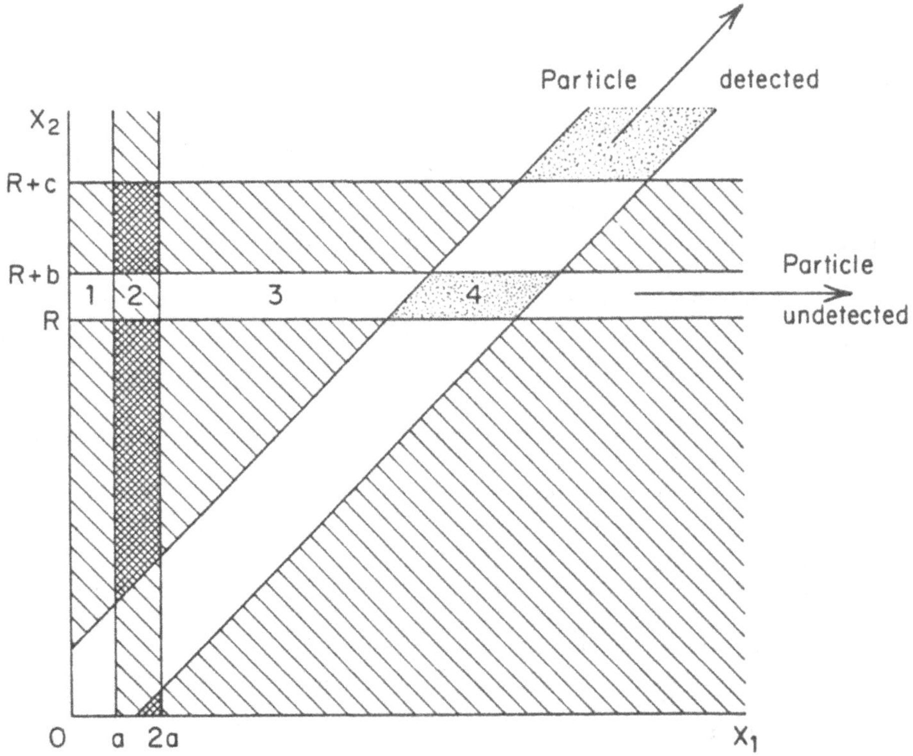

Fig. 2. Topographic map of the potential energy in Eq. (9). Blank: $V=0$; hatched: $V=V_0$; cross-hatched: $V=2V_0$; dotted: $V=-V_0$. See text for details.

Initially, the wavefunction is localized in region 1, *inside* both potential wells, i. e., the atom is excited and the detector is in its "ready" state. Tunnelling then proceeds via region 2 - this is the radioactive decay - and then the ejected particle freely travels through region 3 until it reaches the detector, in region 4. It will then continue toward larger x_1, either detected or undetected (the efficiency of the detector depends on the value of $c-b$ vs. that of d).

Although this model is grossly oversimplified, it has all the features to describe how the decay of an unstable particle triggers the decay of a detector prepared in a metastable state. Their interaction is continuous and yet it is not subject to any Zeno effect.

ACKNOWLEDGMENTS

This work was supported by the Gerard Swope Fund. The "atom+detector" model of Fig. 2 was concocted during the Udine Conference on Information Complexity and Control in Quantum Physics. I am grateful to CISM for hospitality.

REFERENCES

1. Fleming, G.N.: Nuovo Cimento A16 (1973), 232.
2. Finkelstein, D.: The Physics of Logic, in: Paradigms and Paradoxes (Ed. R.C. Colodry), Univ. Pittsburgh Press, 1971, Vol. V; reprinted in: Logico-Algebraic Approach to Quantum Mechanics (Ed. C.A. Hooker) Reidel, Dordrecht, 1975, Vol. II, pp. 141-160.
3. Peres, A.: Am. J. Phys. 52 (1984), 644.
4. Misra, B. and E.C.G. Sudarshan: J. Math. Phys. 18 (1977), 756.
5. Chiu, C.B., E.C.G. Sudarshan and B. Misra: Phys. Rev. D16 (1977), 520.
6. Peres, A.: Am. J. Phys. 48 (1980), 931.
7. Singh, I. and M.A.B. Whitaker: Am. J. Phys. 50 (1982), 882.
8. Cajori, F.: Am. Math. Mon. 22 (1915), 1, 292.
9. Finding appropriate references, in this book and elsewhere, is left as an exercise for the reader.
10. Horwitz, L.P. and E. Katznelson: Phys. Rev. Lett. 49 (1982), 1804.
11. Curie, I. and F, Joliot: Compt. Rend. Acad. Sci. 198 (1934), 254.
12. Peres, A.: Found. Phys. 14 (1984), 1131.
13. von Neumann, J.: Mathematical Foundations of Quantum Mechanics, Princeton Univ. Press, 1955 (translated from: Mathematische Grundlagen der Quantenmechanik, Springer, Berlin 1932).
14. Wheeler, J.A. and W.H. Zurek: Quantum Theory and Measurement, Princeton Univ. Press, 1983.
15. Gerlach, W. and O. Stern: Z. Phys. 8 (1922), 110; 9 (1922), 349.
16. Caves, C.M. *et al.*: Rev. Mod. Phys. 52 (1980), 341.
17. Peres, A. and W.K.Wootters: Phys. Rev. D32 (1985), 1968.
18. Peres, A.: Am. J. Phys. 48 (1980), 552.
19. Peres, A.: When is a Quantum Measurement?, Am. J. Phys. 54 (1986), in press.

ON THE DESCRIPTION OF QUANTUM DISSIPATIVE PROCESSES

Maurice Courbage
Laboratoire de Probabilités, Paris Cedex, France

ABSTRACT

The description of dissipative processes in quantum mechanics here presented is based on the existence of a time operator canonically conjugate to the Liouville operator. The entropy can be defined as a functional of this time operator and a new dissipative semi-group of time evolution can be constructed. This formalism seems well adapted to the description of the decay phenomena and the measurement process.

I. INTRODUCTION

The Schrödinger equation is symmetric with respect to the time inversion. It was Pauli who introduced the idea of a stochastic evolution of the states of quantum systems caused by the collisions à la Boltzmann in the case of dilute gases. The Pauli equation [1] describes the transition of a many-body system between eigenstates of a free hamiltonian caused by an interaction λV, where λ is the coupling constant. The probability $P_n(t)$ of finding the system in the eigenstate $|n\rangle$ of H_0 corresponding to the eigenvalue E_n, when the state of the system is given by the density matrix ρ_t at time t, is defined by :

$$P_n(t) = \langle n|\rho_t|n\rangle \tag{I.1}$$

The Pauli master equation is a Markovian equation describing the evolution of the probabilities $P_n(t)$, namely :

$$\frac{d}{dt} P_n(t) = \sum_m W_{n\,m} P_m(t) - W_{mn} P_n(t) \tag{I.2}$$

where the transition rate $W_{n\,m}$ is given in terms of the potential V. This equation has been derived by a time-dependent perturbation theory under other stochastic hypotheses. Pauli has proved a theorem-$\mathcal{H}$ using this equation.

The non-equilibrium statistical mechanics aims to derive such an irreversible equation and others from the deterministic equations of quantum mechanics by removing as long as possible the ad-hoc stochastic hypotheses. Thus, the theory of Master equations elaborated by Van Hove [2] and later on by Prigogine and Resibois [3] have permitted to derive the Pauli equation on the general ground of the Liouville equation for density matrices in the so-called weak coupling limit.

However, the general problem of the derivation of irreversible equations from the dynamics becomesvery difficult if we would analyse the problem of the correlations built in the course of evolution under the effect of multiple particles interactions [4]. Especially, one cannot in general remove the contribution of these correlations, neglected in the Pauli's equation, in the definition of the non-equilibrium entropy. A general approach to non-equilibrium statistical mechanics [5][6] has led the Brussels school to develop a point of vue aiming to distinguish sharply between the irreversible description and the reversible one, the second being an idealization of the first. This means that the object of the irreversible description has to be distinguished from that of the reversible one. Instead of the wave function that evolves symmetrically in the future and past, we should consider objects that evolve asymmetrically under the dynamics . When such objects exist the dynamical system is called intrinsically irreversible. Such objects have been displayed for classical unstable systems ; these are the contracting fibers. A description in terms of the contracting fibers induces a change of representation in which the evolution is Markovian and yields an $\mathcal{H}$-theorem ([7],[8],[9],[10],[11]). In this scheme, the problem is not to recover a markovian semi-group of time evolution for a contracted description through a limiting process (e.g. rescaling). One considersthe whole dynamical group $U_t : \rho_0 \to U_t \rho_0$ acting on initial probability distribution ρ_0 and realizesa change of representation by a linear operator Λ depending on the asymmetric properties of the evolution U_t. This transformation is such that the entropy of the new state $\Lambda U_t \rho = \tilde{\rho}_t$ is monotonically increasing with time :

$$- \int \tilde{\rho}_t \log \tilde{\rho}_t \, d\mu \nearrow \tag{I.3}$$

while this quantity is constant for $\rho_t = U_t \rho_0$.

It is clear that such a change in quantum mechanics is meaningfull only when the interactions are such that the wave function is no more a physically observable object. A well know example is the measuring process of some quantum observable where the interaction between the system and the apparatus yields a so-called uncontrollable evolution and leads to the wave reduction. Generally, the mixed states are introduced to express our ignorance of the initial conditions. Nevertheless, their evolution under the Schrödinger equation is reversible and is given by :

$$\rho_t = U_t \rho_0 = e^{-itH} \rho_0 e^{itH} \tag{I.4}$$

where H is the hamiltonian. This evolution moreover preserves the pure states. Now the irreversible evolution in quantum mechanics are not so, and the wave reduction in the measurment process does not preserve the pure states. Here also, as in classical mechanics, the expression $Tr(\rho_t \log \rho_t)$ remains constant under the evolution (I.4).

As outlined above, the transition to the irreversible description will be formulated by a linear operator Λ acting on some suitable space of states $\mathcal{L}$ such that :

1) $\Lambda\rho_t = \tilde{\rho}_t$ evolves under a Markovian semi-group, i.e.,

$$\tilde{\rho}_t = W_t \tilde{\rho}_0 \tag{I.5}$$

where $\tilde{\rho}_0 = \Lambda\rho_0$, for any $\rho_0 \in \mathcal{L}$

this means namely : a) $W_t\rho \geqslant 0$ if $\rho \geqslant 0$, b) $Tr(W_t\rho) = Tr(\rho) = 1$,

c) $W_{t+t'} = W_t W_{t'}$; $t,t' \geqslant 0$.

2) $\Omega(\tilde{\rho}_t) = Tr(\tilde{\rho}_t^2)$ is monotonically decreasing for any ρ_0 that is not invariant under U_t.

Here we shall report some results toward a solution of the above problems. We follow and develop mainly several papers of Misra, Prigogine, the author and Lockhart [12],[13],[14],[15]. Some details omitted in these papers are given here, while others, omitted here, can be found there.

In order to simplify the formulation, the space of states will be the so-called Liouville space, i.e. the space of density matrices endowed with the scalar product :

$$\langle \rho,\rho' \rangle = Tr(\rho^*\rho') \tag{I.6}$$

These states evolve under the group of operators U_t defined by (I.4). Its infinitesimal generator is the Liouville-Von Neumann operator L :

$$L\rho = H\rho - \rho H \equiv [H,\rho] \tag{I.7}$$

U_t can be written in the exponential form : $U_t = e^{-itL}$. In general, H is unbounded and L too. Then L cannot be defined for any $\rho \in \mathcal{L}$, its domain $D(L)$ is defined as $\{\rho \in \mathcal{L} : L\rho \in \mathcal{L}\}$.

The functional $\Omega(\tilde{\rho}_t)$ can be written by using the scalar product :

$$\Omega(\tilde{\rho}_t) = \langle \Lambda\rho_t,\Lambda\rho_t \rangle = \langle \rho_t,\Lambda^*\Lambda\rho_t \rangle \tag{I.8}$$

In order to get an additivity property for the entropy, it can be related to Ω by : $S(\tilde{\rho}_t) = -\log\Omega(\tilde{\rho}_t)$. The property 2) of Ω means that the Liouville operator L admits a Liapounov variable. This is defined as a non-negative bounded linear operator M on $\mathcal{L}$ such that $\langle\rho_t, M\rho_t\rangle$ is decreasing with time :

$$\frac{d}{dt} \langle e^{-itL}\rho, Me^{-itL}\rho\rangle = \langle e^{-itL}\rho, i[L,M]e^{-itL}\rho\rangle \leqslant 0 \qquad (I.9)$$

for any $\rho \in \mathcal{D}(L)$. In order to avoid ambiguities coming from the unboundedness of L, M will be defined as a non-negative operator such that : $M\mathcal{D}(L) \subset \mathcal{D}(L)$ and $-i[L,M] = D$ is a non-negative operator on $\mathcal{L}$. D describes the entropy production. It is also required that M and D commute as a macroscopic operators.

Before the examination of necessary and sufficient conditions for the existence of an M operator with non-vanishing D it can be of interest to show a quantum analogue of Poincaré theorem stating that the entropy cannot be a dynamical variable. Here this means that M cannot be described by an operator on the Hilbert space of the wave functions . We shall prove the following theorem :

Theorem 1.

Let M *be a bounded self-adjoint operator on a Hilbert space* $\mathcal{H}$ *and* H *a non-negative self-adjoint operator in* $\mathcal{H}$ *satisfying :*

i) $M\mathcal{D}(H) \subseteq \mathcal{D}(H)$, *ii)* $D = -i[H,M]$ *is a non-negative operator acting on* $\mathcal{D}(H)$, *iii)* $MD\psi = DM\psi$, $\forall\psi \in \mathcal{D}(H)$, *then* $D = 0$.

Proof :

Formally, the proof is very easy. As in (I.9) we have for any $\psi \in \mathcal{D}(H)$:

$$\frac{d}{dt} \langle e^{iMt}\psi, He^{iMt}\psi\rangle = \langle e^{iMt}\psi, i[H,M]e^{iMt}\psi\rangle = -\langle\psi, e^{-iMt}D\, e^{iMt}\psi\rangle \qquad (I.10)$$

Now, from the commutation of M and D it follows that $e^{-iMt}D\, e^{iMt} = D$. Integrating both sides of (I.10) yields :

$$\langle e^{iMt}\psi, H\, e^{iMt}\psi\rangle = \langle\psi, H\psi\rangle - t\langle\psi, D\psi\rangle \qquad (I.11)$$

Because $H \geqslant 0$, the right hand side is positive and then

$$\langle\psi, H\psi\rangle \geqslant t\langle\psi, D\psi\rangle$$

or $\langle\psi, D\psi\rangle \leqslant \frac{1}{t}\langle\psi, H\psi\rangle \to 0$ as $t \to \infty$ that is, $D \leqslant 0$.

As $D \geqslant 0$, then $D \equiv 0$. In the formula (I.10) the point is to show that $e^{iMt}\psi \in \mathcal{D}(H)$ and that $H\, e^{iMt}\psi$ is differentiable with a derivative given by $iHM\, e^{iMt}\psi$. We leave this to the appendix A.

This theorem is no more valid if one goes to the Liouville space, for in this case the generator of the evolution, L, is no more semi-bounded and its spectrum extends from $-\infty$ to $+\infty$. However, one does not expect M to exist for any dynamical system. As a matter of fact, M cannot exist if H has a discrete spectrum, that is, for a finite quantum system. To see this, let ρ be an eigenvalue of L : $L\rho = \alpha\rho$ then :

$$<e^{-itL}\rho, M\, e^{-itL}\rho> = <\rho, M\rho> \tag{I.12}$$

now taking the derivative of this identity we have $<\rho, D\rho> = 0$. If H has purely discrete spectrum then the set of eigenvectors of L will span $\mathcal{L}$ and then $D \equiv 0$.

Misra has shown that [12] : $<\rho, D\rho> \neq 0$ implies that ρ belongs to the absolutely continuous subspace of L. This is only possible if H has an absolutely continuous spectrum. Moreover, the above theorem shows that L cannot be a bounded operator, and this is possible if H is not bounded. From these arguments it follows that a necessary condition for the existence of an entropy operator is that the hamiltonian has an unbounded absolutely continuous part. We shall see that this is also a sufficient condition.

II. EXISTENCE OF ENTROPY AND TIME OPERATORS IN QUANTUM MECHANICS

The construction of a non trivial entropy operator has been first done for a class of ergodic classical systems by B. Misra [12]. The main requirement of this construction is the existence of a canonically conjugate operator T to L, i.e. T is a self-adjoint operator such that

$$i[L,T] = I \tag{II.1}$$

Let us recall here this construction. From the commutation relation (II.1) it follows that :

$$i[L,f(T)] = f'(T) \tag{II.2}$$

for any differentiable function f. Here f(T) is defined by the functional calculus, i.e.

$$<\rho', f(T)\rho> = \int_{\mathbb{R}} f(\lambda)\, d<\rho', F(\lambda)\rho> \tag{II.3}$$

where $F(\lambda)$ is the family of spectral projections of T. One obtains a Liapounov varibale $M = f(T)$ if f is a positive and decreasing function.

A necessary and sufficient condition for the existence of a self-adjoint T operator canonically conjugate to L is that L has a Lebesgue spectrum with uniform multiplicity extending over the entire real line. Let us give here some mathematical details (see e.g. [16] or [17]). The spectral representation of the self-adjoint operator L is given by a unitary operator $V : \rho \to V\rho = f = (f_1(\lambda), f_2(\lambda), \ldots), f_n(\lambda) \in L^2(\mathbb{R}, d\mu_n)$ and $\sum_{i=1}^{\infty} \int |f_i(\lambda)|^2\, d\mu_i(\lambda) < +\infty$, such that :

$$VL\rho = VLV^{-1}f = (\lambda f_1(\lambda), \lambda f_2(\lambda), \ldots) \tag{II.4}$$

the sequence of measures μ_n can be ordered such that $\mu_1 >> \mu_2 >> \ldots$ where $\mu >> \nu$ means that ν is absolutely continuous with respect to μ. μ is said to be equivalent to ν if $\nu >> \mu$ and $\mu >> \nu$. Now, the spectrum of L is of uniform multiplicity if all μ_n are equivalent. If moreover μ_1 is equivalent to the Lebesgue measure on $\mathbb{R}$ then L has a Lebesgue spectrum with uniform multiplicity extending over $\mathbb{R}$. In general the support of μ_1 is a part of R and this support is the spectrum of L,$\sigma(L)$. If μ_1 is absolutely continuous with respect to the Lebesgue measure, then L has absolutely continuous spectrum.

If L has uniform Lebesgue spectrum, the operator T is given in the spectral representation of L by

$$VT\rho = (i \frac{d}{d\lambda} f_1(\lambda), i \frac{d}{d\lambda} f_2(\lambda), \ldots) \tag{II.5}$$

This is the Schrödinger representation of the commutation relation (II.2). As well known, the spectral representation of T is given by the Fourier transform of that of L :

$$\hat{f} = (\hat{f}_1(\tau), \hat{f}_2(\tau), \ldots) \tag{II.6}$$

$$\hat{f}_1(\tau) = \frac{1}{\sqrt{2\pi}} \int_{-\infty}^{+\infty} e^{i\tau\lambda} f_1(\lambda) d\lambda \quad , \ldots \tag{II.7}$$

$$(\widehat{VTV^{-1}})f = (\tau\hat{f}_1(\tau), \tau\hat{f}_2(\tau), \ldots) \tag{II.8}$$

Let us give an equivalent representation to the spectral representation for an arbitrary self-adjoint operator H in a Hilbert space , the so-called decomposition of $\mathcal{H}$ in a direct integral relatively to H [18]. From the spectral representation we have the family of densities of μ_{K+1} with respect to μ_k :

$$\mu_{K+1}(E) = \int_E \omega_{K+1}(\lambda) d\mu_K \qquad k = 1,2,\ldots \tag{II.9}$$

Let Q_k be the support of μ_K and $\eta_k = \mathbb{R} - Q_K$ be the maximal set of λ such that $\omega_K(\lambda) = 0$. It is clear that $\eta_{K+1} \supseteq \eta_K$. Let $n(\lambda)$ be the maximum of K such $\omega_K(\lambda) \neq 0$, $n(\lambda)$ defines the multiplicity of λ in the spectrum of H. To each $f = (f_1(\lambda), \ldots)$ and to each $\lambda \in \sigma(H)$ associate the sequence $a_k(\lambda)$:

$$a_k(\lambda) = f_k(\lambda)\sqrt{\omega_K(\lambda)} \tag{II.10}$$

Then $a_k(\lambda)$ is a sequence having the following properties :

1) $a_k(\lambda) = 0$ for $k > n(\lambda)$ (II.11)

2) $\sum_i \int_{\sigma(H)} |f_i(\lambda)|^2 \omega_i(\lambda) d\mu_1 = \sum_i \int_{\sigma(H)} |a_i(\lambda)|^2 d\mu_1(\lambda) < +\infty$ (II.12)

Then :

$$\sum_{i=1}^{\infty} |a_i(\lambda)|^2 < +\infty \quad \mu_1 \text{ - a.e.} \tag{II.13}$$

The space of all sequence $a_K(\lambda)$ with properties (II.11) and (II.13) is a Hilbert space, denoted $\mathcal{H}_\lambda$, and the space of all μ_1-mesurable function $a_i(\lambda)$ verifying moreover (II.12) is called the direct integral of $\mathcal{H}_\lambda$ with respect to μ_1 $\int^{\oplus} \mathcal{H}_\lambda \, d\mu_1(\lambda)$.

It is now clear by the spectral representation that H is said to have a uniform multiplicity if and only if $n(\lambda) = m$ for μ_1-almost all $\lambda \in \sigma(H)$, then, H is unitarily equivalent to the multiplication by λ on the direct sum of m copies of $L^2(\mathbb{R}, d\mu_1)$.

We come now to the construction of the time operator in the Liouville space. The relation between the spectra of H and L has been studied by Spohn [19]. Here we shall consider the problem from the point of view of the decomposition of $\mathcal{L}$ into a direct integral relatively to L.

If H has a simple spectrum, that is, the multiplicity of the spectrum of H is one, then H is unitarily equivalent to the multiplication by λ on $L^2(\sigma(H), d\lambda)$. Here we work in this case. As well known, any Hilbert-Schmidt operator ρ on $L^2(\sigma, d\lambda)$ can be identified with a Lebesgue square integrable Kernel $\rho(\lambda, \lambda')$ via the following formula :

$$(\rho\psi)(\lambda) = \int_{\sigma(H)} \rho(\lambda, \lambda')\psi(\lambda')d\lambda' \tag{II.14}$$

Then $(L\rho)(\lambda, \lambda') = (\lambda - \lambda')\rho(\lambda, \lambda')$. To simplify the demonstration, we suppose that $\sigma(H) = R^+$. Then the change of variable :

$$\phi : (\lambda, \lambda') \leftrightarrow \left(\nu = \lambda - \lambda', \ \nu' = \frac{\lambda + \lambda'}{2} - \left|\frac{\lambda - \lambda'}{2}\right|\right)$$

is a one to one Lebesgue measure preserving map from $\mathbb{R}^+ \times \mathbb{R}^+$ into $R \times R^+$ such that the operator induced by ϕ :

$$(V\rho)(\nu, \nu') = \rho\left(\nu' + \frac{|\nu|}{2} - \frac{\nu}{2}, \ \nu' + \frac{|\nu|}{2} + \frac{\nu}{2}\right)$$

$$= \begin{cases} \rho(\nu', \nu' + \nu), & \nu > 0 \\ \rho(\nu' - \nu, \nu'), & \nu < 0 \end{cases} \tag{II.15}$$

maps unitarily $L^2(R^+ \times R^+, d\lambda d\lambda)$ onto $L^2(R \times R^+, d\nu d\nu')$ and :

$$(VL\rho)(\nu, \nu') = \nu (V\rho)(\nu, \nu')$$

This gives the decomposition of $\mathcal{L}$ into a direct integral relatively to L. Taking now a basis $e_i(\nu')$ of $L^2(R^+, d\nu')$ any function $f \in L^2(R \times R^+, d\nu d\nu')$ can be developed as follows :

$$f(\nu,\nu') = \sum_{i=1}^{\infty} a_i(\nu)e_i(\nu') \tag{II.16}$$

$$a_i(\nu) = \int_0^{\infty} f(\nu,\nu')\overline{e_i(\nu')}d\nu' \tag{II.17}$$

the sequence $\{a_i(\nu)\}_{i=1}^{\infty}$ gives the desired decomposition. It follows then that L has Lebesgue uniform spectrum with countable multiplicity and therefore the time operator exists.

The above result can be generalized to hamiltonians which are not simple. We will present this generalization in the appendix B. So we have shown the following theorem :

THEOREM 2.

If H *has an absolutely continuous spectrum extending on an interval* $[a,\infty[$, *then the Liouville von-Neumann operator has a countable uniform Lebesgue spectrum.*

In the spectral representation of L, the time operator is defined by :

$$(Tf)(\nu,\nu') = i\frac{d}{d\nu} f(\nu,\nu') \tag{II.18}$$

The time operator is an observable on the same foot of the other physical quantities as required by the fourth uncertainty relation. To be consistent with the rule giving the expectation values of quantum observables, we define the mean value of T in a state ρ by

$$\langle T\rangle_\rho = \langle \rho^{\frac{1}{2}}, T\rho^{\frac{1}{2}}\rangle = Tr(\rho^{\frac{1}{2}}.T(\rho^{\frac{1}{2}})) \tag{II.19}$$

In fact, the mean value of an observable θ in a state ρ is given by :

$$\langle\theta\rangle_\rho = Tr(\rho\theta) = Tr(\rho^{\frac{1}{2}}.\theta.\rho^{\frac{1}{2}}) = \langle\rho^{\frac{1}{2}},\hat{\theta}(\rho^{\frac{1}{2}})$$

where $\hat{\theta}$ is the operator in the Liouville space defined by

$$\hat{\theta}(\rho) = \theta.\rho \tag{II.20}$$

To see the meaning of T, recall that the commutation relation (II.1) is equivalent to the Weyl relation :

$$e^{iLt}\, T\, e^{-iLt} = T + tI \tag{II.21}$$

Then, from (II.19) we have for a state $\rho_t = e^{-itL}\rho$:

$$\langle T\rangle_{\rho_t} = \langle T\rangle_{\rho_{t_0}} + (t-t_0) \tag{II.22}$$

Now, we shall show that the commutation relation (II.1) entails the uncertainty relation between time and energy. As well known [20], the proofs of this relation were rather controversial because of the impossibility to derive it by the arguments used for the momentum and position, due to the non existence of a time operator in $\mathcal{H}$.

To prove the fourth uncertainty relation, we recall that the commutation (II.1) entails that :

$$(\Delta L)_\rho (\Delta T)_\rho \geq \frac{1}{2} \qquad \text{(II.23)}$$

where $(\Delta L)_\rho$ and $(\Delta T)_\rho$ are given by :

$$(\Delta L)_\rho = \sqrt{<\rho^{\frac{1}{2}}, L^2\rho^{\frac{1}{2}}> - <\rho^{\frac{1}{2}}, L\rho^{\frac{1}{2}}>^2} \qquad \text{(II.24)}$$

$$(\Delta T)_\rho = \sqrt{<\rho^{\frac{1}{2}}, T^2\rho^{\frac{1}{2}}> - <\rho^{\frac{1}{2}}, T\rho^{\frac{1}{2}}>^2} \qquad \text{(II.25)}$$

we compute $(\Delta L)_\rho$ in terms of $(\Delta E)_\rho$ using the following definition of L :

$$L\rho = \frac{1}{\hbar}[H\rho - \rho H], \qquad \hbar = \frac{h}{2\pi}$$

and $(\Delta E)_\rho$ is given by $Tr(\rho H^2) - (Tr(\rho H))^2$.

From the definition of L we see that :

$$<\rho^{\frac{1}{2}}, L\rho^{\frac{1}{2}}> = 0 \qquad \text{(II.26)}$$

then

$$(\Delta L)^2_\rho = \frac{2}{\hbar^2}[Tr(\rho H^2) - Tr(H\rho^{\frac{1}{2}}H\rho^{\frac{1}{2}})]$$

$$= \frac{2}{\hbar^2}[(\Delta E)^2_\rho + (Tr(\rho H))^2 - Tr(H\rho^{\frac{1}{2}}H\rho^{\frac{1}{2}})]$$

Using $Tr(\rho) = 1$, we have for any real x :

$$Tr(x\rho^{\frac{1}{2}} + \frac{1}{2}\rho^{1/4}H\rho^{1/4})^2 = x^2 + x\,Tr(\rho H) + \frac{1}{4}Tr(H\rho^{\frac{1}{2}}H\rho^{\frac{1}{2}})$$

This is a positive quadratic form in x, which implies that :

$$Tr((H\rho))^2 \leqslant Tr(H\rho^{\frac{1}{2}}H\rho^{\frac{1}{2}}) \qquad \text{(II.27)}$$

that is $$(\Delta L)^2_\rho \leqslant \frac{2}{\hbar^2}(\Delta E)^2_\rho \qquad \text{(II.28)}$$

substituting this inequality in (II.23) we get the fourth uncertainty relation :

$$(\Delta E)_\rho (\Delta T)_\rho \geqslant \frac{\hbar}{\sqrt{2}}(\Delta L)_\rho (\Delta T)_\rho \geqslant \frac{\hbar}{2\sqrt{2}} \qquad \text{(II.29)}$$

Remark

Here above the time operator has been interpreted as an aging of the states. The uncertainty relation (II.29) suggests that it can as well be interpreted as the time occurrence of specified events such as time of arrival of a particle at a given position or time of decay of the system. The time of occurrence of such events fluctuates and then we may speak of the probability of the occurrence of the event in a time interval $\Delta =]t_1,t_2]$. This time is therefore an observable and it can consistently be given by a time operator T' verifying the commutation relation :

$$[L,T'] = iI \tag{II.30}$$

$$\Leftrightarrow \quad e^{iLt}\, T'\, e^{-iLt} = T - tI \tag{II.31}$$

This operator can be interpreted in accordance with the rules of quantum mechanics : if $F'(\lambda)$ is the family of spectral projections of T', then $||F'(\Delta)\rho^{\frac{1}{2}}||^2 = P(\Delta,\rho)$ is the probability of the occurrence of the event, or the decay, in the time interval $\Delta =]t_1,t_2]$. This suggests to define an initially "undecayed" state by :

$$||F'(\lambda)\rho^{\frac{1}{2}}||^2 = 0 \qquad \lambda \leq 0$$

That is $\rho = (1-F'_0)\rho$. The expectation value of this time decay in ρ is $\langle T'\rangle_\rho$ (see (II.19)) and its relation to the time parameter t under the Schrödinger representation : $\rho_t = e^{-itL}\rho_0$ is given by the Weyl relation (II.31) :

$$\langle T'\rangle_{\rho_t} = \langle T'\rangle_{\rho_0} - t \tag{II.32}$$

The usual treatment of this question faced difficulties due to the non existence of canonically conjugate operator of H [21],[22]. In [22], Misra and Sudarshan have raised the question of a quantum expression for the probability that a system prepared in the undecayed state ρ (i.e. $\rho \in M$, the subspace of the undecayed states) is find to decay sometime during the interval $\Delta =]0,t]$. They pointed that such an expression has to be distinguished from the probability, at the time instant t, for finding the system "decayed" when initially it was in the state ρ :

$$p(t) = 1 - \mathrm{Tr}(E\rho_t) \tag{II.33}$$

where E is the projection onto the subspace M of the undecayed states. The first one is monotonically increasing, which is not the case of (II.33). The quantity $P(\Delta,\rho)$ is a candidate for the first kind of probability.

We have seen in § I, that time and entropy observables cannot be represented by operators in $\mathcal{H}$. The difference between these operators and the other observables can be seen in considering the Liouville space $\mathcal{L}$ as the Hilbert-space of a representation of the algebra $\mathfrak{A}$ of the bounded self-adjoint operators on $\mathcal{H}$, representing the observables of the quantum description. In fact, (II.20) defines a representation $\pi : \theta \in \mathfrak{A} \to \Theta = \pi(\theta)$, where the operator Θ belongs to the set of bounded operators on $\mathcal{L}$, $\mathcal{B}(\mathcal{L})$. We shall show now that the entropy operator M cannot belong to $\pi(\mathfrak{A})$.

The important point here liesin the transition from the Liouville-von-Neumann operator to the irreversible description. As explained in §I, this transition has been formulated as a change of representation $\rho_t \to \tilde{\rho}_t = \Lambda\rho_t$

verifying 1) and 2). This leads necessarily to the Liapounov variable $M = \Lambda^*\Lambda$.

Conversely, if M is a Liapounov variable M such that $M^{\frac{1}{2}}$ maps states into states, then it defines a transformation $\Lambda = M^{\frac{1}{2}}$ which is a candidate to the change of representation yielding the dissipative semi-group W_t (I.5) :

$$\Lambda\rho_t = W_t\Lambda\rho$$

if Λ *and* W_t *maps states into states.* In that case, one can characterizes in this representation the unstable states.

It is known that the unstable states have exponential decay rate. It can be shown that the class of the undecaying states characterized here above by the time operator T' have such a rate under the irreversible evolution. In fact, a comparaison between T' (II.30) and T (II.1) shows that we can take T' = -T. Thus, if $F(\lambda)$ is the family of spectral projections of T, i.e.

$$T = \int_{-\infty}^{+\infty} \lambda dF(\lambda) \tag{II.34}$$

then the spectral projections of T' are given by $F'(\lambda) = 1 - F(-\lambda)$. This entails that $1 - F'(0) = F(0)$. The unstable states are positive density matrices in the subspace on which F(0) projects ; i.e., $\rho_0 = F_0\rho_0$. To compute the rate of decrease of $||\Lambda U_t\rho_0||$ we take, following a canonical procedure of construction [7], Λ as a decreasing function of T : let $h(\lambda)$ a bounded positive function such that $h' < 0$ and $h(\lambda) \to 0, \lambda \to \infty$, then Λ has the spectral form :

$$\Lambda = \int_{-\infty}^{+\infty} h(\lambda)dF(\lambda) \tag{II.35}$$

Let U_t denote the unitary group : $U_t = e^{-itL}$, then :

$$||\Lambda U_t\rho_0||^2 = ||U_{-t}\Lambda U_t\rho_0||^2 \tag{II.36}$$

Now the operator $U_t^{-1}\Lambda U_t$ has the spectral form :

$$U_t^{-1}\Lambda U_t = \int_{\mathbb{R}} h(\lambda)d\, U_t^{-1}\, F_\lambda\, U_t$$

$$= \int_{\mathbb{R}} h(\lambda)d\, F_{\lambda-t} = \int_{\mathbb{R}} h(\lambda+t)dF_\lambda \tag{II.37}$$

here we have used the relation $U_tF_\lambda U_{-t} = F_{\lambda+t}$ which is equivalent to the Weyl relation (II.21). Inserting this identity into (II.36) we get :

$$||\Lambda U_t\rho_0||^2 = \int_{-\infty}^{0} h^2(\lambda+t)d\,||F(\lambda)\rho_0||^2 \tag{II.38}$$

The right hand side of this relation converges to zero as $t \to \infty$. If we take an $h(\lambda)$ of the form :

$$h(\lambda) = \frac{e^{-\lambda/\tau}}{1 + e^{-\lambda/\tau}} \quad \text{for } \lambda < 0$$

Then (II.38) yields :

$$||\Lambda U_t \rho_0||^2 = e^{-2t/\tau} \int_{-\infty}^{0} \frac{d||F(\lambda)\rho_0||^2}{(e^{\lambda/\tau} + e^{-t/\tau})^2}$$

$$< e^{-2t/\tau} \int_{-\infty}^{0} e^{-2\lambda/\tau} \, d||F(\lambda)\rho_0||^2$$

Then for ρ_0 such that $\int_{-\infty}^{0} e^{-2\lambda/\tau} \, d||F(\lambda)\rho_0||^2 < +\infty$, the rate of decay of $||\Lambda U_t \rho_0||$ is exponential with uniform bound :

$$||\Lambda U_t \rho|| \leq K_\rho \, e^{-t/\tau}$$

III. STRUCTURE OF TIME AND ENTROPY OPERATORS

An important feature of irreversible evolutions is that they does not make any distinction between pure states and mixtures and W_t cannot preserve the set of pure states. The class of operators that preserve the pure states is slightly more general than $\pi(\mathfrak{A})$. It corresponds to the *factorizable operators* that act on ρ as a multiplication on the right and on the left by an operator on $\mathcal{H}$:

$$\rho \in \mathcal{L} \to A\rho B \quad , \quad A,B \in \mathcal{B}(\mathcal{H}) \tag{III.1}$$

We shall show here that M, wherever it exists, cannot be of the form (III.1).

THEOREM 3.

If M *is a factorizable Liapounov variable that preserves hermicity (i.e.* $(M\rho)^* = M\rho$ *if* $\rho^* = \rho$*), then* $D = 0$.

Proof : This theorem is proved in several steps :

I) *If* M *is a hermitian factorizable operator in* $\mathcal{L}$ *then* M *has the following form :*

$$M\rho = \tilde{A}\rho\tilde{B}, \quad \tilde{A} = \tilde{A}^*, \quad \tilde{B} = \tilde{B}^* \tag{III.2}$$

where $\tilde{A}, \tilde{B} \in \mathcal{B}(\mathcal{H})$.

Proof : Let $M\rho = A\rho B$ and $M = M^*$ then

$$\begin{aligned} \langle \rho, M\rho' \rangle &= \mathrm{Tr}(\rho^* A \rho' B) \\ &= \mathrm{Tr}(B\rho^* A\rho') \\ &= \mathrm{Tr}((A^*\rho B^*)^*\rho') \end{aligned}$$

$$= \langle A^* \rho B^*, \rho' \rangle$$

On the other hand $\langle \rho, M\rho' \rangle = \langle M\rho, \rho' \rangle$ then it follows that :

$$M\rho = A^* \rho B^* = A \rho B \tag{III.3}$$

Taking now $\rho = |\phi\rangle\langle\psi|$ in this relation we get :

$$|A^*\phi\rangle\langle B\psi| = |A\phi\rangle\langle B^*\psi|$$

Now by acting on ψ the above equation yields :

$$A^*\phi = A\phi \,.\, \frac{\langle B^*\psi|\psi\rangle}{\langle B\psi|\psi\rangle}$$

which implies that $A\phi = CA^*\phi$ where C is a constant independent of ϕ. Then $A = CA^*$ and this implies that $|C| = 1$. Similarly $B = dB^*$ with $|d| = 1$. Inserting these two relations into (III.3) we get : $Cd = 1$. Denote $C = e^{i\theta}$, $d = e^{-i\theta}$, $\tilde{A} = e^{-i\theta/2}A$, $\tilde{B} = e^{i\theta/2}B$, it follows that :

$$M\rho = \tilde{A}\rho\tilde{B}$$
$$\tilde{A}^* = e^{i\theta/2}A^* = e^{-i\theta/2}A = \tilde{A}$$
$$\tilde{B}^* = e^{-i\theta/2}B^* = e^{i\theta/2}B = \tilde{B}$$

II) *If* M *is a non-negative, hermiticity preserving operator in* $\mathcal{L}$ *then there exists a non-negative operator* $M_1 \in \mathcal{B}(\mathcal{H})$ *such that :*

$$M\rho = M_1 \rho M_1 \tag{III.4}$$

Proof :

By I), $M\rho = \tilde{A}\rho\tilde{B}$. Now, for any hermitian $\rho \in \mathcal{L}$ we have, on account of the hermiticity preservation :

$$\tilde{A}\rho\tilde{B} = \tilde{B}\rho\tilde{A}$$

$\tilde{A}$ and $\tilde{B}$ being self-adjoint. This relation extends to any $\rho \in \mathcal{L}$ in view of the decomposition into hermitian and antihermitian part :

$$\rho = \rho_1 + i\rho_2 \qquad \text{with} \quad \rho_1 = \rho_1^*, \quad \rho_2 = \rho_2^*.$$

Using now the same argument as in I), we have :

$$\tilde{A} = C\tilde{B}$$

The positivity of M implies that :

$$Tr(|\phi\rangle\langle\phi|.M(|\phi\rangle\langle\phi|)) = C\langle\phi,\tilde{B}\phi\rangle^2 \geqslant 0$$

and C is a positive constant. Then :

$$M\rho = C\tilde{B}\rho\tilde{B} = C^{\frac{1}{2}}\tilde{B}\rho C^{\frac{1}{2}}\tilde{B}$$

Now taking $M_1 = C^{\frac{1}{2}}\tilde{B}$ we get for any $\psi, \phi \in \mathcal{H}$:

$$<(|\phi><\psi|),M(|\phi><\psi|)> = <\phi,M_1\phi><\psi_1 M_1\psi> \geqslant 0$$

That is, M_1 has a definite sign. This achieves the proof.

III) We give now the main ideas of the proof of the theorem, avoiding all the algebraic calculations that are straitforward (see [13]) : Let M be a Liapounov variable, then in view of II), there exists a non negative operator A such that $M\rho = A\rho A$. It follows that :

i) D acts on elements of the form $|\psi><\phi|$ as :

$$D(|\phi><\psi|) = |D_1\phi><A\psi| + |A\phi><D_1\psi| \qquad \text{(III.5)}$$

where D_1 is defined by $D_1 \equiv -i[H,A]$

ii) D_1 is a non negative operator on $\mathcal{H}$.

iii) There exists a constant C such that $-i[D_1,A] = CA^2$.

Either $C \neq 0 \Rightarrow A = 0 \Rightarrow M = 0$ and $D = 0$

Or $C = 0 \Rightarrow [D_1,A] = 0$ and then A is a Liapounov variable for H. The theorem I implies that $D_1 = 0$. Then from (III.5), $D|\phi><\psi| = 0$ for any ϕ and ψ in the domain of H. As pointed above, H and L are unbounded operators with dense domains. D_1 is defined on $\mathcal{D}(H)$ and D on $\mathcal{D}(L)$. Above, we only considered operators of the form $|\phi><\psi|, \psi, \phi \in \mathcal{D}(H)$. To extend the result, one needs a relation between $\mathcal{D}(L)$ and $\mathcal{D}(H)$. This is given by a theorem proved in [23] which states that :

$$\rho \in \mathcal{D}(L) \Leftrightarrow \rho\mathcal{D}(H) \subset \mathcal{D}(H) \text{ and } H\rho - \rho H \text{ extends to Hilbert-Schmidt operator}$$

then $\mathcal{D}(D)$ contains the set $\{|\phi><\psi|, \phi, \psi \in \mathcal{D}(H)\}$ which is dense in $\mathcal{D}(L)$. As D is null on this set, then $D\rho = 0 \ \forall \rho \in \mathcal{D}(L)$ because D is non-negative closable operator.

As a result of this theorem, it follows that M cannot preserve the pure states.

<u>THEOREM 4</u>. *If* M *is a Liapounov variable that preserves hermiticity and purity then* D = 0.

<u>Proof</u> : A pure state is a density matrix of the form

$\rho_\phi = |\phi><\phi|, \phi \in \mathcal{H}$ and $||\phi|| = 1$. Let us first show that M preserves positivity, that is, $\rho \geqslant 0 \Rightarrow M\rho \geqslant 0$. In fact, any $\rho \geqslant 0$, $\rho \in \mathcal{L}$, has the spectral form :

$$\rho = \sum_{i=1}^{\infty} \lambda_i |\psi_i><\psi_i| \qquad \text{(III.6)}$$

$\lambda_i \geqslant 0$, where $\{\psi_i\}$ is an orthonormal system. The series here above is convergent in the Hilbert-Schmidt norm. From the purity preservation it follows that :

$$M(|\psi_i\rangle\langle\psi_i|) = |\phi_i\rangle\langle\phi_i|$$

As M is continuous, it follows from (III.6) that $M\rho$ is non-negative. M is then positivity preserving. This entails that it can only have one of the three forms [34] :

i) M is factorizable.

ii) $M\rho = A'\rho^* A'^*$ with A' an antilinear operator on $\mathcal{H}$.

iii) $M\rho = \mathrm{Tr}(\rho B)|\psi\rangle\langle\psi|$

with B a fixed positive operator and ψ a fixed vector. Case i) implies that D = 0. The case ii) implies that $M^2\rho = A'^2\rho A'^2$,

$$D' = -i[L,M^2] = 2DM = 2M^{\frac{1}{2}}DM^{\frac{1}{2}} \geq 0$$

and $\quad [D',M^2] = 2[DM,M^2] = 0.$

Then M^2 is factorizable Liapounov variable, thus D' = 0. As $M\rho > 0$ for any ρ not belonging to the null space of L, then D = 0. Finally, iii) implies that M is an operator of rank one and has therefore a discrete spectrum which is also impossible as can be shown by using the same argument showing that M is not compatible with a pure discrete spectrum of L (see (I.12) and use $\langle e^{itM}\rho, L\, e^{itM}\rho\rangle$).

IV. ON THE QUANTUM MEASUREMENT AND IRREVERSIBILITY

The quantum measurement theory is concerned with the problem of the change of the state of a quantum system under the measurement of some observable.

Let us consider, for simplicity, an observable which has the set of measured values $a_1, a_2, \ldots$, then by the quantum mechanics, it corresponds to it a hermitian operator A given by the spectral representation

$$A = \sum_{n=1}^{\infty} a_n E_n \tag{IV.1}$$

If the system is in a coherent superposition $\psi = \sum_n \psi_n$, where $\psi_n = E_n\psi$, then after the measurement, the system is in a mixture of states :

$$\rho' = \sum_n |\psi_n\rangle\langle\psi_n| \tag{IV.2}$$

and the probability of finding a_i in this state is $||\psi_n||^2$, that is, $\mathrm{Tr}(\rho E_n)$. More generally, if the state of the system before measurement is ρ, then after measurement it will be ρ' :

$$\rho' = \sum_{n=1}^{\infty} E_n \rho E_n \qquad (IV.3)$$

The aim of the measurement theory is to explain the transition from ρ to ρ' as a result of an interaction between the quantum system and a "macroscopic apparatus". Von-Neumann has given an analysis of this problem in terms of a spin measurement. The observable to be measured is the z-component σ_z acting on C^2 with eigenvalues ± 1 corresponding to $\psi_+ = \binom{1}{0}$ and $\psi_- = \binom{0}{1}$. Before the measurement the apparatus is in a neutral state ϕ_0 and under the measurement the states of the apparatus becomes $\phi_\pm$. Thus the coupled system S+A makes the transition :

$$\psi_\pm \otimes \phi_0 \rightarrow \psi_\pm \otimes \phi_\pm \qquad (IV.4)$$

Let the system before the measurement be in a state $C_+\psi_+ + C_-\psi_-$. If the transition where unitary as in the Schrödinger picture, one arrives to the coherent superposition $C_+\psi_+ \otimes \phi_+ + C_-\psi_- \otimes \phi_-$ and this leads to a macroscopic uncertainty. However, in a series of identical experiments, the relative frequency of spin up and down will be $|C_+|^2$ and $|C_-|^2$, then the state of the system has to be described by a mixture :

$$\rho' = |C_+|^2 |\psi_+ \otimes \phi_+\rangle\langle\psi_+ \otimes \phi_+| + |C_-|^2 |\psi_- \otimes \phi_-\rangle\langle\psi_- \otimes \phi_-| \qquad (IV.5)$$

The substitution of the coherent superposition $(C_+\psi_+ + C_-\psi_-) \otimes \phi_0$ by ρ' is called the "reduction of the wave paquet". (For a discussion of this problem see Jauch [24]).

HEPP [25] has proposed a model in which the spin interacts with an infinite macroscopic system in such a way that as a result of an infinitely long time interaction between S and A the coherent superposition is transformed into a mixture. More precisely, Hepp presented a model in which a coherent superposition becomes indistinguishable from a mixture(*). To formulate this, Hepp uses the algebraic description of quantum mechanics where quasi local observables form a C*-algebra $\mathfrak{A}$ with identity 1, and states $\mathcal{S}$ are linear and positive functionals over $\mathcal{S}$ such that $\omega(1) = 1$, $\forall \omega \in \mathcal{S}$. If one considers a representation $\pi(\mathfrak{A})$ of $\mathfrak{A}$, in some Hilbert space $\mathcal{H}_\pi$, then a vector state $\psi \in \mathcal{H}_\pi$ defines $\omega_\psi \in \mathcal{S}$ by :

$$\omega_\psi(A) = \langle\psi, \pi(A)\psi\rangle, \qquad ||\psi|| = 1 \qquad (IV.6)$$

Let ψ_1 and ψ_2 be normalized vectors in $\mathcal{H}_\pi$ and consider ω_ψ corresponding to the vector $\psi = C_1\psi_1 + C_2\psi_2$, $|C_1|^2 + |C_2|^2 = 1$, such that :

(*) This notion is discussed in [24].

$$\langle\psi_1, \pi(A)\psi_2\rangle = 0 \qquad \forall\ A \in \mathfrak{A} \tag{IV.7}$$

Then :

$$\omega_\psi(A) = |C_1|^2\omega_1(A) + |C_2|^2\omega_2(A) \tag{IV.8}$$

where $\omega_1 = \omega_{\psi_1}$ and $\omega_2 = \omega_{\psi_2}$.

ω_ψ is then undistinguishable from the mixture $|C_1|^2\omega_1 + |C_2|^2\omega_2$ with respect to the algebra $\mathfrak{A}$. The states ω_1 and ω_2 such that (IV.7) is verified for any representation π, in which they are represented by vectors ψ_1 and ψ_2, are called *disjoint*. In Hepp opinion, the solution of the wave packet reduction can be performed if under a family of evolution operators S_t^1 on ψ_1 : $\psi_{1,t} = S_t^1\psi_1$ and S_t^2 on ψ_2 ; $\psi_{2,t} = S_t^2\psi_2$ the states $\omega_{1,t}$ and $\omega_{2,t}$ become in the limit $t \to \infty$, disjoint. That is :

$$\begin{aligned} \omega_{1,t}(A) &= \langle\psi_{1,t}, \pi(A)\psi_{1,t}\rangle \to \omega_1(A) \\ \omega_{2,t}(A) &= \langle\psi_{2,t}, \pi(A)\omega_{2,t}\rangle \to \omega_2(A) \end{aligned} \tag{IV.9}$$

with ω_1 and ω_2 disjoint. In that case one can show that [25] :

$$\langle\psi_{1,t}, \pi(A)\psi_{2,t}\rangle \to 0, \qquad t \to \infty \tag{IV.10}$$

$\forall A \in \mathfrak{A}$.

Now given an automorphism group of evolution, α_t, on $\mathfrak{A}$, it can be implemented by a group of unitary operators U_t on $\mathcal{H}_\pi$, such that :

$$\pi(\alpha_t A) = U_t\pi(A)U_{-t} \tag{IV.11}$$

One would such Liouvillian evolutions (i.e. $\alpha_t A = e^{itL}A$), called automorphisms of $\mathfrak{A}$, to lead to disjointness (IV.9) by taking $S_t^1 = S_t^2 = U_{-t}$. Yet, this is impossible [25]. Hepp suggests to use non automorphic evolutions in several models of apparatus leading to disjointness for $t \to \infty$. However, the infinitely long time necessary for the reduction (IV.10) and the possibility of construction of "time reversal" observable inspired J.S. Bell [28] objection as to the decay of (IV.10). For any t, he constructed an observable z_t such that :

$$\langle\psi_{1,t}, \pi(z_t)\psi_{2,t}\rangle \neq 0 \tag{IV.12}$$

In order to eliminate such unphysical observables, Lockhart and Misra have constructed [15] an irreversible model of measurement inspired from the X-ray edge model of Hepp and have obtained a group of automorphisms on the representation $\pi(\mathfrak{A})$ different from (IV.11). The transition from this group to a dissipative semi-group is realized by a Projection operator E_0. Such a transition has been used in classical systems that are K-flows [10] as follows :

Let $(\Gamma,\mathcal{B})$ be a measurable space with a flow S_t leaving the measure invariant, then $(\Gamma, ,S_t,\mu)$ is said a dynamical K-system if there exists a partition ξ_0 of Γ formed by measurable subsets from the σ-algebra $\mathcal{B}$, such that

i) $S_t\xi_0 = \xi_t$ is finer than ξ_0 for $t > 0$

ii) ξ_t tends to the partition of Γ into points as $t \to \infty$

iii) ξ_t tends to the partition formed by one element Γ for $t \to -\infty$.

This flow induces on the space of measurable functions a group of operators U_t :

$$(U_t f)(\omega) = f(S_{-t}\omega) = f_t(\omega)$$

and this group is a group of unitary operators when it acts on L^2_μ. If we take the "coarse grained" evolution with respect to the partition ξ_0, that is, the projection E_0 onto the subspace $L^2_\mu(\xi_0)$, of the functions that are constant on the fibers of the partition ξ_0, then U_t is converted into a semi-group W_t, $t > 0$, by :

$$E_0 U_t f = W_t E_0 f \tag{IV.13}$$

where W_t is a *contraction semi-group of operators associated to a Markov process in* $(\Gamma,\mathcal{B})$ *; such that* :

$$||W_t E_0 f||^2 \searrow_{t\to\infty} 0 \tag{IV.14}$$

The projection operator E_0 onto $L^2_\mu(\xi_0)$ is a "coarse-graining" operation, also called the conditional expectation in probability theory. For a countable partition $\mathcal{P} = (P_1,P_2,\ldots,P_n,\ldots)$ it is given by :

$$(E_{\mathcal{P}} f)(\omega) = \frac{1}{\mu(P_i)} \int_{P_i} f(x)d\mu(x)$$

for any $\omega \in P_i$ [11]. Note that the conditional expectation is always defined with respect to a given measure μ.

In quantum mechanics, the conditional expectation can also be defined (see e.g. [26]). The K-flow structure can be characterized using the family of the conditional expectation E_t onto $L^2_\mu(\xi_t)$ then :

$$E_t = U_t E_0 U_{-t} \tag{IV.15}$$

$$E_t \nearrow_{t\to\infty} \mathbb{1}_{L^2_\mu} \tag{IV.16}$$

$$E_t \searrow_{t\to-\infty} \qquad E_{-\infty} = |1\rangle\langle 1| \tag{IV.17}$$

where $|1\rangle\langle 1|$ is the projection onto constant functions, i.e. :

$$E_{-\infty} f = \int_{\Gamma} f \, d\mu .$$

For non commutative algebra of observables such a K-systems has been already considered [27]. Let us now come back to the X-ray edge model. It is a model of spin coupled to an infinite Fermi system. The algebra of observables is $\mathcal{B}(C^2) \otimes \mathfrak{A}_A$, $\mathfrak{A}_A$ is the algebra of the canonical anti-commutation relations.

Lockhart and Misra have considered some state ω on $\mathcal{B}(C^2) \otimes \mathfrak{A}_A$ and the Gelfand-Neimark-Segal representation $\pi_\omega(\mathcal{B}(C^2) \otimes \mathfrak{A}_A)$. Under some automorphisms group α_t on the von-Neumann algebra $\mathfrak{M}$ generated by $\pi_\omega(B(C^2) \otimes \mathfrak{A}_A)$ they have constructed a projection E_0 onto some sub-algebra $\mathfrak{M}_0 \subset \mathfrak{M}$ having properties analoguous to (IV.15)-(IV.17), giving rise to a positivity preserving semi-group W_t (see (IV.13)) and to disjointness property (IV.10). The essential point is that the observables z_t with non-vanishing off-diagonal elements at time t, cannot belongs to the sub-algebra of the physical observables $\mathfrak{M}_0$. We do not give here the construction of α_t and E_0, details can be found in [15]. Although the model is not built on a hamiltonian system, its main difference with respect to Hepp theory consists in the restriction of the observables to a subset of $\mathfrak{M}$ non symmetric with respect to time inversion (time asymmetry is also discussed in [29],[30]).

Remarks : The time evolution in Lockhart and Misra work is taken in the Heisenberg representation as acting on the observables. The dissipative semi-group that they construct is only acting on the observables and not on states as made in the previous formulations of the non-unitary transformation Λ given in §I. Then, the approach outlined in this section is not relevant to the problem of the entropy as a functional of the state. Moreover, the reduction is not obtained under the irreversible evolution in a finite time but asymptotically.

V. CONCLUSION

The approach to the problem of the irreversibility here presented is based on the existence of a time operator in quantum mechanics. Two conditions are necessary : first the passage to mixed states (the Liouville space), secondly an absolutely continuous spectrum of H carried on an unbounded subset of $\mathbb{R}$. In that case, one may construct a dissipative semigroup of time evolution related to the unitary group of the Liouville equation by a bounded operator of similarity Λ. The main problem that is open to future work is the positivity preserving of Λ. In classical mechanics the positivity preserving property is due to hyperbolicity of the dynamics. The Λ transformation maps a Dirac measure into a measure carried by a contracting fiber and Λ is highly delocalizing (see [32],[31],[30]). In quantum mechanics too, Λ is not purity preserving. Nevertheless, the analysis of delocalization caused by Λ is not well known, for, this depends on the version of Λ (defined up to unitary equivalence in §II) that is positivity preserving. This problem has also been considered from the point de vue of the subdynamics theory (see [33]). A further study of this point may help to understand the problem of the wave packet reduction in the measurement process where the mixture arise as a result of the irreversible evolution. In any case this representation seems well adapted to the description of the unstable state as it provides in the same time an answer to the computation of probabilities which cannot be computed in usual quantum mechanics, but can be here fitted in a generalized class of states that also display an exponentially decreasing rate.

APPENDIX A

Let us first show that $e^{iMt}\psi \in \mathcal{D}(H)$ if $\psi \in \mathcal{D}(H)$.

As M is bounded, e^{iMt} is given by the uniformly convergent series :

$$e^{iMt} = \sum_{p=0}^{\infty} \frac{(it)^p M^p}{p!} \tag{A.1}$$

Define ψ_n by :

$$\psi_n = \sum_{0\leqslant p\leqslant n} \frac{(it)^p}{p!} M^p \tag{A.2}$$

$\psi_n \in \mathcal{D}(H)$ for $M\mathcal{D}(H) \subset \mathcal{D}(H)$. By the closedness of H, if $H\psi_n$ and ψ_n are convergent then $\lim_{n\to\infty} \psi_n = e^{iMt}\psi$ belongs to $\mathcal{D}(L)$. It suffices to show that $H\psi_n$ is a convergent sequence. Now, the relation $[H,M]\psi = iD\psi$ and the commutation of M and D imply :

$$[HM^p - M^pH]\psi = ipM^{p-1}D\psi \tag{A.3}$$

Then :

$$H\psi_n = \sum_{p\leqslant n} \frac{(it)^p}{p!} M^pH\psi - t(\sum_{p=0}^{n-1} \frac{(it)^p}{p!} M^p)D\psi \tag{A.4}$$

the uniform convergence of (A.1) gives the result. Moreover (A.4) implies that for any $\psi \in \mathcal{D}(H)$:

$$H\, e^{iMt}\psi = e^{iMt}H\psi - t\, e^{iMt}D\psi \tag{A.5}$$

this implies the differentiability of $H\, e^{iMt}\psi$ when $\psi \in \mathcal{D}(H)$ and (A.5) yields :

$$\frac{d}{dt} H\, e^{iMt}\psi = iHM\, e^{iMt}\psi.$$

APPENDIX B

Let H be a hamiltonian with degenerate absolutely continuous spectrum, then H is unitarily equivalent to the multiplication by λ on $\mathcal{H} = \bigoplus_{k=1}^{N} L^2(Q_k, d\lambda)$ where $Q_1 = \sigma(H)$ and Q_k is the support of $d\mu_k$, N is finite or infinite. From now on, we identify H with this spectral representation. We have first the following :

Proposition : Let $\mathcal{H} = \bigoplus_{k=1}^{N} L^2(Q_K, d\lambda)$. Then, ρ is a Hilbert-Schmidt operator on $\mathcal{H}$ if and only if there is a family of functions $K_{ij}(\lambda,\lambda') \in L^2(Q_i \times Q_j, d\lambda d\lambda')$ with

$$(\rho\psi)_i(\lambda) = \sum_j \int_{Q_i} K_{ij}(\lambda,\lambda')\psi_j(\lambda')d\lambda' \qquad (B.1)$$

$$\sum_{ij} \int_{Q_i \times Q_j} |K_{ij}(\lambda,\lambda')|^2 d\lambda d\lambda' = ||\rho||^2 \qquad (B.2)$$

Proof : Let E_i be the family of the orthogonal projections onto $L^2(Q_i, d\lambda)$. Then any ρ can be decomposed into :

$$\rho = \sum_{ij} E_i \rho E_j \qquad (B.3)$$

$E_i \rho E_j$ is an orthogonal family in $\mathcal{L}$, i.e.

$$\langle E_i\rho E_j, E_{i'}\rho E_{j'}\rangle = Tr(E_j \rho^* E_i E_{i'} \rho E_{j'})$$
$$= \delta_{ii'}\delta_{jj'} ||E_i \rho E_j||^2 \qquad (B.4)$$

Now it is well known (see e.g. [17] p. 210) that $E_i\rho E_j$ is Hilbert-Schmidt if and only if there is $K_{ij}(\lambda,\lambda') \in L^2(Q_i \times Q_j, d\lambda d\lambda')$ such that for any $\psi \in E_j\mathcal{H} = L^2(Q_j, d\lambda')$ we have :

$$(E_i \rho E_j \psi)(\lambda) = \int_{Q_j} K_{i,j}(\lambda,\lambda')\psi(\lambda')d\lambda' \qquad (B.5)$$

$$\int_{Q_i \times Q_j} |K_{ij}(\lambda,\lambda')|^2 d\lambda d\lambda' = ||E_i \rho E_j||^2. \qquad (B.6)$$

Taking the decomposition of any $\psi \in \mathcal{H}$ we get :

$$(\rho\psi)_i(\lambda) = (E_i\rho\psi)(\lambda) = \sum_j (E_i \rho E_j \psi)(\lambda)$$

and (B.1) follows from (B.5).(B.2) follows from the orthogonality of $E_i\rho E_j$ and (B.6) :

$$||\rho||^2 = \sum_{ij} ||E_i\rho E_j||^2 = \sum_{ij} \int_{Q_i \times Q_j} |K_{ij}(\lambda,\lambda')|^2 d\lambda d\lambda'$$

Theorem : (with the above notations and hypotheses). If $\sigma(H)$ is unbounded, then the Liouville operator has countable uniform Lebesgue spectrum.

Proof : Let us consider the identification of any Hilbert-Schmidt with $\{\rho_{i,j}(\lambda,\lambda')\}$ as above. Then clearly $L\rho$ is represented by $\{(\lambda-\lambda')\rho_{ij}(\lambda,\lambda')\}$. As $Q_1 \supset Q_2 \supset \ldots$ we define the spectral representation of L by the unitary mapping $\{V_{ij}\}$

$$V_{ij} : \rho_{ij}(\lambda,\lambda') \to (V_{ij}\rho_{ij})(\nu,\nu')$$

where ν and ν' are given as in (II.15). As $d\mu_1$ is the dominating measure then $Q_1 \times Q_1 \supset Q_i \times Q_j$ for any ij. Let $\mathcal{L}_{ij}$ be the subspaces of $\rho \in \mathcal{L}$ s.t. $\rho = E_i \rho E_j$, these are invariant subspaces and L restricted to $\mathcal{L}_{11}$ has countable uniform Lebesgue spectrum. This means that there are infinitely $\rho_n \in \mathcal{L}_{11}$ such that $\{\overline{f(L)\rho_n, f \in C(\mathbb{R})}\}$ are invariant orthogonal subspaces of $\mathcal{L}$ and the measures $\langle E(\Delta)\rho_n,\rho_n\rangle$ are equivalent to the Lebesgue measure on $\mathbb{R}$, here $\{E(\Delta)\}$ are the spectral projections of L. From this it necessarily follows that L has countable uniform Lebesgue spectrum.

* * *

The author expresses his gratitude to the members of the Brussels group for many fruitfull discussions, especially with B. Misra and I. Prigogine.

REFERENCES

[1] Pauli, W. : In Probleme der modernen physik, P. Debye ed. Leipzig (1928)

[2] van Hove, L. : Physica 21, 362 (1955), 23, 441 (1957)

[3] Prigogine, I. and Résibois, P. : Physica, 27, 629 (1961)

[4] Prigogine, I. : Non equilibrium Statistical Mechanics Wiley-Inter-Science, N.Y. (1962)

[5] Balescu, R. : Equilibrium and non equilibrium Statistical Mechanics, Wiley, N.Y. (1975)

[6] Résibois, P. and De Leener, M. : Classical Kinetic Theory of Fluids, Wiley, N.Y. (1977)

[7] Misra, B., Prigogine, I. and Courbage, M. : Physica A 98, 1 (1979)

[8] Goldstein, S., Misra, B. and Courbage, M. : J. Stat. Phys. 25, 111 (1981)

[9] Goldstein, S., Penrose, O. : J. Stat. Phys. 24, 325 (1981)

[10] Misra, B. and Prigogine, I. : Suppl. Progr. Theor. Phys. 69, 101 (1980)

[11] Courbage, M. : in "Dynamical Systems and Microphysics" eds A. Avez and A. Blaquière, Academic Press, 1982

[12] Misra, B. : Proc. Natl. Acad. Sci. USA, 75, 1627 (1978)

[13] Misra, B., Prigogine, I. and Courbage, M. : Proc. Natl. Acad. Sci. USA, 76, 4768 (1979)

[14] Courbage, M. : Lett. Math. Phys. 4, 425 (1980)

[15] Lockhart, C.M. and Misra, B. : Physica 136A, 47 (1986)

[16] Kato, T. : Perturbation theory for linear operators, Springer 1966

[17] Reed, M. and Simon, B. : Methods of Modern Mathematical Physics I, Academic Press 1970

[18] Naimark, M.A. : Normed Rings, P. Noordhoff, Groningen

[19] Spohn, H.: J. Math. Phys. 17, 59 (1976)

[20] Jammer, M. : The Philosophy of Quantum Mechanics, Wiley

[21] Allcock, G.R. : Ann. Phys. (N.Y.) 53, 253 (1969)

[22] Misra, B. and Sudarsham, E.C.G. : J. Math. Phys. 18, 756 (1977)

[23] Courbage, M. : J. Math. Phys. 23, 646 (1982)

[24] Jauch, J.M. : Helv. Phys. Acta 37, 193 (1964)

[25] Hepp, J. : Helv. Phys. Acta 45, 237 (1972)

[26] Takesaki, M. : J. Functional Anal. 9, 306 (1972)

[27] Emch, G.G. : Com. Math. Phys. 49, 191 (1976)

[28] Bell, J.S. : Helv. Phys. Acta 48, 93 (1975)

[29] Misra, B. and Prigogine, I. : In "Long Time Prediction in Dynamics" Eds. C.W. Horton, Jr., L.E. Reichl, A.G. Szebehely, Wiley, 1983

[30] Courbage, M. and Prigogine, I. : Proc. Natl. Acad. Sci. USA 80, 2412 (1980), and Courbage, M. : Physica 122A, 459 (1983)

[31] Martinez, S. and Tirapegui, E. : J. Stat. Phys. 25, 111 (1981)

[32] Misra, B. and Prigogine, I. : Lett. Math. Phys. 7, 421 (1983)

[33] Prigogine, I. and George, C. : Proc. Natl. Acad. Sci. USA, 80, 4590 (1983)

[34] Davies, E.B. : Quantum Theory of Open Systems, Academic Press, N.Y. (1976), Theorem 3.1, p. 21.

QUANTIZATION OF THE KICKED ROTATOR WITH DISSIPATION

T. Dittrich and R. Graham
Universität Essen, Federal Republic of Germany

ABSTRACT

The effect of dissipation on a quantum system exhibiting chaos in its classical limit is studied by coupling the kicked quantum rotator to a reservoir with angular momentum exchange. A master equation is derived which maps the density matrix from one kick to the subsequent one. The limits of $\hbar \to 0$ and of vanishing dissipation reproduce the classical kicked damped rotator and the kicked quantum rotator, respectively. In the semi-classical limit the quantum map reduces to a classical map with quantum mechanically determined classical noise terms. For sufficiently small dissipation quantum mechanical interference effects render the Wigner distribution negative in some parts of phase space and prevent its interpretation in classical terms.

1. INTRODUCTION

The theory of chaos in dynamical systems, unlike other recent achievements in theoretical physics, has developped as a purely classical discipline. This becomes most obvious in the central concept of a strange attractor: Its characteristic feature is to exhibit structure without limit, however fine phase space is resolved [1]. The impossibility of an infinite resolution in phase space, on the other hand, is a constituent idea of quantum mechanics. Thus the question arises how the theory of chaos in dynamical systems can be understood as the classical limit of a more complete quantum mechanical theory and in which respects a modification of the classical picture of chaotic behavior will be required [2-16].

Early investigations into the field of quantum chaos, numerical as well as analytical, have indeed shown that simple Hamiltonian systems in a quantum mechanical description do not exhibit stochastic dynamics, even if their classical counterparts do. The reason for this qualitatively different behavior lies in quantum interference effects. Therefore, it seems plausible that taking dissipation into account in the quantum description would again alter the situation by partially destroying quantum interferences, and lead back into the direction of classical chaos.

A straightforward way to study this question further is to supply a well-known Hamiltonian quantum system with some mechanism for dissipation in a quantum mechanical formulation, such that, with the continuous dynamics replaced by discrete time-steps, a deterministic map with Jacobian $J<1$ will appear in the classical limit [17]. This method has already been applied to two types of systems, namely those corresponding to the Kaplan-Yorke map [17] and to the Hénon map [18], respectively, as their classical counterparts. In principle, as was discussed in [17,18], the quantization of such classical maps is not unique in several respects. However, regardless of this fact some results have been found which are independent of the quantization procedure chosen. In particular, it was found that in the semi-classical limit dissipative quantum maps reduce to the classical maps with additional Gaussian noise terms determined by the quantum theory.

The present paper is devoted to an investigation of a third example, namely the standard map [19]

$$\begin{aligned} p' &= p - K\sin\varphi, \\ \varphi' &= \varphi + p', \end{aligned} \tag{1.1}$$

which we will approach along similar lines. Eq. (1.1) corresponds to a kicked rotator or, equivalently, a periodically driven pendulum. Quantized versions of eq. (1.1) have been studied first [3] and perhaps most thoroughly (cf. [4,8,9-16]) of all quantum systems with non-integrable classical counterparts.

Including dissipation in the standard map it is generalized to the form

$$p' = \lambda p - K\sin\varphi, \qquad \varphi' = \varphi + p', \tag{1.2}$$

with $0<\lambda<1$, which is known as Zaslavski's map [20]. Our main goal, in the following, will be to include dissipation in the quantum model, by coupling the quantized kicked rotator to a reservoir which is able to absorb (and at finite temperature to emit) single quanta of the angular momentum of the rotator.

In the present paper we derive the quantum map and study certain limits analytically. We shall begin in section 2 by recalling, in a brief overview, the principal steps of the approach chosen, then develop the elements of our formalism for the kicked damped rotator in section 3 and conclude in section 4 with a discussion of the conservative, dissipative, semi-classical and classical limits.

An extended version of this paper will appear in Zeitschrift für Physik B (Condensed Matter).

2. METHODS

In the present paper, we restrict our investigations, as it has become standard in the field, to a discrete time version of the dynamics. Accepting the natural disadvantage that no information on the evolution of the system between time sections can be obtained, we gain the freedom to decompose the discrete map deliberately into subsequent sub-maps (they may be thought as acting at intermediate "dummy instants" which need not have physical reality). These components can then be treated separately, which is particularly helpful in the process of quantization, to be recombined, in the end, in arbitrary order.

As regards the kicked damped rotator, it is most convenient to choose free propagation, kicks, and damping as three components that constitute the complete map. While quantizing the conservative sub-maps is a straightforward application of the Heisenberg quantization program (which is not unique, in this case, however, [2,3]), it will be our main objective to incorporate a quantum mechanical mechanism of dissipation compatible with the phenomenology of damping [17]. In either case, a source of ambiguity comes into play in that, generally, for each discrete time version of a classical system there is an infinite class of quantum systems which reduce to that same system in the classical limit. The question, in how far our results are affected by this ambiguity, can best be answered by comparing different elaborated examples of dissipative quantum maps.

Due to its cylindrical phase space, the angular momentum of the unperturbed quantum rotator has equidistant discrete eigenvalues $p_l = \hbar l$ and it is convenient to use the representation of the density matrix in the corresponding eigenstates for the discussion of the quantum map.

$$\hat{p}|l\rangle = \hbar l|l\rangle, \quad \langle\varphi|l\rangle = (2\pi)^{-1/2} e^{il\varphi}, \quad l \in \mathbb{Z}, \varphi \in [-\pi, \pi] \tag{2.1}$$

Generally, a single time-step of the system is completely described by a summation kernel $G(l', m'|l, m)$

$$\langle l'|\varrho'|m\rangle = \sum_{l,m} G(l', m'|l, m)\langle l|\varrho|m\rangle. \tag{2.2}$$

To remain consistent with the general properties of the quantum statistical operator it has to satisfy a number of formal conditions, i.e. normalization, hermiticity, positivity and an uncertainty relation.

We shall determine the kernels for free propagation and kicks directly from the respective classical Hamiltonians by evaluating the time transformation operators for discrete time steps.

In the quantization of the locally dissipative component of the map, we follow the procedure of [17]. Compared to the conservative parts of the map an additional step is involved, since it is first necessary to specify a microscopic model which gives back the classical map with dissipation on the level of averaged equations. The kernel is then obtained from the quantum mechanical model Hamiltonian by deriving a continuous time master equation for the density operator and solving it for a single time step with general initial conditions. Consequently, the classical map comes into play only indirectly as the basis of a consistency condition for the master equation.

Finally, the three kernels are combined to form the complete quantum map. The consistency of the quantum map is checked by going to the limits of vanishing and of maximum dissipation and, in particular, to the classical limit. We also determine the leading quantum corrections which we call semi-classical limit.

In order to enable a direct comparison with the classical map, it is preferable to switch from the momentum representation of the density matrix to the Wigner function. Its definition, for a cylindrical phase space, reads

$$W(\varphi, p) = \frac{1}{4\pi^2} \int_{-\infty}^{\infty} d\xi \, \exp\left(-i\frac{p\xi}{2\pi}\right) \left\langle \left(\varphi + \frac{\hbar\xi}{2}\right)(\mathrm{mod}\, 2\pi) \middle| \varrho \middle| \left(\varphi - \frac{\hbar\xi}{2}\right)(\mathrm{mod}\, 2\pi) \right\rangle \tag{2.3}$$

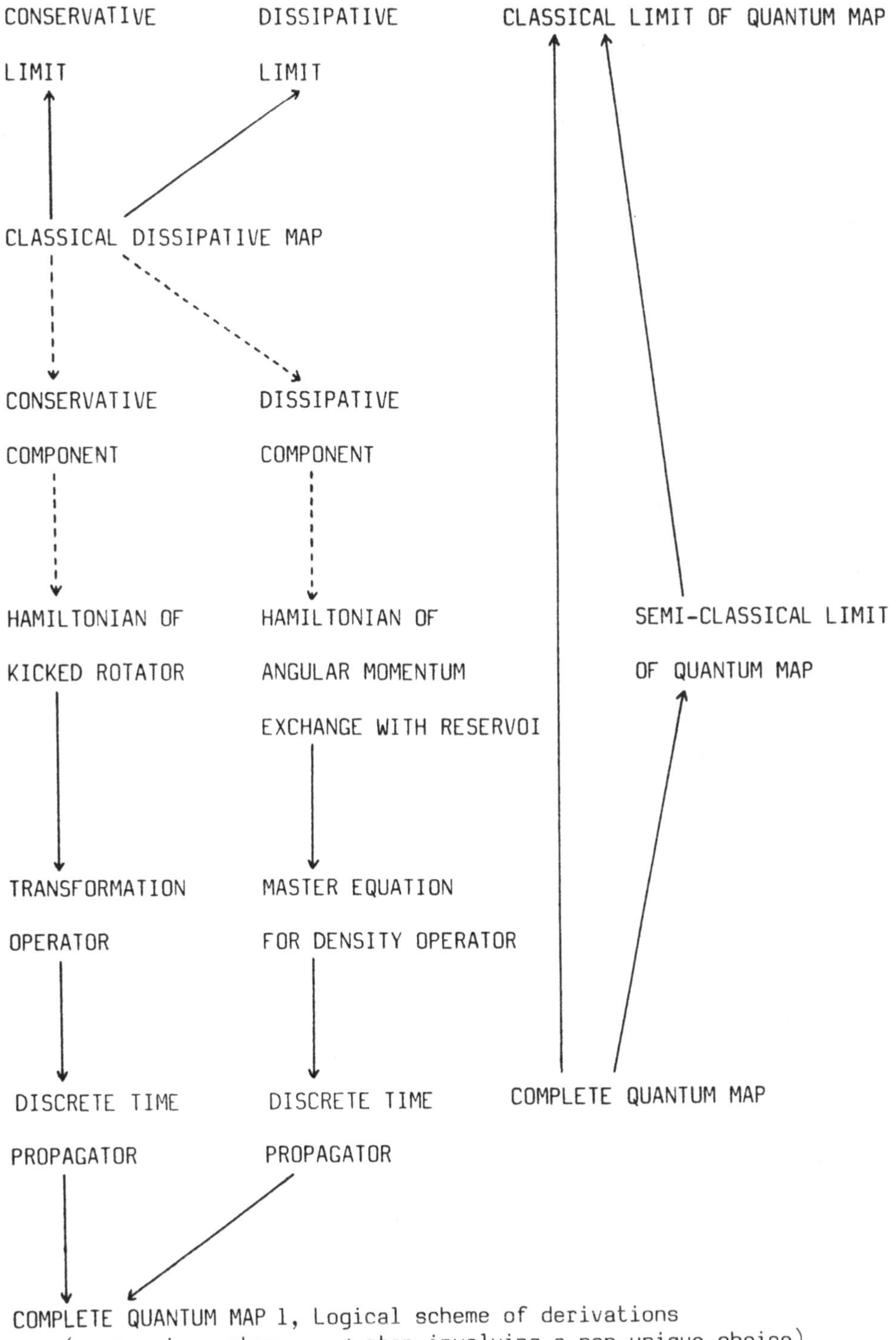

1, Logical scheme of derivations (⟶ unique step, --→ step involving a non-unique choice)

and translates readily into a transformation rule which connects the Wigner function propagator with the time-step generator of the density matrix

$$G_w(l',\varphi'|l,\varphi)=\frac{1}{2\pi}\sum_{m',n'}\delta_{l',m'+n'}\exp(i(m'-n')\varphi')\sum_{m,n}\delta_{l,m+n}\exp(-i(m-n)\varphi)G(m',n'|m,n). \quad (2.4)$$

3. CONSTRUCTION OF THE QUANTUM MAP

We now split the classical map into two parts. The first part is a dissipative step D: $(\varphi,p)\longmapsto(\varphi'',p'')$ which we choose as

$$\begin{aligned}\varphi''&=\varphi,\\ p''&=\lambda p,\end{aligned} \quad (3.1)$$

with $\varphi\in[-\pi,\pi[, p\in]-\infty,\infty[$. The second part is a conservative step which is given by the standard map (1.1) C: $(\varphi'',p'')\mapsto(\varphi',p')$

$$\begin{aligned}\varphi'&=\varphi''+p',\\ p'&=p''-K\sin\varphi'',\end{aligned} \quad (3.2)$$

with $\varphi''\in[-\pi,\pi[, p''\in]-\infty,\infty[$.

Let us first deal with the conservative part C. It can be considered as the stroboscopic map of a periodically kicked rotator [3] with the time-dependent Hamiltonian

$$\begin{aligned}H_S&=H_F(p)+H_K(\varphi,t),\\ H_F(p)&=\frac{p^2}{2},\\ H_K(\varphi,t)&=V(\varphi)\sum_{n=-\infty}^{\infty}\delta(t-n),\ V(\varphi)=-K\cos\varphi.\end{aligned} \quad (3.3)$$

Here we have chosen the period of the kicks as the unit of time and the ratio of the rotator's moment of inertia and the period of the kicks as the unit of angular momentum. The parameter K characterizes the strength of a kick. In our units it is simultaneously an energy, an angular momentum and an action. Integrating the canonical equations of motion following from (3.3) between two times t'', $t'=t''+1$ immediately preceding two subsequent kicks with initial conditions $\varphi(t'')=\varphi''$, $p(t'')=p''$ the map (3.2) is obtained with $\varphi(t')=\varphi'$, $p(t')=p'$.

The quantization of the dynamics generated by (3.3) is easily achieved by using the unitary time transformation operator

$$U_C(t+1,t)=U_F(t+1-\varepsilon,t+\varepsilon)\,U_K(t+\varepsilon,t-\varepsilon). \quad (3.4)$$

U_F and U_K describe the free propagation of the rotator and a kick, respectively:

$$U_F(t+1-\varepsilon, t+\varepsilon) = \exp(-i\frac{p^2}{2\hbar}), \tag{3.5}$$
$$U_K(t+\varepsilon, t-\varepsilon) = \sum_{r=-\infty}^{\infty} b_r(\tfrac{K}{\hbar}) \exp(ir\varphi), \quad b_r(x) := i^r J_r(x) = \frac{1}{2\pi}\int_{-\pi}^{\pi} d\varphi \exp(-ir\varphi)\exp(ix\cos\varphi).$$

Here $J_r(x)$ is the Bessel function of first kind. In the representation of the angular momentum eigenstates $|l\rangle$, defined by (2.1), we arrive at the following propagators for the density matrix

$$G_F(l',m'|l,m) = \delta_{l',l}\,\delta_{m',m} \exp(-i\tfrac{\hbar}{2}(l^2-m^2)), \tag{3.6}$$
$$G_K(l',m'|l,m) = b_{l'-l}(\tfrac{K}{\hbar})\, b^*_{m'-m}(\tfrac{K}{\hbar}).$$

Note that there are two dimensionless parameters now, namely $\hbar$ and $K/\hbar$, which measure, respectively, the unit of angular momentum, and the kick strength in units of $\hbar$.

Let us now turn to the dissipative part (3.1) of the map. We follow ref. [17] and consider the kicked rotator as being coupled weakly to a big reservoir of angular momentum. The reservoir is assumed to be at zero temperature and therefore to act as a sink of angular momentum. Due to the absorption of angular momentum by the reservoir, the state of vanishing angular momentum is preferred by the coupling to the reservoir. In the following we summarize the procedure given in [17]. The weak-coupling Hamiltonian is most conveniently expressed in angular-momentum ladder operators

$$\Gamma_k := \sum_l c_{k,l}\, |l-k\rangle\langle l|, \tag{3.7}$$
$$\Gamma_k^+ := \sum_l c^*_{k,l}\, |l\rangle\langle l-k| = \sum_l c_{k,l+k}\, |l+k\rangle\langle l|,$$

where the complex coefficients $c_{k,l}$ will be determined below.

In this notation, a Hamiltonian with the desired structure reads

$$H = H_S + H_I + H_R,$$
$$H_S = \frac{p^2}{2} + V(\varphi)\sum_{n=-\infty}^{\infty}\delta(t-n), \tag{3.8}$$
$$H_I = \sum_k g_k\,(\Gamma_k^+ R_k + \Gamma_k R_k^+),$$
$$H_R = \sum_k \Omega_k R_k^+ R_k .$$

The reservoir operators R_k, R_k^+ are assumed to commute at equal time with the Γ_k, Γ_k^+ throughout. Taking the reservoir to be Markoffian, i.e. instantaneously responding, and neglecting finite temperature effects, i.e. $T \ll \hbar\omega_k$ with ω_k the eigenvalues of H_S , we obtain from (3.8) by first-order time-dependent perturbation theory the master equation for the density operator

$$\frac{\partial\varrho}{\partial t} = \sum_k \gamma_k\left([\Gamma_k, \varrho\Gamma_k^+] + [\Gamma_k\varrho, \Gamma_k^+]\right), \tag{3.9}$$

where $\gamma_k = \pi g_k^2 D(\Omega_k)/\hbar$ and $D(\Omega_k)$ is the density of states of the reservoir.

The formal conditions mentioned in section 2 are satisfied by (3.9) in the general form given. We restrict the coefficients by requiring the angular momentum component of (3.1) to be valid on average, i.e. $\langle p'\rangle = \lambda\langle p\rangle$. If we assume the reservoir to absorb only single momentum quanta and not to emit into positive or absorb from negative momentum states, and require symmetry of the interaction with respect to the origin of the momentum axis, we obtain

$$\langle l|\dot{\varrho}|m\rangle = |\ln\lambda| \Big[\Theta_{l,0}\Theta_{m,0}((l+1)(m+1))^{1/2}\langle l+1|\varrho|m+1\rangle + \\ +\Theta_{0,l}\Theta_{0,m}((l-1)(m-1))^{1/2}\langle l-1|\varrho|m-1\rangle - \tfrac{1}{2}(|l|+|m|)\langle l|\varrho|m\rangle\Big] \tag{3.10}$$

as the final form of the master equation. Here we defined $\Theta_{l,m} := \begin{cases} 1 \text{ if } l\geq m \\ 0 \text{ if } l<m \end{cases}$.

The solution of eq. (3.10) is found by standard methods and can be combined by convolution with eq. (3.6). If we choose the order of the components according to the classical map (1.2), such that the damping acts first, then the kick and finally the free propagation, we arrive at the kernel

$$G(l',m'|l,m) = \exp\left(-i\frac{\hbar}{2}(l'^2-m'^2)\right)\lambda^{\frac{1}{2}(|l|+|m|)} \cdot \\ \cdot\Big[b_{l'-l}\left(\tfrac{K}{\hbar}\right)b^*_{m'-m}\left(\tfrac{K}{\hbar}\right) + \\ +\Theta_{l\cdot m,0}\sum_{j=1}^{\min(|l|,|m|)}\left(\binom{|l|}{j}\binom{|m|}{j}\right)^{1/2}\left(\frac{1-\lambda}{\lambda}\right)^j b_{l'-l+\frac{l}{|l|}j}\left(\tfrac{K}{\hbar}\right)b^*_{m'-m+\frac{m}{|m|}j}\left(\tfrac{K}{\hbar}\right). \tag{3.11}$$

Interpreting the index j as the angular momentum loss, we may call the j=0-term the elastic and the sum over j 0 the inelastic contributions to a particular element of the density matrix, where each term, in turn, consists of two Bessel function factors which stem from the conservative component of the map, and a combinatorial weight factor which is contributed by the mechanism of dissipation.

4. LIMITING CASES

The limits of vanishing and of infinite dissipation are readily obtained from (3.11):

In the conservative limit $\lambda=1$, the kicked quantum rotator, as introduced by Casati et.al. [3] is recovered

$$G(l',m'|l,m) = \exp\left(-i\frac{\hbar}{2}(l'^2-m'^2)\right)b_{l'-l}\left(\tfrac{K}{\hbar}\right)b^*_{m'-m}\left(\tfrac{K}{\hbar}\right) \tag{4.1}$$

In the dissipative limit $\lambda=0$, the system relaxes to a steady state already after a single time-step, so that any dependence on the initial state is lost:

$$G(l',m'|l,m) = \exp\left(-i\frac{\hbar}{2}(l'^2-m'^2)\right)\delta_{|l|,|m|}\, b_{l'}\left(\tfrac{K}{\hbar}\right)b^*_{m'}\left(\tfrac{K}{\hbar}\right), \\ \langle l|\varrho|m\rangle = \exp\left(-i\frac{\hbar}{2}(l'^2-m'^2)\right)b_l\left(\tfrac{K}{\hbar}\right)b^*_m\left(\tfrac{K}{\hbar}\right). \tag{4.2}$$

Let us now consider the limiting case $\hbar \to 0$. In this limit the angular momentum can be considered as a continuous classical variable, if we provide for some smoothing of this variable on scales of order $O(\hbar)$. In the following we are interested, not only in the classical limit itself, but also in the leading correction to it, for the case that $\lambda \gg O(\hbar)$ and $(1-\lambda) \gg O(\hbar)$. Therefore, we assume that $\hbar$ is small but finite and introduce approximations in (3.11) which are based on this small parameter. In particular the binomial distributions in (3.11) are replaced by Gaussians.

Transforming to the propagator of the Wigner function by means of (2.4) we obtain

$$\begin{aligned} G_W(l',\varphi'|l,\varphi) = \tfrac{1}{2}\Big[& Ga^{(\mathrm{mod}\,2\pi)}\big(\varphi'-\varphi-\tfrac{\hbar}{2}l', \tfrac{1-\lambda}{2\lambda|l|}\big) + \\ & + (-1)^{l}\, Ga^{(\mathrm{mod}\,2\pi)}\big(\varphi'-\varphi-\tfrac{\hbar}{2}l'-\pi, \tfrac{1-\lambda}{2\lambda|l|}\big)\Big] \cdot \\ & \cdot \Big[Ga\big(\tfrac{1}{2}l'-\tfrac{1}{2}\lambda l+\tfrac{K}{\hbar}\sin(\varphi'-\tfrac{\hbar}{2}l'), \tfrac{1}{2}\lambda(1-\lambda)|l|\big) \\ & + (-1)^{l+l'}\, Ga\big(\tfrac{1}{2}l'-\tfrac{1}{2}\lambda l+\tfrac{K}{\hbar}\sin(\varphi'-\tfrac{\hbar}{2}l'-\pi), \tfrac{1}{2}\lambda(1-\lambda)|l|\big)\Big]. \end{aligned} \tag{4.3}$$

(Here $Ga^{(\mathrm{mod}\,2\pi)}(x,\Delta)$ denotes a Gaussian of width Δ convoluted with a periodic delta function of period 2π and $Ga(x,\Delta)$ denotes an ordinary Gaussian.)

The oscillatory dependence of this propagator on l and l' shows that the classical limit is not approached uniformly in the angular momentum but requires some coarse graining. Averaging over two neighbouring values of these momentum indices and replacing them by the classical variables $p=\frac{\hbar}{2}l$, $p'=\frac{\hbar}{2}l'$

$$G_W(p',\varphi'|p,\varphi) = Ga^{(\mathrm{mod}\,2\pi)}\big(\varphi'-\varphi, \hbar\tfrac{1-\lambda}{4\lambda|p|}\big)\, Ga\big(p'-\lambda p+K\sin(\varphi'-p'), \hbar\lambda(1-\lambda)|p|\big). \tag{4.4}$$

This propagator is equivalent to a stochastic map of the form

$$\begin{aligned} p' &= \lambda p - K\sin(\varphi'-p') + \eta' \\ \varphi' &= \varphi + p' + \psi' \end{aligned} \tag{4.5}$$

with the Gaussian stochastic variables η, ψ which satisfy

$$\begin{aligned} &\langle \eta_n \rangle = \langle \psi_n \rangle = 0, \\ &\langle \eta_n \eta_m \rangle = \hbar\lambda(1-\lambda)|p|\,\delta_{n,m}, \\ &\langle \psi_n \psi_m \rangle = \hbar\tfrac{1-\lambda}{4\lambda|p|}\,\delta_{n,m}, \\ &\langle \eta_n \psi_m \rangle = 0. \end{aligned} \tag{4.6}$$

The noise terms in (4.5), though describable in a classical framework as a stochastic process, are of quantum mechanical origin and vanish in the classical limit $\hbar \to 0$. These terms are therefore the leading quantum corrections under the assumption made for . We may consider the stochastic map (4.5), (4.6) again in the limits $\lambda = 1$ and $\lambda = 0$. In the conservative limit the noise terms disappear, i.e. the leading quantum correction is entirely associated with the dissipative mechanism. In the dissipative limit η associated with angular momentum vanishes, while the noise term associated with the phase diverges, in accordance with the uncertainty principle.

Going to the next-to-leading order in the semi-classical limit leads to a replacement of the Gaussians in (4.3) by Airy functions, thus yielding quasi-probability distributions that are negative in some parts of phase space and can no longer be interpreted in classical terms.

REFERENCES

1. R. Shaw, Z. Naturforsch. 36a,80 (1981)
2. M.V. Berry, N.L. Balazs, M. Tabor, A.Voros, Ann. Phys. (N.Y.) 122,26 (1979)
3. G. Casati, B.V. Chirikov, F.M. Izraelev, J. Ford, Lecture Notes in Physics, Vol. 93,334, Springer, Berlin 1979
4. F.M. Izraelev, D.L. Shepelyanski, Teor. Mat. Fiz. 49,117 (1980) (Teor. Math. Phys. 43,553 (1980)
5. H. Hannay, M.V. Berry, Physica 1D,267 (1980)
6. J.S. Hutchinson, R.E.Wyatt, Chem. Phys. Lett. 72,378 (1980)
7. H.J. Korsch, M.V. Berry, Physica 3D,627 (1981)
8. G.M. Zaslavsky, Phys. Rep. 80,157 (1981)
9. T. Hogg, B.V. Huberman, Phys. Rev. Lett. 48,711 (1982)
10. T. Hogg, B.V. Huberman, Phys. Rev. A28,22 (1983)
11. S. Fishman, D.R. Grempel, R.E. Prange, Phys. Rev. Lett. 49,509 (1982)
12. D.L. Shepelyansky, Physica 8D,208 (1983)
13. D.R. Grempel, R.E. Prange, S. Fishman, Phys. Rev. A29,1639 (1984)
14. D.R. Grempel, S. Fishman, R.E. Prange, Phys. Rev. Lett. 53,1212 (1984)
15. G. Casati, I. Guarneri, Comm. Math. Phys. 95,121 (1984)
16. E. Ott, T.M. Antonsen, J.D. Hanson, Phys. Rev. Lett. 53,2187 (1984)
17. R. Graham, Z. Phys. B59,75 (1985)
18. R. Graham, T. Tél, Z. Phys. B60,127 (1985)
19. B.V. Chirikov, Phys. Rep. 52,263 (1979)
20. G.M. Zaslavski, Phys. Lett. 69A,145 (1978); G.M. Zaslavski, Kh.-R.Ya. Rachko, Zh. Eksp. Teor. Fiz. 76,2052 (1979) (Sov. Phys. JETP 49,1039 (1979))

PART IV

QUANTUM MECHANICAL CONTROL SYSTEMS

HAMILTONIAN AND QUANTUM MECHANICAL CONTROL SYSTEMS

Aryan J. van der Schaft
Twente University of Technology, Enschede, the Netherlands

Abstract

After a brief introduction to the notions of controllability and observability for classical control systems, these notions are elaborated for Hamiltonian control systems. It is shown that to every Hamiltonian control system there corresponds a Lie algebra of functions under the Poisson bracket, which completely characterizes the Hamiltonian system from an input-output point of view. Special attention is paid to the case that this Lie algebra is finite-dimensional, leading to an application of the theory of coadjoint representations. It is shown how Hamiltonian control systems can be quantized to quantum mechanical control systems, which are themselves Hamiltonian con trol systems on a Hilbert space. Some relations between the classical and quantized control system are being discussed. Finally the question is addressed which control systems are actually Hamiltonian, and so can be quantized. This leads to a generalization of the Helmholtz conditions for the inverse problem in classical mechanics.

0. Introduction

Recently there have been some attempts striving at a control theory for quantum mechanical control systems. For some references see [1,2,3,4,5] and the contributions to these proceedings. In this paper we shall not deal with the problem of quantum mechanical control as such, but elaborate on the connection between classical and quantum mechanical control systems. It is well known that a transition from classical Hamiltonian systems (without controls) to quantum systems (without controls) is provided by the theory of **quantization**, as originally proposed by Dirac. Inspired by [3] we shall argue in section 4 of this paper that such a theory can be also developed for the quantization of a special class of **control** systems, namely the **Hamiltonian control** systems, to quantum mechanical control systems. This is done after a quick review in sections 1, 2 and 3 of some basic structural results concerning classical (Hamiltonian) control systems, which seem to be of primary interest for the quantization procedure. Finally in section 5 we discuss the necessary and sufficient conditions in order that a general control system can be formulated as a Hamiltonian control system, and so can be quantized. These recently derived conditions ([6]) are based upon the notion of **self-adjointness** of a control system, and yield a variational characterization of Hamiltonian control systems.

1. Classical control systems

In this paper we shall restrict ourselves to nonlinear control systems

$$\dot{x} = g_0(x) + \sum_{j=1}^{m} u_j g_j(x) \qquad x(0) = x_0, \qquad x \in M \tag{1.1a}$$

$$y_j = H_j(x) \qquad j = 1,\dots,p \qquad u = (u_1,\dots,u_m) \in \Omega \subset \mathbb{R}^m \tag{1.1b}$$

Here M denotes the n-dimensional **state space** which is assumed to be a connected manifold and $\Omega \subset \mathbb{R}^m$ is the control space which for simplicity is taken to be a subset of $\mathbb{R}^m$ containing 0 in its interior. The time-functions $u_j(t)$, $j = 1,\dots,m$, in the right-hand side of the differential equations (1.1a) are assumed to belong to a suitable class of functions called the **admissible** controls. For instance we can take the locally L^1 functions. At least the piecewise constant functions have to be admissible. We shall throughout assume that all vectorfields $g_0 + \sum_{j=1}^{m} u_j g_j$ for each constant $u \in \Omega$ are **complete.** This means that for every piecewise constant control u the solution of the differential equations (1.1.a) is defined for any $t \in \mathbb{R}$. Finally the H_j, $j = 1,\dots,p$, are functions from the state space M to $\mathbb{R}$. The outputs $y_j = H_j(x)$ represent the part of the state that we can observe or measure, or alternatively the part we are interested in. In order to simplify our considerations we shall assume that all data in (1.1), i.e. $M, g_0, \ldots, g_m, H_1, \ldots, H_p$ are (real)-**analytic**. We remark that without too much difficulties the theory can be extended to analytic control systems where the inputs enter in a nonlinear way, c.f. [7,6]

$$\dot{x} = F(x,u) \qquad x(0) = x_0 \qquad x \in M \tag{1.2a}$$

$$y_j = h_j(x,u) \qquad j = 1,\ldots,p \qquad u = (u_1,\ldots,u_m) \in \Omega \subset \mathbb{R}^m \tag{1.2b}$$

The general starting point of system and control theory is rather different than in the theory of differential equations and dynamical systems. In the latter case one is primarily interested in the behavior of the differential equations (1.1a) or (1.2a) for a **given** time-function u(t), while in control theory one studies the behavior of the system for **all** (admissible) controls, and especially the **input-output behavior** of the system. Formally the input-output map of (1.1) is given as the map from the control functions $u : [0,T] \to \Omega$ to the output functions $y : [0,T] \to \mathbb{R}^p$, where $y_j = H_j(x(t))$ and $x(t) = x(t,x_0,u(\cdot))$ is the solution of 1.1a) for $u(\cdot)$ and T is such that this solution is defined for all $0 \le t \le T$. Furthermore in control theory one looks for mechanisms (mainly of a **feedback** nature) to adjust or even prescribe the total behavior of the system.

Probably the two most fundamental notions in the theory of control systems are **controllability** and **observability** as introduced by Kalman in the early sixties. A system (1.1) is called controllable if for any two points x_0, $x_1 \in M$ there exists a time $T < \infty$ and an admissible control u such that the solution $x(T,x_0,u(\cdot))$ of (1.1a) starting from $x(0) = x_0$ reaches x_1 in time T. It is called observable if for any two different states $x_1 \neq x_2 \in M$ there exists a time $T < \infty$ and an admissible control u such that for some $j \in \{1,\ldots,m\}$ $H_j(x(T,x_1, u(\cdot)) \neq H_j(x(T,x_2,u(\cdot))$, i.e. the output functions are different.

Apart from their interest per se, both properties play a fundamental role in the solution of control and synthesis problems. For nonlinear systems (1.1) both properties can be characterized to a large extent in the following **algebraic**, and therefore "computable" way. Denote by L the **Lie algebra** generated by the vectorfields $g_0, g_1, \ldots, g_m$ under taking Lie brackets of vectorfields. Let L_0 be the **ideal** in L generated by $g_1,\ldots,g_m$. Then the system is called **accessible** if $\dim L(x) = \dim M$ for any $x \in M$, and **strongly accessible** if $\dim L_0(x) = \dim M$ for any $x \in M$, where $L(x) = \text{span}_{\mathbb{R}}\{f(x) | f \in L\}$ and $L_0(x)$ analogously. Strong accessibility means that the set of points which can be reached at any time $T > 0$ from any point $x_0 \in M$ by choosing different controls in (1.1a) contains a non-empty interior, and accessibility means the same for the set of points which can be reached at arbitrary time $t \geq 0$ with $t \leq T$ for any $T > 0$ [8]. Hence both are a weak form of controllability. Furthermore denote by H the linear space of functions on M of the form $f(H_j) = L_f H_j$, with $f \in L$ and $j = 1,\dots,m$. The system is **observable** if H distinguishes points in M, i.e. for every $x_1 \neq x_2$ there exists an $H \in H$ such that $H(x_1 \neq H(x_2)$. H is called the **observation space.** The system is called **weakly observable** if H only distinguishes nearby points in M. If the system is accessible then necessarily $\dim dH(x)$ is constant for all $x \in M$, where $dH(x) = \text{span}_{\mathbb{R}}\{dH(x) | H \in H\}$ and hence in this case the system is weakly observable if and only if $\dim dH(x) = \dim M$ for each $x \in M$([9]).

A system which is accessible and observable is called **minimal**, and if it is accessible and weakly observable **quasi-minimal.** Minimality intuitively means that the state space M can **not** be reduced without changing the input-output map of the system. Roughly speaking, if the system is **not** accessible we may **leave out** the states which cannot be reached from x_0, and if the system is **not** observable we may **identify** the states which can not be distinguished, and this will not yield a different input-output behavior. On the other hand minimal systems are **uniquely determined** by their input-output map. Let

$$\begin{aligned} \dot{x}^1 &= g_0^1(x^1) + \sum_{j=1}^{m} u_j g_j^1(x^1), \quad x^1(0) = x_0^1, \quad x^1 \in M^1 \\ y_j &= H_j^1(x^1) \quad j = 1,\dots,p \end{aligned} \tag{1.3}$$

and

$$\begin{aligned} \dot{x}^2 &= g_0^2(x^2) + \sum_{j=1}^{m} u_j g_j^2(x^2), \quad x^2(0) = x_0^2, \quad x^2 \in M^2 \\ y_j &= H_j^2(x^2) \quad j = 1,\dots,p \end{aligned} \tag{1.4}$$

be two systems with the **same** input-output map. Then if both systems are minimal there exists a unique (analytic) diffeomorphism $\phi : M^1 \to M^2$ with $\phi(x_0^1) = x_0^2$, which transforms the equations (1.3) into (1.4) ([10]) and if both systems are quasi-minimal then ϕ exists locally around x_0^1 and x_0^2 ([11]). This shows that the state of (quasi-) minimal system, is not something which is more or less arbitrary, but which is up to coordinate transformations uniquely determined by its input-output map.

2. Classical Hamiltonian control systems

The definition of a Hamiltonian control system can be traced back to the Euler-Lagrange equations with external forces. Consider a (conservative) mechanical system with n degrees of freedom $q_1,\dots,q_n$ with Lagrangian $L(q_1,\dots,q_n,\dot{q}_1,\dots,\dot{q}_n)$. The dynamics of motion are given by

$$\frac{d}{dt}\left(\frac{\partial L}{\partial \dot{q}_i}\right) - \frac{\partial L}{\partial q_i} = F_i \qquad i = 1,\dots,n \tag{2.1}$$

where $F = (F_1,\dots,F_n)$ is the vector of generalized external forces. These forces can be interpreted in several ways, from external disturbances to controls. We shall only deal with this last case by assuming that some force components are zero and that the others are arbitrary controls. Cast into the Hamiltonian framework we then obtain

$$\begin{aligned} \dot{q}_i &= \frac{\partial H_0}{\partial p_i} && i = 1,\dots,n \\ \dot{p}_i &= -\frac{\partial H_0}{\partial q_i} + u_i && i = 1,\dots,m \\ \dot{p}_i &= -\frac{\partial H_0}{\partial q_i} && i = m+1,\dots,n \end{aligned} \tag{2.2}$$

with $p_i = \frac{\partial L}{\partial \dot{q}_i}$ and H_0 the internal Hamiltonian. In **arbitrary coordinates** x we therefore obtain ([7,12])

$$\dot{x} = X_{H_0}(x) - \sum_{j=1}^{m} u_j X_{H_j}(x),\ x \in M,\ x(0) = x_0 \tag{2.3a}$$

where M is a symplectic manifold with symplectic form ω and X_{H_i}, $i = 0,1,\dots,m$, are the Hamiltonian vectorfields with Hamiltonian functions H_i defined by $\omega(X_{H_i},-) = -dH_i$. By Darboux's theorem there exist coordinates (q,p) (called **canonical**) such that locally

$$\omega = \sum_{i=1}^{m} dp_i \wedge dq_i,$$

and if we take the (interaction) Hamiltonians H_j equal to q_j, $j = 1,\dots,m$, then we just recover (2.2). The **natural outputs** for equations (2.3a) are

$$y_j = H_j(x) \qquad j = 1,\dots,m \tag{2.3b}$$

i.e., given the external excitations $u_1X_{H_1},\dots,u_mX_{H_m}$ the natural outputs are the "displacements" $H_1,\dots,H_m$, caused by these excitations at the same line of action. For completeness we mention that a general analytic control system (1.2) is called Hamiltonian if it is of the form

$$\dot{x} = X_{H(x,u)} \qquad x \in M,\ x(0) = x_0 \tag{2.4a}$$

where the Hamiltonian H(x,u) is an arbitrary (analytic) function of the state x and control parameters u. In case H(x,u) is of the form

$$H_0(x) - \sum_{j=1}^{m} u_jH_j(x)$$

we recover (2.3a). The natural outputs in this case are given as ([7,13])

$$y_j = -\frac{\partial H}{\partial u_j}(x,u) \qquad j = 1,\dots,m \tag{2.4b}$$

Since the state space M of (2.3) (or (2.4)) is a symplectic manifold we can define the **Poisson bracket**

$$\{F,G\} = X_F(G) = \omega(X_F,X_G) \qquad F,G : M \to \mathbb{R} \tag{2.5}$$

which in canonical coordinates is the familiar expression

$$\{F,G\} = \sum_{i=1}^{n} \left(\frac{\partial F}{\partial p_i}\frac{\partial G}{\partial q_i} - \frac{\partial F}{\partial q_i}\frac{\partial G}{\partial p_i}\right) \tag{2.6}$$

Recall that the observation space $\mathcal{H}$ of a nonlinear system (1.1) is spanned by all functions $f(H_j)$, with $f \in L$. Since $X_F(G) = \{F,G\}$ and because of the identity

$$[X_F,X_G] = X_{\{F,G\}} \tag{2.7}$$

we conclude that for a Hamiltonian system (2.3) $\mathcal{H}$ is spanned by the functions

$$\{F_1, \{F_2, \{\cdots \{F_k,H_j\}\cdots\}\}\} \tag{2.8}$$

with F_r, $r = 1,\ldots,k$, equal to H_i, $i = 0,1,\ldots,m$. In particular $\mathcal{H}$ is the **ideal** generated by $H_1,\ldots,H_m$ in the Lie algebra (under Poisson bracket) generated by $H_0,H_1,\ldots,H_m$. Furthermore it follows from this characterization of $\mathcal{H}$ as a Lie algebra and (2.7) that the mapping

$$H \longrightarrow X_H \tag{2.9}$$

is an isomorphism from $\mathcal{H}$ modulo constant functions to the Lie algebra $\mathcal{L}_0$. Hence ([7,12])

Proposition 2.1. Let (2.3) be a Hamiltonian system. Then
a. The system is strongly accessible and weakly observable $\leftrightarrow$ dim $d\mathcal{H}(x)$ = dim M, $\forall x \in M$ $\leftrightarrow$ the system is quasi-minimal.
b. The system is strongly accessible and observable $\leftrightarrow$ dim $d\mathcal{H}(x)$ = dim M, $\forall x \in M$, and $\mathcal{H}$ distinguishes points in M $\leftrightarrow$ the system is minimal.

In section 1 we indicated that non-minimal systems can be reduced to (quasi-)minimal systems with the same input-output map. In the Hamiltonian case this reduced system is again Hamiltonian as can be seen as follows. Let (2.3) be a Hamiltonian system on (M,ω). Assume that $X_H(x_0) \in \mathcal{L}_0(x_0)$ (otherwise we have to take recourse to time-varying Hamiltonian control systems, c.f. [14]). Take the maximal integral manifold N of the Lie algebra $\mathcal{L}_0$ through x_0. This is in a sense the "controllable" part of the system. Then factor out by the kernel of the codistribution $d\mathcal{H}(x)$ (this can always be done globally!), to obtain a manifold $\bar{M}$. Since $\mathcal{L}_0$ and $\mathcal{H}(\mathrm{mod}\mathbb{R})$ are isomorphic by the mapping (2.9) it follows that $\bar{M}$ is a **symplectic** manifold with symplectic form $\bar{\omega}$, and that the system (2.3) on (M,ω) projects to a Hamiltonian system on $(\bar{M},\bar{\omega})$. Furthermore by construction the observation space $\bar{\mathcal{H}}$ of this reduced system satisfies dim $d\bar{\mathcal{H}}(\bar{x})$ = dim $\bar{M}$, for all $\bar{x} \in \bar{M}$, and so the system is quasi-minimal.

As we mentioned in the preceding section two (quasi-)minimal systems with the same input-output map are necessarily (locally) diffeomorphic. In the Hamiltonian case the equivalence mapping is even a symplectomorphism, i.e. locally a canonical transformation, c.f. [7]. It also follows that the internal Hamiltonians (energies) of both systems are equal up to a constant. Hence the input-output map of a (quasi-) minimal Hamiltonian system not only determines the state, but also the canonical structure of the state and the internal energy of the system.

All these results can be extended to general Hamiltonian systems (2.4) ([7]). The main instrument is again the observation space $\mathcal{H}$, which in this case is the linear space of functions of (x,u) containing $\frac{\partial H}{\partial u_1},\ldots,\frac{\partial H}{\partial u_m}$ and invariant w.r.t. Poisson brackets with H(x,u) and differentiations to $u_1,\ldots,u_m$, and which is again a Lie algebra under Poisson bracket ([7]).

Apart from their controllability and observability duality Hamiltonian control systems have many other nice features (c.f. [7]). In section 4 we shall show that Hamiltonian control systems may be **quantized** to quantum mechanical control systems, which are themselves Hamiltonian control systems on some complex Hilbert space. It can therefore be expected that the theory of classical Hamiltonian control systems plays the same role w.r.t.quantum mechanical control systems, as Hamiltonian differential equations w.r.t. quantum systems.

3. Finite-dimensional observation spaces

In the preceding section we concluded that (quasi-)minimal Hamiltonian systems are up to canonical transformations (locally) completely determined by the input-output map for a certain ground state x_0. On the other hand this input-output map (also in the non-minimal case) is completely determined by the observation space H (even by the **values** of the functions in H in the point x_0). This stresses the importance of H, and motivates the study of the Lie algebraic structure of it. In general H is an infinite-dimensional Lie algebra, and so is hard to study. We shall now consider the special case that H is **finite-dimensional.** So let

$$\dot{x} = X_{H_0}(x) - \sum_{j=1}^{m} u_j X_{H_j}(x),\ x(0) = x_0,\ x \in (M^{2n},\omega) \tag{3.1}$$

$$y_j = H_j(x) \qquad j = 1,\dots,m$$

be a Hamiltonian system, and assume that H, which is the ideal generated by $H_1,\dots,H_m$ within the Lie algebra generated by $H_0,H_1,\dots,H_m$ is an N-dimensional Lie algebra. Let $\phi_1,\dots,\phi_N$ be a basis of H, then there exist constants $c_{ijk} \in \mathbb{R}$ such that

$$\{\phi_i,\phi_j\} = \sum_{k=1}^{N} c_{ijk}\phi_k \tag{3.2}$$

Furthermore c_{ijk} are the structure coefficients of the Lie algebra H and so satisfy the relations

$$c_{ijk} = -c_{jik},\quad i,j,k = 1,\dots,N$$

$$\sum_{s=1}^{N} (c_{ijr}c_{rks} + c_{jks}c_{ris} + c_{kir}c_{rjs}) = 0 \tag{3.3}$$

We can see the elements $\phi_1,\ldots,\phi_N \in H$ not only as (nonlinear) functions on M, but also as linear coordinate functions on the dual Lie algebra $H^* \simeq \mathbb{R}^N$ (since $\phi_1,\ldots,\phi_N$ is a basis of H). The trick will now be to write (3.1) as a subsystem of a **Poisson system** on the state space H^*.
A Poisson system is defined in the same way as a Hamiltonian system, except for the fact that the state space need not be a symplectic manifold but a more general object, namely a **Poisson manifold.** Let M be an N-dimensional manifold. A **Poisson structure** on M is a bracket operation

$$\{F,G\} : M \to \mathbb{R} \qquad \text{for } F,G : M \to \mathbb{R} \tag{3.4}$$

satisfying

$$\{F,G\} = -\{G,F\} \qquad \forall F,G,H : M \to \mathbb{R} \tag{3.5}$$

$$\{F,\{G,H\}\} + \{G,\{H,F\}\} + \{H,\{F,G\}\} = 0$$

M together with its Poisson structure is called a **Poisson manifold.** Given a function $H : M \to \mathbb{R}$ we define the vectorfield X_H on $(M,\{,\})$ as the unique vectorfield satisfying

$$X_H(F) = \{H,F\} \qquad \text{for all } F : M \to \mathbb{R} \tag{3.6}$$

Hence a Poisson structure defines a linear mapping from T_x^*M to T_xM given by $dH(x) \longrightarrow X_H(x)$. The rank of this mapping is called the **rank** of the Poisson structure in $x \in M$. We note that if M is a **symplectic** manifold then the Poisson bracket $\{F,G\} = \omega(X_F,X_G)$, with ω the symplectic form, satisfies (3.5) and so defines a Poisson structure. Furthermore the rank of this Poisson structure is everywhere maximal (= dim M). Conversely if $\{\ ,\ \}$ is a Poisson structure with rank equal to dim M, then there exists a symplectic form on M such that $\{\ ,\ \}$ equals the Poisson bracket.
A **Poisson control system** on a Poisson manifold $(M,\{\ ,\ \})$ is of the form

$$\dot{x} = X_{H_0}(x) - \sum_{j=1}^{m} u_j X_{H_j}(x) \qquad x(0) = x_0,\ x \in M \tag{3.7}$$

$$y_j = H_j(x), \quad j = 1,\ldots,m$$

(Poisson control systems were first used in [15].) If the rank of $\{\ ,\ \}$ is everywhere equal to dim M, then (3.7) is just a Hamiltonian system (2.3). Also for Poisson structures we have the equality (c.f. (2.7))

$$[X_F, X_G] = X_{\{F,G\}} \tag{3.8}$$

Hence the observation space H and the strong accessibility Lie algebra L_0 of (3.7) are homomorphic Lie algebras under the mapping $F \longrightarrow X_F$. However in general we have $L_0(x) \leq \dim dH(x)$ for any $x \in M$, since $\dim L_0(x)$ is always bounded by the rank of the Poisson structure in x. In particular if this rank is less than dim M then the system is never strongly accessible, while it may be (weakly) observable. A Poisson structure $\{\ ,\ \}$ on a vector space V is called **linear** if the bracket of any two linear functions on V is again linear. Let $x_1,\dots,x_N$ be linear coordinate functions on V then this implies that

$$\{x_i, x_j\} = \sum_{k=1}^{N} c_{ijk} x_k \tag{3.9}$$

for certain constants $c_{ijk} \in \mathbb{R}$. Furthermore using the conditions (3.5) it follows that these constants satisfy (3.3) and so are the structure coefficients of a Lie algebra. Since $x_1,\dots,x_N$ can be seen as a basis of V^* this makes V^* into a Lie algebra. In our case we go the other way around. We **start** with a Lie algebra H with basis $\phi_1,\dots,\phi_N$. Viewing $\phi_1,\dots,\phi_N$ as linear coordinate functions on the dual Lie algebra H^* it follows that H^* is a Poisson manifold with linear Poisson structure given by (3.2).

Since $\phi_1,\dots,\phi_N$ are not only coordinates for H^* but also functions on the state space M of the Hamiltonian system (3.1) we can calculate their time-evolution as

$$\frac{d\phi_i}{dt} = \{H_0, \phi_i\} - \sum_{j=1}^{m} u_j \{H_j, \phi_i\} \tag{3.10}$$

$$= \sum_{k=1}^{N} a_{ik}\phi_k - \sum_{k=1}^{m} u_j \sum_{k=1}^{N} b^j_{ik}\phi_k, \qquad i = 1,\dots,N$$

for certain constants $a_{ik}, b^j_{ik} \in \mathbb{R}$. Furthermore since $H_1,\dots,H_m \in H$ we have for certain constants $h^j_k \in \mathbb{R}$

$$y_j = \sum_{k=1}^{N} h^j_k \phi_k \qquad j = 1,\dots,m \tag{3.11}$$

Interpreting ϕ_i as linear coordinates z_i on H^* this yields the following **bilinear** system on H^*

$$\dot{z}_i = \sum_{k=1}^{N} a_{ik} z_k - \sum_{j=1}^{m} u_j \sum_{k=1}^{N} b^j_{ik} z_k, \quad z_i(0) = \phi_i(x_0), \; i = 1,\dots,N$$

$$y_j = \sum_{k=1}^{N} h^j_k \phi_k, \qquad j = 1,\dots,m, \; (z_1,\dots,z_n) \in H^* \tag{3.12}$$

The idea of writing an arbitrary (not necessarily Hamiltonian) control system with finite-dimensional observation space H as a bilinear control system on H^* is due to Hijab [16] and Fliess & Kupka [17]. In this last reference it is shown that the input-output map of the bilinear system (3.12) **equals** the input-output map of the original control system (3.1), and so we can interpret (3.1) as a **subsystem** of (3.12). In our case, H^* is not an arbitrary vector space but a **Poisson** manifold. Furthermore the natural mapping from M to H^* (i.e. $x \to (\phi_1(x),\dots,\phi_{N'}(x))$) is a **Poisson mapping** ([18]) and hence (3.12) is a **Poisson system.** Poisson control systems may be reduced to quasi-minimal control systems in the same way as we did for Hamiltonian systems. As a matter of fact the quasi-minimal system will always be a Poisson system with maximal rank of the Poisson structure, so a Hamiltonian system. In our case the Poisson system (3.12) is automatically **observable**, since $\phi_1,\dots,\phi_N \in H$ are a coordinate system on H^*. So the reduction procedure consists of taking the maximal integral manifold $\bar{M}$ of the strong accessibility algebra L_0 of (3.12) through the point $z_0 = (\phi_1(x_0),\dots,\phi_N(x_0)) \in H^*$. Since L_0 is the image of H under the mapping $F \to X_F$ on H^* it follows that $\bar{M}$ is exactly the symplectic leaf of the linear Poisson structure on H^* through z_0. Said in another way, $\bar{M}$ is the coadjoint orbit through z_0 of the Lie algebra H acting on H^* via its coadjoint representation (c.f. [18]). Summarizing

Theorem 3.1 Let (3.1) be a Hamiltonian system with finite-dimensional observation space H. Then there exists a **minimal** Hamiltonian system with the same input-output map, which has as state space the coadjoint orbit through $z_0 = z(0) = (\phi_1(x_0),\dots,\phi_N(x_0))$ in the dual Lie algebra H^*.

Remark The **dimension** of the coadjoint orbit through z_0 is equal to the rank of the Poisson structure in z_0.

Now let us assume that (3.1) is quasi-minimal. It is known ([19]) that the space of a quasi-minimal system is a **covering space** of the state space of a minimal system with the same imput-output map. Hence Theorem 3.1 in this case implies that M is a covering space of a coadjoint orbit in H^*. This was already proved by Basto Goncalves ([14]) for the case that H is isomorphic to L_0 (in general H **modulo** the constant functions is isomorphic to L_0), and hence there exists a **Poisson action** of the Lie group associated to $H \simeq L_0$ on M. In this case Theorem 3.1 reduces to the famous Kostant-Kirillov-Souriau theorem.

As a special case of Theorem 3.1 let us consider the case that H is finite-dimensional and for any two elements $F_1, F_2 \in H$ the Poisson bracket $\{F_1, F_2\}$ is **constant.** If $\{F_1, F_2\} = 0$ for all $F_1, F_2 \in H$ then clearly a minimal state space is the empty one, while in all other cases we may assume that the function 1 belongs to H and that H has a basis $\phi_1, \ldots, \phi_{2k}, 1$ such that

$$\{\phi_i, \phi_j\} = c_{ij} \in \mathbb{R} \qquad i,j = 1,\ldots,2k \qquad (3.13)$$
$$\text{rank } (c_{ij}) = 2l, \qquad \text{with } l \leq k$$

Furthermore ([20]) it follows that ϕ_i, $i = 1,\ldots,2k$, is either a linear combination of the output functions H_j, or the Poisson bracket with H_0 of such a linear combination. Coadjoint orbits in $H^* = \mathbb{R}^{2k+1}$ are simply $2l$-dimensional affine subspaces. It follows ([20]) that a minimal reduced Hamiltonian system on such a state space is such that the internal Hamiltonian H_0 is quadratic-linear, while the interaction Hamiltonians H_j, $j = 1,\ldots,m$ are linear-constant.

4. Quantization

Classical Hamiltonian control systems are not only of interest per se, but they also form a natural starting point for obtaining **quantum mechanical control systems.** Just like one may try to quantize a Hamiltonian vectorfield $\dot{x} = X_H(x)$, one may quantize a Hamiltonian control system

$$\dot{x} = X_{H_0}(x) - \sum_{j=1}^{m} u_j X_{H_j}(x) \qquad , x \in M \qquad (4.1)$$
$$y_j = H_j(x)$$

or the more general form

$$\dot{x} = X_{H(x,u)} \qquad ,x \in M \tag{4.2}$$

$$y_j = -\frac{\partial H}{\partial u_j}(x,u)$$

into a quantum mechanical system controlled by the (macroscopic) controls $u_1,\ldots,u_m$. The difference is that instead of quantizing a single Hamiltonian H one now has to quantize a **set** of Hamiltonians $H_0,H_1,\ldots,H_m$, or, in the general case, a Hamiltonian H(x,u) depending on the control parameters u. On the other hand in the usual quantization procedure for a Hamiltonian H one starts not by quantizing H, but by quantizing a "complete" set of physical quantities, normally the generalized positions and momenta $q_1,\ldots,q_n,p_1,\ldots,p_n$, and then expressing H into these quantities. Recall that quantization of a set of functions on M is the assignment to every f in this set of a Hermitian operator $\hat{f}$, acting on some complex Hilbert space. Furthermore these operators should satisfy some commutation relations, corresponding to the Poisson bracket relations of the classical functions. If $M = \mathbb{R}^{2n}$ and we choose to quantize the canonical coordinates $q_1,\ldots,q_n,p_1,\ldots,p_n$ then the commutation relations are

$$[\hat{q}_i,\hat{q}_j] = 0,\ [\hat{p}_i,\hat{p}_j] = 0,\ [\hat{q}_i,\hat{p}_j] = i\hbar\delta_{ij}I \tag{4.3}$$

From now on we shall normalize $\hbar$ to 1. Note that to every Hermitian operator $\hat{f}$ there corresponds the skew-Hermitian operator $-i\hat{f}$. Hence (4.3) means that the mapping $f \to -i\hat{f}$ is (up to a factor i) a Lie algebra morphism from the classical functions $q_1,\ldots,q_n,p_1,\ldots,p_n,1$ under Poisson bracket to a Lie subalgebra (under the commutator) of the skew-Hermitian operators. (As a matter of fact a Poisson bracket $\{f,g\}$ is mapped onto $-i[\hat{f},\hat{g}] = i[-i\hat{f},-i\hat{g}]$.)
The usual quantization scheme for $q_1,\ldots,q_n,p_1,\ldots,p_n$ satisfying (4.3) is to take the Hilbert space to be $L^2(\mathbb{R}^n,\mathbb{C})$, and to assign to q_j the operator $\hat{q}_j = q_j$. (Multiplication by q_j) and to p_j the operator $\hat{p}_j = -i\frac{\partial}{\partial q_j}$. This same quantization scheme may be also used for the quantization of some simple Hamiltonian control systems. For example, consider the harmonic oscillator with external force u. The Hamiltonian is given by $H(q,p,u) = \frac{p^2}{2m} + \frac{1}{2}kq^2 - uq$. Quantization yields

$$i\frac{\partial\phi}{\partial t} = -\frac{1}{2m}\frac{\partial^2\phi}{\partial q^2} + \frac{1}{2}kq^2\phi - uq\phi \qquad ,\phi \in L^2(R,C) \tag{4.4}$$

which describes a particle in an oscillator well, subject to a uniform classical external field, whose strength and direction is an arbitrary function of time u(•). This example is also given in [3], which is one of the first papers that deals with quantization of control systems (see also [1,2]). In general if the Hamiltonians $H_0, H_1, \ldots, H_m$ in (4.1) can be quantized to Hermitian operators on some complex Hilbert space H then the quantized system in Schrödinger representation is

$$i\frac{\partial\phi}{\partial t} = \hat{H}_0\phi - \sum_{j=1}^{m} u_j\hat{H}_j\phi \qquad \phi \in H \tag{4.5}$$

The operator $\hat{H}_0$ is the Hamiltonian of the quantum system in isolation, while the second term of (4.5) represents a coupling to m external fields of strength $u_j(t)$, $j = 1,\ldots,m$, through some system observables $\hat{H}_1,\ldots,\hat{H}_m$.

We remark that for a complex Hilbert space H the imaginary part of the Hermitian inner product < , > defines a symplectic form on H. With respect to this symplectic form the skew-adjoint operators $-i\hat{H}_j$ are just the (linear) Hamiltonian vectorfields on H corresponding to the (quadratic) Hamiltonians

$$\langle\hat{H}_j\phi,\phi\rangle = \langle\phi|\hat{H}_j|\phi\rangle \qquad j = 0,1,\ldots,m \tag{4.6}$$

i.e. the **expected values** of the observables $\hat{H}_j$. Therefore (4.5) yields a **Hamiltonian system** on the infinite-dimensional state space H

$$\frac{\partial\phi}{\partial t} = -i\hat{H}_0\phi - \sum_{j=1}^{m} u_j(-i\hat{H}_j)\phi \tag{4.7}$$

$$y_j = \langle\phi|\hat{H}_j|\phi\rangle \qquad j = 1,\ldots,m$$

with macroscopic controls $u_1,\ldots,u_m$ and the outputs $y_1,\ldots,y_m$ equal to the expected (average) values of the observables $\hat{H}_1,\ldots,\hat{H}_m$ (c.f. [7]). Note that since y_j equals the expected value of $\hat{H}_j$ there is no measurement problem. Furthermore, although the act of measurement of $\hat{H}_j$ may introduce disturbances into the system, these disturbances are along the same "channels" as the corresponding input u_j. Because H is a symplectic space the Poisson bracket of the expected values of two observables $\hat{H}_1$ and $\hat{H}_2$ is defined as

$$\{\langle\phi|\hat{H}_1|\phi\rangle\ ,\ \langle\phi|\hat{H}_2|\phi\rangle\} = -\ \langle\phi|[\hat{H}_1,\hat{H}_2]|\phi\rangle \tag{4.8}$$

(where the minus sign comes from the fact that we have to consider the commutator of $-i\hat{H}_1$ and $-i\hat{H}_2$). Hence the observation space of the quantum mechanical control system (4.7) is given as the ideal generated by the expected values of $\hat{H}_1,\dots,\hat{H}_m$ within the Lie algebra generated by the expected values of $\hat{H}_0,\dots,\hat{H}_m$ under this Poisson bracket. It seems of interest to compare this observation space with the observation space of the original system (4.1). In case the Hamiltonians $H_0,H_1,\dots,H_m$ are all quadratic-linear it follows from Ehrenfest's theorem that both observation spaces are equal (since the expected values satisfy the classical equations). Consequently in this case the quantum mechanical control system can never be minimal. Finally let us remark that in order to quantize (4.1) it could be of interest **not** to start with the quantization of $q_1,\dots,q_n,p_1,\dots,p_n$ but to quantize a basis of the observation space of (4.1). For the harmonic oscillator with external force this makes no difference, since the observation space is spanned by q,p and 1. It would be interesting to study the quantization of other finite-dimensional observation spaces as treated in section 3.

Remark In the literature there have been various attempts to quantize Hamiltonian systems with external, non-potential, forces. The above approach however is rather different since we do **not** quantize a Hamiltonian control system for a **given** input (force) function as is usually done, but we quantize the **full** control system. The paper certainly closest in spirit is [3], which was actually a major motivation for writing this section.

5. Variational characterization of Hamiltonian control systems

In the preceding section we showed how classical Hamiltonian control systems may be quantized to quantum mechanical control systems. This raises the question which classical control systems actually can be formulated as **Hamiltonian** control systems and so may be quantized. More precisely, let

$$\dot{x} = g_0(x) + \sum_{j=1}^{m} u_j g_j(x), \qquad x \in M, \qquad x(0) = x_0 \tag{5.1}$$

$$y_j = H_j(x) \qquad\qquad j = 1,\dots,m$$

be a classical control system. The question is, when does there exist a **symplectic form** ω on M such that the vectorfields g_j equal the Hamiltonian vectorfields with Hamiltonian $-H_j$, $j = 1,\dots,m$, and g_0 is a locally Hamiltonian vectorfield with internal Hamiltonian H_0, resulting in the Hamiltonian system

$$\dot{x} = X_{H_0}(x) - \sum_{j=1}^{m} u_j X_{H_j}(x) \qquad x \in (M,\omega), \qquad x(0) = x_0 \tag{5.2}$$

$$y_j = H_j(x)$$

This problem is closely related to the so-called Inverse Problem in Mechanics, which is intensively studied in the literature (e.g. [21]), and can be stated as follows: Given a set of second-order differential equations $R_i(q,\dot{q},\ddot{q}) = 0$, $i = 1,\dots,n$, $q \in \mathbb{R}^n$, when does there exist a Lagrangian $L(q,\dot{q})$ such that

$$R_i(q,\dot{q},\ddot{q}) = \frac{d}{dt}\left(\frac{\partial L}{\partial \dot{q}_i}\right) - \frac{\partial L}{\partial q_i} \tag{5.3}$$

i.e. the equations equal a set of Euler-Lagrange equations. In [22,7] it is shown how this problem can be naturally interpreted as representing the equations

$$R_i(q,\dot{q},\ddot{q}) = u_i \qquad i = 1,\dots,n \tag{5.4}$$

with u_i **external forces** as Euler-Lagrange equations with external forces

$$\frac{d}{dt}\left(\frac{\partial L}{\partial \dot{q}_i}\right) - \frac{\partial L}{\partial q_i} = u_i \qquad i = 1,\dots,n \tag{5.5}$$

In section 2 we already saw that (5.5) yields the Hamiltonian control system

$$\dot{q}_i = \frac{\partial H}{\partial p_i} \qquad i = 1,\dots,n$$

$$\dot{p}_i = -\frac{\partial H}{\partial q_i} + u_i \tag{5.6}$$

$$y_i = q_i$$

with $p_i = \frac{\partial L}{\partial \dot{q}_i}$ and $H(q,p)$ the Legendre transform of L. It already follows from (5.3) that R_i has to be linear in $\ddot{q}$, i.e. we must have

$$R_i(q,\dot{q},\ddot{q}) = \sum_{k=1}^{n} A_{ik}(q,\dot{q})\ddot{q}_k + B_i(q,\dot{q}) \qquad i = 1,\dots,n \tag{5.7}$$

Usually it is also assumed that the matrix with elements $\frac{\partial R_i}{\partial \ddot{q}_j} = A_{ij}(q,\dot{q})$ is non-singular. Therefore in our framework the Inverse Problem is to decide whether the control system

$$
\begin{aligned}
\dot{q}_i &= \dot{q}_i \\
\ddot{q}_i &= -\sum_{k=1}^{n} A^{ik}(q,\dot{q})B_k(q,\dot{q}) + \sum_{k=1}^{n} A^{ik}(q,\dot{q})u_k \qquad i = 1,\dots,n \\
y_i &= q_i
\end{aligned}
\tag{5.8}
$$

with (A^{ik}) the inverse matrix of (A_{ik}) is a Hamiltonian control system (5.6). The basic concept to derive necessary and sufficient conditions for the solvability of the Inverse Problem (the Helmholtz conditions, c.f. [21]) is the notion of self-adjointness of a set of (higher order) differential equantions. Following [6] we shall show how this concept can be extended to general control systems, and so derive necessary and sufficient conditions in order that a control system is Hamiltonian.
Let $u(t) = (u_1(t),\dots,u_m(t))$ be an admissible control for (5.1), giving rise to a solution $x(t)$, $x(0) = x_0$ of the system, and an output $y(t)$ with $y_j(t) = H_j(x(t))$. Then along such a state-input-output trajectory $(x(t),u(t),y(t))$ we define the **variational system** of (5.1) as

$$
\begin{aligned}
\dot{v}(t) &= \frac{\partial g_0}{\partial x}\big(x(t)\big)v(t) + \sum_{j=1}^{m} u_j(t)\frac{\partial g_j}{\partial x}\big(x(t)\big)v(t) \\
&\quad + \sum_{j=1}^{m} u_j^v g_j\big(x(t)\big) \qquad v(0) = v_0 \in \mathbb{R}^k,\ k = \dim M \\
y_j^v(t) &= \frac{\partial H_j}{\partial x}\big(x(t)\big)v(t) \qquad j = 1,\dots,m
\end{aligned}
\tag{5.9}
$$

with variational state $v(t) \in \mathbb{R}^k$, variational inputs $u^v = (u_1^v,\dots,u_m^v)$ and variational outputs $y^v = (y_1^v,\dots,y_m^v)$. This system is called variational because of the following. Let $\big(x(t,\varepsilon),u(t,\varepsilon),y(t,\varepsilon)\big)$, $t \in [0,T]$ be a **family** of state-input-output trajectories of (5.1), parameterized by ε, such that $x(t,0) = x(t)$, $u(t,0) = u(t)$ and $y(t,0) = y(t)$, $t \in [0,T]$. Then the quantities

$$v(t) = \frac{\partial x(t,0)}{\partial \varepsilon} \qquad (= \delta x(t))$$

$$u^v(t) = \frac{\partial u(t,0)}{\partial \varepsilon} \qquad (= \delta u(t)) \tag{5.10}$$

$$y^v(t) = \frac{\partial y(t,0)}{\partial \varepsilon} \qquad (= \delta y(t))$$

satisfy (5.9). Furthermore we define the **adjoint system** along the same trajectory (x(t), u(t), y(t)) as

$$-\dot{p}(t) = \left(\frac{\partial g_0}{\partial x}\right)^T(x(t)) + \sum_{j=1}^{m} u_j(t)\left(\frac{\partial g_j}{\partial x}\right)^T(x(t))p(t)$$

$$+ \sum_{j=1}^{n} u_j^a\left(\frac{\partial H_j}{\partial x}\right)^T(x(t)) \qquad p(0) = p_0 \in R^k \tag{5.11}$$

$$y_j(t) = p^T(t)g_j(x(t)) \qquad j = 1,\dots,m$$

with state $p(t) \in R^k$, inputs $u^a = (u_1^a,\dots,u_m^a)$ and outputs $y^a = (y_1^a,\dots,y_m^a)$. The adjoint system is characterized by the variational system since it is the **unique** system such that

$$\frac{d}{dt}\, p^T(t)v(t) = (u^v(t))^T y^a(t) - (u^a(t))^T y^v(t) \tag{5.12}$$

for any $u^v(t)$ and $u^a(t)$. Both the variational and adjoint system are time-varying **linear** systems. We now say that the system (5.1) is **self-adjoint** if along any state-input-output trajectory the input-output map of the variational system for $v_0 = 0$ equals the input-output map of the adjoint system for $p_0 = 0$, i.e. if $u^a(t)$ equals the variational input $u^v(t)$, **then** the output function $y^a(t)$ should equal the variational output $y^v(t)$.

Theorem 5.1 [6] Let (5.1) be a strongly accessible and observable control system. Then it is a minimal Hamiltonian system if and only if the system is self-adjoint.

We note that the conditions for self-adjointness of (5.1) can be explicitly expressed into conditions on $g_0, g_1, \dots, g_m$ and $H_1, \dots, H_m$. As a matter of fact, for equations (5.8) we exactly recover the Helmholtz conditions. (Note that (5.8) is almost automatically strongly accessible

and observable.)
Self-adjointness can be also expressed in the following appealing way. Let $u(t)$ be an admissible control. A variational input $\delta u(t) = u^v(t)$ with support within $[0,T]$ such that the resulting variational output $\delta y(t) = y^v(t)$ also has support within $[0,T]$ is called **admissible** if for any other control function $u_1(t)$ such that $u(t) = u_1(t)$, $t \in [0,T]$, $\delta y(t)$ still has support within $[0,T]$.

Theorem 5.2 [6] A strongly accessible and observable system (5.1) is self-adjoint (and hence Hamiltonian) if for any piecewise constant input u and for any two admissible piecewise constant variational inputs $\delta_1 u(t)$, $\delta_2 u(t)$ with compact support $[0,T]$ such that the variational outputs $\delta_1 y(t)$, $\delta_2 y(t)$ also have support within $[0,T]$, the following identity holds

$$\int_0^\infty [\delta_1^T u(t)\delta_2 y(t) - \delta_2^T u(t)\delta_1 y(t)]dt = 0 \tag{5.12}$$

This last expression can be considered as the dynamical generalization of the symmetry condition for a **static** input-output system $y_j = G_j(u)$, $j = 1,\dots,m$ to be a static Hamiltonian system given by a potential V, i.e. $y_j = \frac{\partial V}{\partial u_j}(u)$, $j = 1,\dots,m$. The (reciprocity) condition in this case is that for any two infinitesimal inputs $\delta_1 u$, $\delta_2 u$ in a point u, the corresponding infinitesimal outputs $\delta_1 y$, $\delta_2 y$ satisfy

$$\delta_1^T u\ \delta_2 y - \delta_2^T u\ \delta_1 y = 0 \tag{5.13}$$

REFERENCES

1. A.G. Butkovskii and Yu.I. Samoilenko: Control of quantum systems, Automat. Remote Cont. 40 (1979), 485-502.
2. A.G. Butkovskii and Yu.I. Samoilenko: Control of quantum systems II Automat. Remote Cont. 40 (1979), 629-645.
3. T.J. Tarn, G. Huang and J.W. Clark: Modelling of quantum mechanical control systems, Math. Modelling, 1 (1980), 109-121.
4. J.W. Clark, Ong, T.J. Tarn, G. Huang: Quantum Nondemolition filters, Math. Syst. Theory, 18 (1985), 33-55.
5. V.P. Belavkin: Optimal linear randomized filtration of quantum boson signals, Probl. of Control and Inf. Theory, 3 (1974), 47-62.

6. P.E. Crouch and A.J. van der Schaft: Variational characterization of Hamiltonian Systems, to appear.
7. A.J. van der Schaft: System theoretic descriptions of physical systems, CWI Tract 3, CWI, Amsterdam 1984.
8. H.J. Sussmann and V. Jurdjevic: Controllability of nonlinear systems, Journ. Diff. Eqns. 12 (1972), 95-116.
9. R. Hermann and A.J. Krener: Nonlinear controllability and observability, IEEE Trans. Aut. Cont., AC-22 (1977), 728-740.
10. H.J. Sussmann: Existence and uniqueness of minimal realizations of nonlinear systems, Math. Syst. Theory, 10 (1977), 263-284.
11. J. Basto Goncalves: Nonlinear observability and duality, Syst. and Contr. Letters, 4 (1984), 97-101.
12. A.J. van der Schaft: Controllability and observability for affine nonlinear Hamiltonian systems, IEEE Trans. Aut. Cont. AC-27 (1982), 490-492.
13. R.W. Brockett: Control theory and analytical mechanics, in: Geometric control theory (Eds. C. Martin and R. Hermann), Vol VII of Lie Groups: History, Frontiers and Applications, Math. Sci. Press, Brookline 1977.
14. J. Basto Goncalves: Realization theory for Hamiltonian systems, to appear in SIAM Journ. Cont. and Opt. 1986.
15. P.S.Krishnaprasad: Lie-Poisson structures and Dual-Spin Spacecraft, Report. Dept. of Elect. Eng, Univ. of Maryland, 1983.
16. O.B. Hijab: Minimum energy estimation, Ph. D. thesis, Berkeley, 1980.
17. M. Fliess and I. Kupka: A finiteness criterion for nonlinear input-output differential systems, SIAM J. Contr. and Opt. 21 (1983), 721-728.
18. A. Weinstein: The local structure of Poisson manifolds, J. Diff. Geom, 18 (1983), 523-557.
19. P.E. Crouch: Dynamical realizations of finite Volterra series, SIAM J. Contr. and Opt., 19 (1981), 177-202.
20. A.J. van der Schaft: Linearization of Hamiltonian and gradient systems, IMA J. Math. Cont & Inf. 1 (1984), 185-198.
21. R.M. Santilli: Foundations of theoretical mechanics I, Springer, New York 1978.
22. F. Takens: Variational and conservative systems, Report ZW-7603, Univ. of Groningen, 1976.

QUANTUM STATISTICS FOR SYSTEMS INTERACTING WITH A COHERENT ELECTROMAGNETIC FIELD

A. Alaoui
Mohammed V University, Rabat, Marocco

ABSTRACT

The adiabatical principle (in the meaning of Ehrenfest) plays a fundamental role in quantum and classical statistics. Starting from the hypothesis that adiabatic invariant states are the most probable states at the equilibrium of a set of quantum systems, we show that it is possible to elaborate a statistical scheme for non-conservative systems.

In the case of quantum systems interacting with a coherent field, the Floquet's theorem allows us to determine the adiabatic states and then to give the statistical scheme explicitly. The application of this theory to NMR shows a good agreement with experimental facts.

1. INTRODUCTION

The main idea of the present work can be summed up in the following question : does it make sense to speak about a thermodynamical equilbrium of a set of quantum systems interacting with a strong e.m field ? and if so, how to describe this equilibrium state.

The basic concepts of quantum statistics are usually defined for conservative systems only. However several phenomenons in physics, and particularly the NMR, show that an equilibrium state exists for systems interacting with radiation. At least for coherent radiation.

Usually, to describe a statistical set of non-conservative systems, the representation is modified in order to obtain a conservative case. The problem is that the choice of the transformation intended to eliminate time-dependance is purely formal, and, in any case, is not based on a fundamental physical criterium [1][2][3].

As matter of fact the number of transformations which permit the reduction of the wave equation to a new one having a constant hamiltonian,is infinite [4]. What is then the privilege of those that give results in good agreement with experimental facts ?

We think that this privilege is due to the fact that these transformations determine the adiabatic invariant states (in the meaning of Ehrenfest) of these non-conservative systems. We know that from the point of view of thermodynamics, the stationnary states have a particular prerogative,because they are the most probable at the equilibrium and, consequently, they correspond to maximum of entropy (as L. De Broglie says,they are more stable than the others). This property is not enclosed in the Schrödinger equation, it is related to the fact that these states are adiabatic invariant [5].

It seems then natural to consider the adiabatic invariant criterium as one of the basic principles of quantum statistics.

In other words, as suggested by G. Lochak [5] this is the criterium that we must use to choose, among all the possible states, the equilibrium states for conservative systems as well as for non-conservative systems.

This hypothesis allowed as to built a statistical theory for an important class of non-conservative systems : systems interacting with a coherent radiation, strong enough to consider that the stationnary states do not exist any more. [6]

This theory, gave us, in a way, the same possibilities as the conservative quantum statistics. In order to show that our theory is well founded, we have compared it with well known other theories and experimental facts. For several reasons, we choose, for this confrontation, the NMR. Since this phenomenon belongs to R.F. spectroscopy, we can consider that the fields are coherent. Furthermore, we do not need any quantization of these fields.

By another way, the relaxation processes play in NMR an important role, but are usually described by methods based on mathematical tricks which hide the physical reality, instead of beeing based on a general principle. The adiabatic invariance property allowed us to give a physical justification to these methods and to regroup them in a general scheme.

The statistical description of systems interacting with coherent radiation, requires the knowledge of the adiabatic states of these systems. The existence of such states has been shown by several authors [7] [8] [9]. and among them G. Lochak who called them : "the permanent states" [10] However, no one of these authors, except the later, has considered these states as the most probable of matter interacting with radiation.

2. THE PERMANENT STATES PROPERTIES

Consider the wave equation of a quantum system having a periodical hamiltonian.

$$(2,1) \qquad i\hbar \frac{d|\psi(t)\rangle}{dt} = H(t)|\psi(t)\rangle \quad , \qquad H(t+\tau) = H(t)$$

The Floquet's theorem [11] stipulates that any fundamental operator V(t) of the equation (2,1) can be written in the form :

$$(2,2) \qquad V(t) = T(t)\, e^{-\frac{i}{\hbar} Rt}$$

Where T(t) is a unitary periodic operator, with period τ, and R an hermitian constant operator. The couple [R,T(t)] is called the Floquet's decomposition of the system. For the same solution V(t), it exists an infinity of such decompositions.
The operator T(t) can be considered as a transformation which reduces (2,1) to equation with a constant hamiltonien R called the reduced hamiltonien.

$$(2,3) \qquad |\psi(t)\rangle = T(t)|\phi(t)\rangle \qquad i\hbar \frac{d|\phi(t)\rangle}{dt} = R\, |\phi(t)\rangle$$

The hamiltonien R defines, in the initial representation, a first integral of the system called the reduced energy :

$$(2,4) \qquad \tilde{R}(t) = T(t)\, RT^{-1}(t) = H(t) - i\hbar \frac{\partial T}{\partial t}(t).\, T^{-1}(t)$$

If $\{|r_k\rangle\}$ are the eigenstates of R with eigenvalues μ_k, we can easily see that we have :

$$(2,5) \qquad \tilde{R}(t)|\psi_k(t)\rangle = \mu_k|\psi_k(t)\rangle; \text{ with } |\psi_k(t)\rangle = T(t)\, e^{-\frac{i}{\hbar}\mu_k t}|r_k\rangle$$

The states $|\psi_k(t)\rangle$ are the Permanent states of the system [10] . They are solutions of (2,1) and form an orthonormal basis of the states space. We are interested only in the discret part's spectrum of the system, because the linked states alone are adiabatic invariant.
In another, Whatever the choice of the Floquet's decomposition is, the permanent states are univocally defined by the equation (2,1). However their eigenvalues μ_k, which represent the "reduced energy levels", are defined up to $n\hbar\omega$, where n is an integeer and $\omega = \frac{2\pi}{\tau}$. In addition, the

definition of the eigenvalues degeneracy is linked to a characteristic property of the permanent states which is that whatever the Floquet's decomposition may be, It is always possible to built a new one, called principal, verifying the condition :

(2,6) $\mu_m - \mu_p \neq n\hbar\omega \quad \forall\ l,m,n \in N.$

The set of the principal decompositions constitutes an equivalence class (the P. class) and plays a fundamental role in the theory, since it is precisely this class which determines the partition of the adiabatic invariant states $|\psi_k(t)\rangle$.

The degeneracy of an eigenvalue μ_k is then determined with regard to the class P.

Since the order of the class P is infinite, the reduced energy levels μ_k are still defined up to a constant and consequently,we cannot consider a statistical description of the systems. However, it is possible to eliminate this indetermination owing to a "limit postulate" which links up the reduced energy spectrum to the free hamiltonien spectrum , when the external periodic field amplitude vanishes adiabatically. The Floquet's decomposition we define like this is unique and we call it principal and canonical.[12]

Let us now say a few words about the reduction of wave equations like (2,1), but with a hamiltonien depending on time in an arbitrary way. If this equation has an evolution operator U(t), and if S is an arbitrary constant hermitian operator, it exists an infinity of linear, continuous and bounded transformations (as well as their inverse), called Liapounov transformations [13] :

(2,7) $L(t) = U(t)\, e^{-\frac{i}{\hbar} St}$

That allow us to reduce the initial equation to : $i\hbar \dfrac{d|\rho(t)\rangle}{dt} = S|\rho(t)\rangle$

However the eigenstates of the first integral $\tilde{S}(t) = L(t)S\,L^{-1}(t)$ we obtain like this,have no particular physical meaning regarding to thermodynamics. The adiabatical criterium is the only one which privileges these states as it is the case for eigenstates of the reduced energy $\tilde{R}(t)$ defined previously.

3. THE STATISCAL SCHEME

Go on now to the statistical description of a set of systems, interacting with an intense coherent radiation, under the hypothesis that the most probable states at the equilibrium are the adiabatic invariant states of these systems.

We must recall that the adiabatic invariance played a prime role in the early quantum theory [14] and in classical and quantum statistics.

For instance, in classical statistics, this is this property of the phase volume Ω that makes the definition $S = k \log \Omega$ of entropie compatible with the second principle [5]. In quantum statistics, most of thermodynamic quantities are defined with the implicit condition that the stationnary states are diabatic invariant [15]. Without this condition, the definition of entropy given by Von Neumann would not be, here too, compatible with the second principle.

Taking into account all these considerations, it seems natural to generalise this definition of entropy to a statistical set of systems interacting with a coherent radiation.

3.1. Generalisation of entropy

Consider then a statistical set of N identical systems weakly coupled together and submitted to a periodic field with period τ. The hamiltonian of each system is $H(\alpha,t)$, where α is a set of parameeters linked to the systems or the external fields.

The systems are assumed to be in a mixed state described by a density matrix :

$$(3,1) \qquad \rho(\alpha,t) = \sum_k p_k(\alpha) \; |\psi_k(\alpha,t)\rangle\langle\psi_k(\alpha,t)|$$

The permanent state $|\psi_k(\alpha,t)\rangle$ occupied with the probability $p_k(\alpha)$, and the eigenvalues $\mu_k(\alpha)$ are supposed to belong to the principal and canonical Floquet's decomposition.

The hypothesis that the permanent states are the most probable at the equilibrium leads us to adopt for the entropy, the definition [5][15]:

(3,2) $$S(\alpha) = K\ N\ \mathrm{tr}\ \{\rho(\alpha,t)\ \log \rho(\alpha,t)\}$$

We can now determine the equilibrium distribution of the mixture.

3.2. The equilibrium distribution and the variation of energy

Like in the conservative case, this distribution corresponds to the maximum of entropy. Under the conditions :

(3,3) $$\mathrm{tr}\ \{\rho(\alpha,t)\} = 1 \quad ; \quad \langle\tilde{R}\ (\alpha,t)\rangle = \mathrm{cte}$$

and with the help of Lagrange multipliers, we obtain

(3,4) $$\rho_0(\alpha,t) = \xi^{-1}(\alpha)\ e^{-\beta\tilde{R}(\alpha,t)} \ ; \ [\rho_0(\alpha,t)]_{kk} = \xi^{-1}(\alpha)e^{-\mu_k\beta} \ ;$$

$$\xi^{-}\ (\alpha) = \mathrm{tr}\ \{e^{-\beta\mu_k(\alpha)}\}$$

The equilibrium density matrix is diagonal in the basis $\{|\psi_k(\alpha,t)\rangle\}$ and β is a parameeter fixed by the averrage value of the reduced energy Let's remark that in the reduced representation we have :

(3,4') $$\rho_0^{\star}(\alpha) = T(\alpha,t)\rho_0(\alpha,t)\ T^{-1}(\alpha,t) = \xi^{-1}(\alpha)\ e^{-\beta R(\alpha)}$$

which shows that the systems are distributed on the eigenstates $\{|r_k\rangle\}$ of the constant hamiltonian $R(\alpha)$. However, let's recall that the knowledge of any constant hamiltonian is not sufficent to determine the equilibrium states. Here, the eigenstates of $R(\alpha)$ have a physical meaning because in the laboratory frame they are the only ones that are adiabatic invariants.

Suppose now that an infinitly slow modification of the parameeters α makes the mixture pass from the equilibrium state defined by the values (α,β), to an infinitly close equilibrium state defined by $(\alpha+d\alpha,\beta+d\beta)$. During this quasi-static transformation, the variation of the averrage value $\bar{R}(\alpha) = N\langle\tilde{R}(\alpha,t)\rangle$ is given by :

(3,5) $$d\bar{R}(\alpha) = N \sum_k \rho_{kk}(\alpha)\, d\mu_k(\alpha) + N\sum_k \mu_k(\alpha) d\rho_{kk}(\alpha)$$

The first term corresponds to the variation of the reduced energy levels without any involvement of the non-conservative forces responsible for the equilibrium. We read it as the averrage work received from the external forces and call it "reduced work" : $d\bar{W}(\alpha)$. The second one represents a redistribution of the systems on the permanent states while the external parameters remain constant. We read it as the heat received by the mixture from an external source and call it the "averrage reduced heat" : $d\bar{Q}(\alpha)$. By another way, just like the energy of a conservative system, $\tilde{R}(\alpha,t)$ is the only one first integral whose eigenstates are the adiabatic invariant states of the system [6]. Moreover $\tilde{R}(\alpha,t)$ is an additive quantity [15]. Thus, it seems natural to consider $\bar{R}(\alpha)$ as the internal energy of the statistical set ; we will say "reduced internal energy" ; and consequently to read the relation (3,5) as the first principle of thermodynamics :

(3,6) $$d\bar{R}(\alpha) = d\bar{W}(\alpha) + d\bar{Q}(\alpha)$$

Consider now the variation of entropy. It is easy to show that during any quasi-static transformation we have :

(3,7) $$dS(\alpha) = k\beta\, d\bar{Q}(\alpha)$$

Imagine that during this transformation the mixture, remaining submitted to the periodic field, is thermally isolated from out side [2][3]. We will have then $dS(\alpha) = 0$ which indicates that entropy defined in this scheme, has the same properties as the entropy defined in classical thermodynamics, provided that we can identify the parameeter β to $1/kT$ where T is the equilibrium temperature.
The relation (3,7) represents then the second principle. This interpretation of β is confirmed by the confrontation with experimental facts [15].

Let's recall now that the total $H(\alpha,t)$ is not a first integral of the system and that its averrage value is not constant at the equilibrium. What is then the variation of $\overline{H}(\alpha,t)$ when the parameeter α is modified infinitly slowly ? Starting from the expression (2,4), we can write :

$$(3,8)\quad d\overline{H}(\alpha,t)=d\overline{W}(\alpha)+d\overline{Q}(\alpha)+d\overline{F}(\alpha,t) \quad ; \quad \overline{F}(\alpha,t) = i\hbar N < \frac{\partial T}{\partial t}(\alpha,t)T^{-1}(\alpha,t)>$$

Suppose that during this transformation : $d\overline{R}(\alpha) = 0$ ("pseudo-isolated mixture") (3,8) becomes : $d\overline{H}(\alpha,t) = d\overline{F}(\alpha,t)$

Then, the only one energy exchange that can occur here has an electromagnetic nature, and the variation of this electromagnetic interaction energy is given by $d\overline{F}(\alpha,t)$. In NMR with a strong R.F. field, we used this result to calculate the energy absorbed by a sample at the equilibrium at a temperature T and we showed that this energy is equal to zero. In the usual theories [3] this is explained by the fact that the quantum transitions between stationnary states are saturated due to the strong R.F. field.

From our point of view, when the R.F. field is strong, the stationnary states do not exist any more, and this "saturation" is due to the fact that the systems constituting the sample are distributed and relaxe on the permanent states.

4. APPLICATION TO NMR

4.1. The saturated Bloch equation

Consider a sample consisting of N identical nuclear spin $\vec{I}$ having a magnetic moment $\vec{m}$, weakly coupled together and submitted to a magnetic static field $\vec{h}_0$ and a strong coherent R.F. field. $\vec{h}_1(t)$

This spin system is supposed to be in a mixed state discribed by a density matrix ρ.

In order to describe the evolution of the system under the effect of the relaxation processes and the interaction with the fields, we proposed an equation of motion that we called "saturated Bloch equation" :

$$(4,1)\qquad i\hbar \frac{d\rho(t)}{dt} = [H(t), \rho(t)] - i\hbar \frac{\rho(t) - \rho_0(t)}{\theta}$$

where $H(t) = H_0+\omega_1 H_1(t)$ represents the total hamiltonian of one spin. The part H_0 describes the interaction with the static field $\vec{h}_0$ and $H_1(t)$ the interaction with the strong R.F. field $\vec{h}_1(t)$ whose amplitude is proportional to ω_1.

The last term of (4,1) gives a phenomenological description of the spin system relaxation.We suppose that one relaxation time θ is enough (gas, liquids [3]). This equation has the same form as the usual equations [3] [17], however the fundamental and new hypothesis here is that the equilibrium density matrix $\rho_0(t)$ is equae to :

$$(4,2) \qquad \rho_0 = \xi^{-1}(T)\, e^{-\tilde{R}(t)/kT} \quad ; \ \xi = \mathrm{Tr}\ \{e^{-\tilde{R}(t)/kT}\}$$

where $\tilde{R}(t)$ is the reduced energy of one spin (Boltzmann statistics) and T the temperature of the spin system environnement considered as a thermostat. It means that we assume the permanent states of the spins to be the equilibrium states.

Let's show now, very briefly, that this equation, for various intensity of the R.F. field $\vec{h}_1(t)$ leads to the usual descriptions of thermal relaxation.

i) if the intensity of $\vec{h}_1(t)$ is very weak, we can consider, according to the limit postulate, that $\tilde{R}(t) \cong H_0$ and consequently $\rho_0(t) \propto e^{-H_0/kT}$. The equation obtained, in this case, from (4,1) is the well known Bloch equation [17] which gives a correct description of solids and liquids in weak R.F. field.

ii) Suppose now that the R.F. field intensity is comparable to the static field intensity and that its variation in time is very slow. In this case the relation (2,4) shows that $\tilde{R}(t) \cong H(t)$ and from (4,2) we have $\rho_0(t) \propto e^{-H(t)/kT}$. The equation that we obtain from (4,1) is the modifified Bloch equation introduced by Garnstens [18],who proposed this hypothesis for $\rho_0(t)$ called "instantaneous quasi-stationnary equilibrium". In fact the physical justification of this hypothesis is that the eigenstates of H(t) are the adiabatic invariant states of the spin system.

iii) Finally, when the R.F. field intensity is strong enough to have the saturation condition $|\omega_1|\theta \gg 1$, we must use the expression (4,2) for $\rho_0(t)$. In order to determine the magnetisation of the spin system we solved,in this case, the equation (4,1) for different R.F.field polarisations [15] . We obtained results in agreement with experimental facts and we verified the validity of our statistical scheme.

4.2. A general spin temperature theory

The description of NMR in solids submitted to a high R.F. field presents many difficulties due to the strong interactions between the nuclear spins. The Bloch equations, in this case, are not valid anymore [3][17].

However, when the R.F. field is rotating A.G. Redfield proposed the hypothesis of a spin temperature in the rotating frame [1][2][3] . This theory, in agreement with experimental facts, is based on the use canonical transformation(the rotating frame) intended to eliminate the time dependance of the spin system hamiltonian and,consequently, to do statistics in a conservatice case.

We are going to show, through a general spin temperature theory based on our statistical scheme, that the transformation to a rotating frame is precisely the transformation which determines the invariant adiabatic states of the spin system [4].

Consider then a solid sample consisting of N identical nuclear spins $\vec{I}$ strongly coupled toge ther, submitted to a magnetic field $\vec{h}_0$ and to a coherent R.F. field having any polarisation and an intensity close to the saturation level.

Let's recall that in solids we must consider two relaxation times T_1 et T_2 ($T_1 \gg T_2$). T_1 is the spin-lattice relaxation time and T_2 the spin-spin relaxation time [3][17]. Suppose now that the phenomenons we are going to describe last a time t verifying the condition $T_2 < t < T_1$. In this way we can consider the nuclear spins as isolated from the lattice. The total hamiltonian of the system is then:

$$(4,3) \qquad H_S(t) = H_0 + \omega_1 H_1(t) + H_D \quad ; \quad H_S(t+\tau) = H(t)$$

where $\tau$is the period of $\vec{h}_1(t)$. The term H_D is the spin-spin interactions hamiltonien. Without going into details [3][15], let's say that these between spins are so important that we must consider the whole sample

as a unique system.
Since $H_s(t)$ is periodic in time, we can use the Floquet's theorem to find a decomposition $[R_s, T_s(t)]$ and consequently a reduced energy $R_s(t)$, whose eigenstates are the adiabatic invariant states of the system. According to our statistical scheme we say that, after a time T_2, nuclear spins reach an equilibrium characterized by a spin temperature T_s and the density matrix :

$$(4,4)\quad \rho_s(t) = \xi^{-1}(T)e^{-R_s(t)/kT_s} \quad ; \; \xi(T) = \text{tr}\,\{e^{-R_s(t)/kT_s}\}$$

The determination of a Floquet's decomposition $[R_s, T_s(t)]$ is not easy at all in the general case [15]. Suppose here that the R.F field $\vec{h}_1(t)$ is rotating around oz with an angular velocity ω. We obtain then the following decomposition :

$$(4,5)\qquad T_s(t) = e^{-i\omega I_z t} \quad ; \quad R_s = (\omega_0 - \omega)I_z + \omega_1 I_x + H_D'$$

where ω_0 is proportionnal to the amplitude of $\vec{h}_0$ and H_D' is the secular part of H_D(the part which commutes with Iz). Finally, we can say that $T_s(t)$ is the transformation to the frame rotating with the field $\vec{h}_1(t)$ and R_s is the spin system's hamiltonian in this frame. It is precisely the hamiltonian invoked by Redfield [1].
The physical justification of Redfield's theory is then that, fortuitously, the rotating frame representation determines the adiabatic invariant states of the spin system.
Let's remark that if we want to determine the magnetization of the spin system after a time $t \gg T_1$, we must take into account the interaction with the lattice. [2] [3] [15]. This interaction, with regard to spin-spin interactions, is weak and his effect is to change the spin temperature T_s without changing the equilibrium states (eigenstates of $R_s(t)$).
Notice also that in NMR, double irradiation technics (two coherent fields) are used very often. However the resonance frequencies which appear in these phenomenons are not correctly explained, in usual theories, as

quantum transitions frequencies.

In the framework of the statistical scheme presented here, we gave an equation of motion to describe the double irradiation phenomenons, and we showed that the resonance frequencies are in fact the frequencies of quantum transitions, induced by the weak coherent field, between permanent states due to the strong R.F field [15].

To a conclusion let's say that, in our opinion, the invariant adiabatic states are not restrained to periodic hamiltonians and could may be exist for other hamiltonians depending on time in a different way [5].

REFERENCES

1. Redfield, A.G. Phys. Review, V 98, N6, (1955)
2. Goldmann, M : Spin temperature and nuclear magnetic resonance in solids Clarendon Press, Oxford 1970.
3. Abragam, A. Les principes du magnetisme nucléaire - PUF. Paris 1961
4. Alaoui, A. Lochak, G. : C.R.A.S. 280 (1975), p.589
5. Lochak, G. : Ann, Fond. Louis de Broglie, 1, n°2, p 56, 1976
7. Grichkowsky, D. Phys. Review 7, N6 (1973) p 2096.
8. Shirley, J.H. : Phys. Review. 138 B, (1965), p 979.
9. Sambe, H. : Phys. Review A, 7, (1973) p 2203.
10. Lochak, G, C.R.A.S. Série B, 272, (1971), p 1281.
11. Coddington, E. Levinson, N. : Theory of ordinary differential equations, Mc Graw Hill, N.Y. 1955.
12. Alaoui,A, Lochak, G. : Ann. Fond. Louis de Broglie, V2, n 2, (1977) p 87
13. Riesz, F. Nagy, Sz. : Leçons d'analyse fonctionnelle, Paris, Gauthier-Villars, 1972.
14. Brillouin, L. : L'atome de Bohr. Ed du journal de Physique. (1936) Paris.
15. Alaoui, A : Thèse d'Etat (1982) RABAT.
16. Castaing, R. : Thermodynamique statistique, MASSON, 1970.
17. Bloch, F. : Phys. Review, V 70, N 7, 8, (1946)
18. Garstens, M.A. Kaplan, J.I. : Phys. Review, V 99, N 2, (1955).

NON-DEMOLITION MEASUREMENT AND CONTROL IN QUANTUM DYNAMICAL SYSTEMS

Viacheslav Belavkin
Miem, Moscow, U.S.S.R.

ABSTRACT

A multi-stage version of the theory of quantum-mechanical measurements and quantum-statistical decisions applied to the non-demolition control problem for quantum objects is developed. It is shown that in Gaussian case of quantum one-dimensional linear Markovian dynamical system with a quantum linear transmission line optimal quantum multi-stage decision rule consists of classical linear optimal control strategy and quantum optimal filtering procedure, the latter contains the optimal quantum coherent measurement on the output of the line and the recursive processing by Kalman-Busy filter. All the results are illustrated by an example of the optimal problem solution for a quantum one-dimensional linear oscillator on the input of a quantum wave transmission line.

1. INTRODUCTION

Perspective use of quantum devices for communication and control and constantly growing demands on increase of observation and control precision stimulate theoretical investigation and mathematical modelling of controlled dynamic systems, which are described on quantum-mechanical level from general statistical positions. Mathematical description given in [1] of statistical processes of control and observation in open quantum systems, the stochastic theory of which was intensively constructed during the last decade in [3-8] allows to apply a lot of concepts and methods of statistical theory of control to quantum controlled objects. Note that the deterministic statement of standard problems of optimal observation and control in quantum case is in general due to the uncertainty principle impossible, except the problem of programming control of a closed quantum-mechanical system without measurements and feed-back, considered in [9]. In particular, the solution obtained in [10,11] for the problem of optimal observation of the linear Markovian system leads to the non-singular Kalman-Busy filter, the estimate error of which is never equal to zero.

Here we describe the solution of a statistical problem of optimal control of quantum Markovian system under assumption, that the output observable process is given by non-direct continuous measurements which non-demolish a given quantum Markovian subsystem (quantum controlled object). Such non-demolition measurements in linear boson Markovian systems we described in [11] for the problem of optimal filtration, where we proved that the optimal continuous measurements of canonical observables can be realized by a coherent Markov filter described by a quantum stochastic linear equation [12]. Note that these optimal measurements are indirect and can be reduced from the scheme of repeated non-ideal measurements in initially Hamiltonian system by the limit procedure [7].

Here we describe the multi-stage problem of optimal non-demolition measurements in application to the optimal mean-square control of a quantum Markovian oscillator with linear matched channels and Gaussian noises and give its solution in discrete time using the multi-stage generalization [13,14] of quantum statistical decision theory [15]. The optimal filtering and control strategy are also found in continuous case by the limit procedure $\Delta t \to 0$. In the next chapter we consider an interesting example physically motivated in [11] and give the solution of the corresponding problem of optimal damping of a quantum active or dissipative oscillator based on non-demolition filtration. The rigorous treatment and proving of this result is given in the 3-th and 4-th chapters for the general one-dimensional case of a quantum open linear Markovian system.

2. CONTROLLED QUANTUM LINEAR OSCILLATOR WITH QUANTUM TRANSMISSION LINE

Let x be an operator of complex amplitude of a quantum oscillator with the Hamiltonian $\Omega \cdot x^* x$, which satisfies the canonical commutation relations with x^* being an adjoint operator

$$[x, x^*] = xx^* - x^*x = \hbar \mathbb{1} \tag{2.1}$$

where $\mathbb{1}$ is the unit operator, $\hbar > 0$ is the Plank constant. Assume that in general case this oscillator is controlled by the complex amplitude u by means of a quantum-mechanical transmission line with wave resistance $\gamma/2$, where the operator of the wave $y(t-s/c)$ travelling from the oscillator into the line is measured. In the simplest case of ideal conjugation between the line and the measuring apparatus, when there is no reflection of the wave travelling from the oscillator, i.e. in case of the matched line, $x(t)$ and $y(t)$ are described by the pair of linear equations [10]

$$dx(t)/dt + \alpha x(t) = \gamma u(t) + v(t), \quad x(0) = x, \tag{2.2}$$

$$y(t) = \bar{\alpha} x(t) - dx(t)/dt = \gamma(x(t) - u(t)) - v(t), \tag{2.3}$$

where, generally speaking, α is a complex number with the fixed real part, $\alpha + \bar{\alpha} = \gamma$, and with arbitrary imaginary part depending on the choice of the representation, $v(t+s/c)$ the amplitude operator of the wave travelling out of line to the oscillator, this operator is responsible for the commutator preservation. Under natural for super-high and optical frequences assumption of narrowness of the frequency band which we deal with the commutators for $v(t)$ in the representation of "rotating waves" have the delta-function form[18]:

$$[v(t), v(t')] = 0, \quad [v(t), v(t')^*] = \gamma \hbar \mathbb{1} \delta(t - t'). \tag{2.4}$$

By integrating equation (2.2) and taking into account that $v(t)$ does not depend on $x(t')$ when $t > t'$, it is easy to verify that the commutator $[x(t), x(t)^*]$ is constant, moreover, $x(t)$ is commutative both with $y(t')$ and $y(t')^*$ when $t > t'$, and the commutators for $y(t)$, $y(t')$, $y(t')^*$ coincide with (2.4). The latter means that the consideration of von Neumann reduction which appears as a result of some quantum measurement of $y(t)$ at previous instants of time $t' < t$ does not affect the future behaviour of $x(t)$, $y(t'')$, $t'' > t$, so that equations (2.2), (2.3) remain unchanged. This fact called the non-demolition principle together with the hypothesis that the quantum process is Markovian which holds for quantum thermal equilibrium states of the wave $v(t)$ in case of narrow band approximation the optimal measurement and control problems for the simplest quantum dynamical system mentioned above.

For instance, let try to choose the optimal measurement of the controlled quantum oscillator (2.2) with transmission line (2.3), so that to minimize its energy $\Omega\langle x^*(t)x(t)\rangle$ at the final instant of time $t=\tau$ by means of the control strategy, the norm $\int_0^\tau |u(t)|^2dt$ of which should not be too great. Assume, that the initial state x is Guassian with the mathematical expectation $\langle x\rangle = Z$ and

$$\langle (x-Z)(x-Z)\rangle = 0, \quad \langle (x-Z)^*(x-Z)\rangle = \hbar\Sigma ,$$

v(t) is quantum Gaussian white noise [19] which is described by the following correlations

$$\langle v(t)\, v(t')\rangle = 0, \qquad \langle v(t)^*\, v(t')\rangle = \hbar\sigma\tilde{\delta}(t-t'),$$

what corresponds to the equilibrium state with the temperature T: $\sigma = \gamma(\exp(\hbar\Omega/kT) - 1)^{-1}$. This problem is characterized by the quality criterion

$$\Omega\langle x(\tau)^*x(\tau)\rangle + \int_0^\tau \langle\theta|u(t)|^2 + \omega(x(t)-u(t))^*(x(t)-u(t))\rangle dt, \quad (2.6)$$

where θ, $\omega \geq 0$ are parameters responsible for the measurement quality: when $\theta = \Omega = 0$ (2.6) corresponds to the problem of pure filtration, when $\omega = 0$, $\theta \neq 0$ it corresponds to the control problem.

It will be shown below (see §3) that the optimal measurement minimising criterion (2.6) is statistically equivalent to the measurement of the stochastical process $Z(t) = \hat{x}(t) + \mathring{x}(t)$ described by the Kalman-Bucy filter:

$$d\hat{x}(t)/dt + \alpha\hat{x}(t) = \gamma u(t) + \varkappa(t)(y(t) - \gamma(\hat{x}(t) - u(t))). \qquad (2.7)$$

Here $\hat{x}(0)=Z$, $\varkappa(t) = (\gamma\Sigma(t) - \sigma)/(\mu + \sigma)$, $\Sigma(t)$ is the solution of the equation

$$d\Sigma(t)/dt = (\sigma - \gamma\Sigma(t))(\mu + \gamma\Sigma(t))/(\mu + \sigma), \quad \Sigma(0) = \Sigma,$$

$$d\mathring{x}(t)/dt + \alpha\mathring{x}(t) = \varkappa(t)(\mathring{v}(t) - \gamma\mathring{x}(t)), \quad \mathring{x}(0) = 0, \qquad (2.8)$$

and $\dot{v}(t)$ is the amplitude operator with commutators

$$[\dot{v}(t),\dot{v}(t')] = 0, \quad [\dot{v}(t),\dot{v}(t')^*] = -\hbar\gamma\tilde{\delta}(t - t')$$

and with correlations of vacuum noise of intensity $\mu = \gamma$ if $\gamma>0$ and $\mu=0$ if $\gamma \leq 0$:

$$\langle\dot{v}(t)\,\dot{v}(t')\rangle=0, \quad \langle\dot{v}(t)^*\dot{v}(t')\rangle=\hbar\mu\tilde{\delta}(t - t'), \qquad (2.9)$$

which change the quantum process $\hat{x}(t)$ into the classical (commutative) diffusion complex process. For instance, such measurement takes place by the heterodyning where [20] $\dot{v}(t)$ stands for a standard wave. In this case the optimal control strategy coincides with the classical: $u(t) = -\lambda(t)Z(t)$, where $\lambda(t)=(\gamma\Omega(t)-\omega)/(\theta+\omega)$, and $\Omega(t)$ is a solution of the equation:

$$-d\Omega(t)/dt=(\omega-\gamma\Omega(t))(\theta+\gamma\Omega(t))/(\theta+\omega),\ \Omega(\tau)=\Omega, \qquad (2.10)$$

which together with (2.8) defines the minimum quantity of losses (2.6):

$$\hbar(\Omega(0)\Sigma+\int_0^\tau(\Omega(t)\sigma+(\gamma\Omega(t)-\omega)^2\Sigma(t)/(\theta+\omega))dt)+\Omega(0)|Z|^2.$$

By setting $\sigma=0$, $\omega=0$, we obtain in particular the solution of the terminal control problem for an oscillator with thermal noise equal to zero. But in this case unlike to the classical one the optimal measurement remains indirect and the equation (2.8) remains regular associated with the white noise in the channel of intensity $|\gamma|\hbar$. Thus to consider the quantum measurement postulates is statistically equivalent to the adding of white noise into the channel of intensity $|\gamma|\hbar$ what excludes the singular case of pure measurement of the amplitude x .

It is interesting to note that in the case of thermal equilibrium when $\gamma>0$, $T>0$ and $\Sigma=(\exp\{h\Omega/kT\}-1)^{-1}$ the optimal amplification coefficient $\varkappa(t)$ equals to zero which means the possibility of optimal control of the quantum oscillator without measurement. It also holds when $\omega = \gamma\Omega$, the solution of equation (2.9) is stationary and optimal feed-back coefficient $\lambda(t)$ equals to zero. But in the contrary case $\gamma < 0$, $T < 0$ which is associated with the active medium of the oscillator (laser) the optimal coefficients $\varkappa(t), \lambda(t)$ are strictly negative and non-zero even for the stationary solution $\Sigma(t)=0$, $\Omega(t)=\theta/|\gamma|$ of equations (2.8), (2.9).

3. QUANTUM GAUSSIAN FILTRATION OF BOSON LINEAR MARKOVIAN OSCILLATOR STATE

We examine a Markovian one-dimensional quantum dynamical system with a channel, described at discrete times t_k by the unital $*$-algebras A_k and B_k generated by the operators $x_k = x(t_k)$ and $y_k = y(t_k)$ respectively, which act in a Hilbert space H, where they satisfy the equations:

$$x_k = \varphi\, x_{k-1} + \beta\, u_{k-1} + v_k \tag{3.1}$$

$$y_k = \gamma\, x_{k-1} + \delta\, u_{k-1} + w_k \tag{3.2}$$

Here $\varphi, \beta, \gamma, \delta$ are some complex parameters, u_k are controls which can accept complex values, $x_0 = x$ is the initial operator in H, which generates an algebra A, and v_k, w_k are the operators acting in the space H and generating some algebras B_k^o. To the Markov hypothesis corresponds the independence of the states on A, B_k^o, k=1,2... for which it is necessary to demand that the operators from A, B_k^o corresponding to different instants of time t_k should commutate. We shall examine the following simultaneous commutation relations for the generating operators x, v_k, w_k:

$$[x, x^*] = \hbar\mathbb{1} \qquad [v_k, v_k^*] = (1-|\varphi|^2)\hbar\mathbb{1}$$
$$[w_k, w_k^*] = (\varepsilon - |\gamma|^2)\hbar\mathbb{1}, \quad [w_k, v_k^*] = -\bar{\varphi}\gamma\hbar\mathbb{1}, \tag{3.3}$$

($\hbar>0$, and $\mathbb{1}$ =the unit in H), supposing the unwritten commutators, including all those corresponding to different instants of time to be equal to zero. Here the choice of the commutator $[w_k, v_k^*]$, responsible for the commutability $[y_k, x_k^*]=0$ is essential, the other nonzero commutators are chosen so that the commutators

$$[x_k, x_k^*] = I\hbar\mathbb{1}, \quad [y_k, y_k^*] = I\varepsilon\hbar\mathbb{1}$$

should be constant. Such a system will be called a discrete linear Markovian quantum partially observable oscillator.

Let describe the states π_k on A_k by the Glauber distributions $p_k(\xi)$, normed with respect to the Lebesque measure $d\xi = d\mathrm{Re}\,\xi\, d\mathrm{Im}\,\xi/\pi\hbar$ on the complex plane $\mathbb{C}\ni\xi$. In the representation described in the appendix of [13], the Markovian morphisms, corresponding to linear equations (3.1), (3.2) transform the distributions $p_{k-1}(\xi)$ into the two-dimensional distributions

$$g_k(\xi,\eta)=\int q_k(\xi-T\xi'-\beta u, \eta-H\xi'-\delta u)p_{k-1}(\xi')d\xi', \tag{3.4}$$

where $q_k(\xi, \eta)$ are (not necessary the Glauber) two-dimensional distributions, which describe the independent states ς_k^o on algebras B_k^o. When $\varepsilon=0$, the operators y_k, y_k^* are mutually measurable, a posteriori Glauber distributions $p_k(\xi|\zeta^{k+1})$ describing states at instants of time t_k with the fixed u_k, $y_k=\eta_k$ are defined recurrently by the formula analogous to the Bayes one

$$p_k(\xi|\zeta^{k+1}) = g_k(\xi, \eta_k|\zeta^k)/r_k(\eta_k|\zeta^k),$$

where $g_k(\xi,\eta|\zeta^k)$ are distributions obtained by substitution of $p_{k-1}(\xi|\zeta^k)$ into (3.4) and

$$r_k(\eta|\zeta^k) = \int g_k(\xi,\eta|\zeta^k)d\xi$$

are the probability distributions, describing the complex

stochastical values η_k, which arise as the results of direct measurements of y_k under fixed $\xi^k=(u_o,\eta_1,u_1,\ldots,\eta_{k-1},u_{k-1})$.

When $\varepsilon \neq 0$, only indirect measurement of y_k is possible which is described, for instance, by the B_k-valued measure

$$b_k(d\eta) = \cdot m(\eta - y_k)\cdot d\eta \tag{3.5}$$

In this case in order to calculate a posteriori Glauber [17] distribution one should change $q_k(\xi,\eta)$ in formula (3.4) for the distribution

$$q_k'(\xi,\eta)=\int m_k(\eta-\eta')q_k(\xi,\eta')d\eta'. \tag{3.6}$$

THEOREM 1. Let the initial state π of the quantum oscillator be described by the Glauber distribution of Gaussian type

$$p(\xi)= \exp\{-|\xi-z|^2/\hbar\Sigma\}/\Sigma, \tag{3.7}$$

the distributions $q_k(\xi,\eta)$, describing the transitions (3.4) at instants t_k, be also Gaussian

$$q_k(\xi,\eta)=\exp\{-(\vartheta|\xi|^2+2\mathrm{Re}\,v\xi\bar{\eta}+\sigma|\eta|^2)/\hbar(\sigma\vartheta-|v|^2)\}/(\sigma\vartheta-|v|^2), \tag{3.8}$$

and the measurements be indirect (3.5), described by the Gaussian distributions

$$m_k(\eta)=\exp\{-|\eta|^2/\hbar\mu\}/\mu. \tag{3.9}$$

Then a posteriori states (3.3) in [13] at each instant t_k, $k=1,2\ldots$, are described by the conditional Glauber distributions also Gaussian

$$p_k(\xi|\xi^{k+1})=\exp\{-|\xi-z_k|^2/\hbar\Sigma_k\}/\Sigma_k, \tag{3.10}$$

where z_k, $_k$ are defined by the recurrent equations of the complex Kalman filter:

$$z_k=\varphi z_{k-1}+\beta u_{k-1}+\varkappa_k(\eta_k-\gamma z_{k-1}-\delta u_{k-1}), \quad z_o=z, \tag{3.11}$$

$$\Sigma_k = |\varphi|^2 \Sigma_{k-1} + \sigma - |\varkappa_k|^2 \Psi_k, \quad \Sigma_o = \Sigma, \tag{3.12}$$

where

$$\varkappa_k = (\varphi \bar{\gamma} \Sigma_{k-1} - v)/\Psi_k, \quad \Psi_k = |\gamma|^2 \Sigma_{k-1} + \nu', \quad \nu' = \nu + \mu.$$

PROOF of the theorem, due to the chosen representation is quasi-classic, is an exact copy of the appropriate classic proof(see, for instance, [16]). One should only take into account that distributions (3.6) are also Gaussian (3.8) after changing the parameters ν for $\nu' = \nu + \mu$. By substituting in (3.4) $q(\xi,\eta)$ by $q'(\xi,\eta)$ and $p_{k-1}(\xi)$ by the conditional distribution $p_{k-1}(\xi|\varsigma^k)$ of type (3.10), we obtain

$$g'_k(\xi,\eta|\varsigma^k) = p_k(\xi|\varsigma^{k+1}) r'_k(\eta|\varsigma^k), \tag{3.13}$$

where $p_k(\xi|\varsigma^{k+1})$ is the Gaussian distribution (3.10) with the parameters (3.11), (3.12) and

$$r'_k(\eta|\varsigma^k) = \exp\{-|\eta - \gamma z_{k-1}|^2/\hbar\Psi_k\}/\Psi_k. \tag{3.14}$$

Note, that distinct from the classical case the correlation of distributions (3.8), (3.9) should not only be non-negative definite but also should satisfy the Geisenberg uncertainty principle

$$\begin{pmatrix} \sigma & -v \\ -v & \nu \end{pmatrix} + \begin{pmatrix} 1-|\varphi|^2 & \varphi\bar{\gamma} \\ \gamma\bar{\varphi} & \varepsilon - |\gamma|^2 \end{pmatrix} \geqslant 0, \quad \mu \geqq \varepsilon \quad ,$$

excluding the case $\mu = 0$ when $\varepsilon > 0$ in accordance with the inequality.

As shown in the next paragraph, a posteriori mathematical expectation z_k with $\mu = \max(0,\varepsilon)$ appear to be the optimal estimates $u_k^o = z_k$ of the operators x_k with respect to the squared quality criterion $c_k(u_k) =: |x_k - u_k|^2$: which minimize the error $\hbar\Sigma_k$. In the commutative case $[x_k, x_k^*] = 0$ this optimality was proved, and as well in linear approximation.

Note, that instead of calculating z_k by means of the recurrent formula (3.11) using the results $(\eta_1,\ldots,\eta_k)$ of the indirect measurement (3.5) one may obtain z_k at once as the results of the measurement described by the B_k-valued measures:

$$b_k(\xi^k,dz) = \cdot n_k(z-\hat{x}_k)\cdot dz, \tag{3.15}$$

where

$$n_k(z) = m_k(z/|\varkappa_k|)/|\varkappa_k|^2,$$

and

$$\hat{x}_k = \varphi\ z_{k-1} + \beta u_{k-1} + \varkappa_k(y_k - \gamma z_{k-1}) \tag{3.16}$$

is an operator, depending only on the last values z_{k-1}, u_k, and independent of the preceding measurement and the control results.

When time is continuous the quantum oscillator (3.1), (3.2) is described by the quantum stochastic differential equations

$$dx(t)+\alpha x(t)dt = \beta u(t)dt + v(dt), \tag{3.17}$$

$$y(dt) = \gamma x(t)dt + \delta u(t)dt + w(dt), \tag{3.18}$$

i.e. by equations (3.1), (3.2) with $x(t_k)=x_k$, $y(\Delta t_k)=y_k$,

$$\varphi \simeq 1-\alpha\cdot\Delta,\ \beta\simeq\beta\Delta\ ,\ \gamma\simeq\gamma\Delta\ ,\ \varepsilon\simeq\varepsilon\Delta\ ,$$

where $|\Delta t_k| = t_k - t_{k-1} = \Delta$ tends to zero. In addition to that the commutation relations (3.3) change in the following way

$$[x,x^*] = \hbar\mathbb{1}, \qquad [v(dt),v(dt)^*] = (\bar{\alpha}+\alpha)\hbar dt\,\mathbb{1},$$

$$[w(dt),w(dt)^*] = \varepsilon\hbar dt\,\mathbb{1}, \quad [w(dt),v(dt)^*] = -\gamma\hbar dt\,\mathbb{1},$$

and the other commutators including the corresponding to the different instants of time are equal to zero. Under the assumptions of theorem 1 it is easy to obtain by passing to the limit as $\Delta \to 0$ that when $\sigma \simeq \sigma \cdot \Delta$, $\vartheta \simeq \vartheta\Delta$, $\vartheta \simeq \vartheta\,\Delta$ a posteriori state $\pi(t, \xi^k)$ is described by the Glauber distribution $p(t,\xi|\xi^k)$ of Gaussian type (3.10) with the parameters $z(t)$, $\Sigma(t)$ which correspond to the Kalman-Busy filter:

$$dz(t)+\alpha z(t)dt=\beta u(t)dt+\varkappa(t)(\eta(dt)-\gamma z(t)-\delta\breve{u}(t)dt). \tag{3.19}$$

Here $\varkappa(t)=(\bar{\gamma}\Sigma(t)-v)/\vartheta'$, $\vartheta'=\vartheta+\mu$, $z(0)=z$, $\Sigma(0)=\Sigma$,

$$d\Sigma(t)/dt + (\alpha+\bar{\alpha})\Sigma(t)=\sigma-|\varkappa(t)|^2\vartheta',$$

and $\eta(dt)$ are the results of the corresponding indirect measurements $y(dt)$ which are realized by the measurement of the sum $y(dt) + \mathring{w}(dt)$ where $\mathring{w}(dt)$ is an operator-valued measure, defined by the coefficients ε, μ:

$$[\mathring{w}(dt),\mathring{w}(dt)^*]=-\varepsilon\hbar dt\,\mathbb{1}, \quad \langle\mathring{w}(dt)^*\mathring{w}(dt)\rangle=\mu\hbar dt.$$

As shown in the next paragraph, such "continuous" measurement appears to be also optimal in the squared Gaussian case when $\mu = \max(0, \varepsilon)$.

4. OPTIMAL MEASUREMENT AND CONTROL OF QUANTUM OSCILLATOR

In the following theorem it is not required that the distributions p_0, q_k and m_k should be Gaussian and it is assumed only that they should have the zero mathematical expectations, and the covariations should coicide with the covariations Σ, σ, v, ϑ, μ of the distributions (3.7)-(3.9) respectively, not necessary being of the form (3.11).

THEOREM 2. Let $a_k = \Omega x_k^* x_k$ be a final operator, $\Omega \geq 0$,

$$c_k(u_k)=\omega x_k^* x_k-\vartheta\bar{u}_k x_k-\bar{\vartheta}u_k x_k^*+\vartheta'|u_k|^2,\ \omega\geq 0,\ \vartheta'>0 \tag{4.1}$$

be squared loss operators at the instants of time t_k, $k \in [0,K)$, $u_k = -\lambda_k z_k$ be linear control strategies, where z_k are linear estimates (3.11) based on the results η_k of the indirect measurement (3.5) and

$$\lambda_k = (\varphi \bar{\beta} \Omega_{k+1} - \vartheta)/\Upsilon_k, \tag{4.2}$$

where $\Upsilon_k = |\beta|^2 \Omega_{k+1} + \vartheta'$ and Ω_k satisfies the following equation

$$\Omega_k = |\varphi|^2 \Omega_{k+1} + \omega - |\lambda_k|^2 \Upsilon_k, \quad \Omega_K = \Omega \tag{4.3}$$

Then the operators of future losses (see [13], equ.(4.5), or [14]) are also squared when

$$a_k(\xi^{k+1}) = d_k \mathbb{1} + \Upsilon_k |u_k + \lambda_k z_k|^2 + \Omega_k x_k^* x_k +$$

$$+ \Gamma_k (z_k - x_k)^* (z_k - x_k) - 2\mathrm{Re}\, \Lambda_k (u_k + \lambda_k z_k)^* (z_k - x_k) \tag{4.4}$$

where $d_k = h \sum_{i=k+1}^{K} (\Omega_i \sigma + \Gamma_i (\sigma + 2\mathrm{Re}\, \varkappa_i \bar{v} + \vartheta' |\varkappa_i|^2))$,

$$\Gamma_k = |\lambda_k|^2 \Upsilon_k + |\varphi - \varkappa_{k+1} \gamma|^2 \Gamma_{k+1}, \quad \Gamma_k = 0, \tag{4.5}$$

and

$$\Lambda_k = \varphi \bar{\beta} \Omega_{k+1} - \vartheta$$

PROOF. In the representation $a_K =: \alpha_K(x_K):$, $c_k(u_k) = =: \sigma(x_k, u_k):$, $a_k(\xi^{k+1}) =: \alpha_{k+1}(x_k, \xi^{k+1}):$, $k \in [0,K)$ the recurrent equation (4.5) in [13] has the form $\alpha_k(\xi_k, \xi^{k+1}) =$

$$= \int \alpha_{k+1}(\xi, \xi^{k+1}, \eta, u) q'(\xi - \varphi \xi_k - \beta u_k, \eta - \gamma \xi_k - \delta u_k) d\xi d\eta + \sigma(\xi_k, u_k), \tag{4.6}$$

where

$$u=-\lambda_{k+1}z, \quad z=\varphi z_k+\beta u_k+\varkappa_{k+1}(\eta-\gamma' z_k).$$

We assume that the function $\alpha_k(\xi)$ has the squared form (4.4), in particular, at the instant t_k it has the same form, namely, $\alpha_k(\xi)=\Omega|\xi|^2$. Insering the latter expression into (4.6) and integrating, we obtain, that the function α_{k-1} is of the analogous form, moreover, $\Upsilon_{k-1}=$ $=\nu'+|\beta|^2\Omega_k$ and Ω_k is defined by Ω_{k-1} due to the formula (4.3), Γ_k is defined due to the formula (4.5) and

$$d_k=d_{k+1}+\hbar(\Omega_{k+1}\sigma+\Gamma_{k+1}(\sigma+2\mathrm{Re}\varkappa_{k+1}\bar{v}+\nu'|\varkappa_{k+1}|^2.$$

Summing $\sum d_k-d_{k+1}$ and taking into account $d_K=0$, $\Gamma_K=0$, we obtain (4.4).

LEMMA. We assume, that from the time t_{k+1} the controls $u_{k'}$ are chosen to be linear $u_{k'}=-\lambda_{k'}z_{k'}$ with the coefficients (4.2), and $z_{k'}$, $k'>K$ are calculated according to the results of the subsequent indirect measurement $\eta_{k+1}, \ldots, \eta_{K-1}$ due to the formula (3.11) with the initial condition $z_k=z$. Let also the indirect measurement be described by the Gaussian distributions (3.9) up to the time t_k. Then the operator $\rho_k(\xi^k,\xi)$, defined in [13]; equ. (4.9) , or in [14]; equ.(6) , has the following normal form

$$\rho_k(\xi^k,\xi)=\lambda_k(\xi^k)+:(\Upsilon_k|u+\lambda_k\hat{x}_k|^2+|\varphi-\varkappa_{k+1}\gamma|^2) \Gamma_{k+1}|z-\hat{x}_k|^2)r^{\circ}_k(y_k|\xi^k):, \tag{4.7}$$

where

$$\lambda_k(\xi^k)=:(\Omega_k|\hat{x}_k|^2+\hbar(\Omega_k+\Gamma_k)\Sigma_k\mathbb{1}+d_k\mathbb{1})r^{\circ}_k(y_k|\xi^k):, \tag{4.8}$$

the operator $\hat{x}_k$, defined in (3.16), is linear with respect to y_k, and $r^{\circ}_k(\eta|\xi^k)$ is the distribution (3.14) with the parameters $\nu^{\circ}=\nu+\mu^{\circ}$, where $\mu^{\circ}=\max(0,\varepsilon)$.

Indeed, the operator $\rho_k(\xi^{k+1})$ analogically to any operator of density on B_k is defined by the distribution

$$r_k(\eta, \xi^{k+1}) = \iint \varkappa_k(\xi, \xi^{k+1}) q_k(\xi - \varphi \xi_{k-1} - \beta u_{k-1}, \eta - \gamma \xi_{k-1} - \delta u_{k-1}) p_{k-1}(\xi_{k-1} | \xi^k) d\xi d\xi_{k-1} \tag{4.9}$$

as by a symbol of the contrary order. This symbol is normal when $\varepsilon \leqslant 0$ and antinormal when $\varepsilon < 0$. In the former case inserting the operator symbol (4.4) into (4.9) and integrating with respect to the Gaussian type of the distribution $p_{k-1}(\xi_{k-1} | \xi^k)$ we obtain (4.7), where $r_k^o(\eta | \xi^k)$ coincides with the distribution $r_k(\eta | \xi^k)$ of Gaussian type (3.14) with the parameter $\nu' = \nu$. In the latter case $\varepsilon > 0$ the normal symbol of the operator $\rho_k(\xi^{k+1})$ is obtained from (4.9) by means of the convolution with the distribution (3.9) when $\mu = \varepsilon$, as the result of which the parameter ν increases for ε. In any case $r_k^o(\eta | \xi^k)$ is the normal symbol of the conditional density operator on B_k.

THEOREM 3. Let the quantum oscillator (3.1), (3.2) be described by the Gaussian initial and final distributions (3.7), (3.8) and the quality criterion (see [13], Equ. (4.2) or [14]) be defined by the sum of mean-squared final and transition operators $a_K = \Omega x_K^* x_K$ and $c_k(u_k)$ of form (4.1) respectively. Then the optimal strategy is linear: $u_k = -\lambda_k z_k$, where λ_k is defined by (4.2), and z_k are optimal linear estimates (4.11) based on the results $\{\eta_i\}_{i \leqq k}$ of the coherent measurements (3.5) which are described by the distributions (3.9 with the minimal value of the parameter $\mu = \mu^o$.

PROOF. is reduced to the verification of conditions (4.10), (4.11) in [13], or (5) in [14] for the operator (4.7) and the mentioned above measurement at each instant t_k. As Υ_k,

$\Gamma_k \geq 0$ and the density operator $:r_k^o(y_k|\xi^k):$ is non-negative definite, the differencies $\rho_k(\xi^{k+1}) - \lambda_k(\xi^k)$ are non-negative definite operators. It remains to verify equations (4.13) in [13] for the optimal strategy $\gamma_k^o(\xi^k,\eta) = -\lambda_k z_k$ of the coherent measurements (3.5) or, what is the same, of the measurements (3.15) with the Gaussian distributions $n_k^o(z)$ corresponding to the case $\mu = \mu^o$. Inserting $u = -\lambda_k z$ into (4.7) and considering (4.5), we obtain

$$\rho_k(\xi^k,\eta,\gamma_k^o(\xi^k,\eta)) - \lambda_k(\xi^k) = \Gamma_k : |z-\hat{x}_k|^2 r_k^o(y_k|\xi^k):.$$

Thus, equations (4.13) in [13] with $\varepsilon > 0$ can be written in the form

$$(z-\hat{x}_k)\cdot n_k^o(z-\hat{x}_k)\cdot = 0, \tag{4.10}$$

and the conjugate ones with $\varepsilon < 0$ can be written in the form

$$\cdot n_k^o(z-\hat{x}_k)\cdot(z-\hat{x}_k) = 0. \tag{4.11}$$

As the operators $\cdot n_k^o(z-\hat{x}_k)\cdot$ described by the Gaussian distributions $n_k^o(z)$, which realize the lower bound of the Heisenberg inequality, are proportional to the known coherent projectors, as the operators $\hat{x}_k$ when $\varepsilon > 0$ are proportional to the annihilation operators and when $\varepsilon < 0$ to the creation operators, for which the coherent projectors are the right and the left eigenprojectors respectively, equations (4.10), (4.11) hold. However, in the antinormal case when the coherent projectors are described by the Dirack distributions $\delta(z)$ on $\mathbb{C}$, these equations are evident, for instance, (4.10) has the form

$$\cdot\check{\delta}(z-\hat{x}_k)\cdot(z-\hat{x}_k) = \cdot\,\delta(z-\hat{x}_k)(z-\hat{x}_k)\cdot = 0, \quad \varepsilon < 0.$$

The minimal losses, appropriate to the optimal quantum strategy by analogy to the classical case are defined by the following formula $\alpha_0^0 = \Omega_0 |z|^2 +$

$$+ \hbar(\Omega_0 \Sigma + \sum_{k=1}^{K} (\Omega_k \sigma + \bar{\lambda}_{k-1}(\varphi\bar{\beta}\Omega_k - \nu)\Sigma_{k-1}), \qquad (4.12)$$

where λ_k, Ω_k are defined by (4.2), (4.3), and $\mathcal{H}_k$, Σ_k are defined by (3.11), (3.12) with $\mu = \max(0, \varepsilon)$.

By setting $\omega \simeq \omega \cdot \Delta$, $\vartheta \simeq \vartheta \cdot \Delta$, $\nu' \simeq \nu' \Delta$ in the theorem conditions and passing to the limit as $\Delta \to 0$, it is easy to obtain for continuous time the solution of the optimal control problem for the quantum oscillator (3.17), (3.18) due to the quality criterion

$$\Omega \langle : | x(\tau) |^2 : + \int_0^\tau \langle \omega : | x(t) |^2 : -2\mathrm{Re}\,\nu \bar{u}(t) x(t) + \nu' | u(t) |^2 \, dt.$$

It is evident, that the optimal strategy is of classical [16] form $u(t) = -\lambda(t) z(t)$, where $\lambda(t) = (\bar{\beta}\Omega(t) - \nu)/\nu'$, $\Omega(\tau) = \Omega_0$, $\Omega(t)$ satisfies the equation:

$$-d\Omega(t)/dt + (\alpha + \bar{\alpha})\Omega(t) = \omega - |\lambda(t)|^2 \nu',$$

and $z(t)$ is the optimal estimate $x(t)$ obtained by coherent measurements, corresponding to the case $\mu = \max(0, \varepsilon)$. In particular, when $\beta = \gamma = \varepsilon = \alpha + \bar{\alpha} > 0$, $\nu = \upsilon = \sigma$, $\nu' = \nu + \theta$, $\vartheta = \omega$, we obtain the solution of the optimal control problem for the quantum oscillator connected with the transmission line (2.3) of resistance $\gamma/2$ which was considered as an example in §2. Note, that equations (3.17), (3.18) can be reduced to (2.2), (2.3) due to the generalized derivatives $v(t) = v(dt)/dt$, $y(t) = y(dt)/dt$ which are operator-valued white noises, representing the direct and reverse waves on the input of the oscillator.

ACKNOWLEDGEMENT

The author expresses his warmest thanks to Professor S.Diner and Professor P.Serafini for the hospitality at CISM Udine.

REFERENCES

1. Belavkin,V.P.: To the theory of control in quantum observable systems, Automic and Remove Control, 2(1983).
2. Davies,E.B. and J.T.Lewis : An operational approach to quantum probability, Commun. Math. Phys., 65(1970), 239--260.
3. Accardi, L.: On the non-commutative Markov property, Functional Anal. Appl.,9(1975), 1-8.
4. Belavkin,V.P.: An operational theory of quantum stochastic processes, Proc. of VII-th Conference on Coding and Inform.Trans.Theory, Moscow-Vilnus, 1978.
5. Lindblad,G.: Non-Markovian quantum stochastic processes, Commun.Math.Phys., 65(1979), 281-294.
6. Accardi,L.and Frigerio A. and J.T.Lewis: Quantum stochastic processes, Publ. RIMS Kyoto Univ., 18(1982),97-133.
7. Barchielli,L.Lanz and G.M.Prosperi, Nuovo Cimento, B 72(1982), 79.
8. Belavkin,V.P.: Reconstruction theorem for quantum stochastic process, Theor. and Math.Phys.,62(1985),409-431.
9. Butkovski,A.G. and U.J.Samoilenko: Control of quantum mechanical processes, Nauka, Moscow 1984.
10. Belavkin,V.P.: Optimal quantum filtering of Markov signals, Problems of Control and Inform.Theory, 7(1978), 345-360.
11. Belavkin,V.P.:Optimal filtration of Markovian signals in white quantum noise, Radio Eng. Electron. Phys., 18(1980), 1445-1453.

12. Accardi,L. and K.R. Parthasaraty: Quantum stochastic calculung. Springer LNM (to appear).
13. Belavkin,V.P.: Optimal measurement and control in quantum dynamical systems, Preprint N411, UMK, Toruń, 1979.
14. Belavkin,V.P.:Optimization of quantum observation and control, Lecture Notes in Control and Inform. Sciences, IFIP, Optimization Techniques, Warsaw, 1979, PartI, Springer-Verlag.
15. Holevo,A.S.: Investigations on general statistical decision theory, Proc. of MIAN, CXXIV, Nauka, Moscow, 1976.
16. Aström,K.J.: Introduction to stochastic control theory, Academic Press, New York, 1970.
17. Klauder,J.R. and E.C.D.Sudarshan: Fundamentals of quantum optics. W.A.Benjamin, inc. New York, Amsterdam 1968.
18. Haus,H.A.: Steady-state quantum analysis of linear systems. Proc. IEEE, vol. 58, pp. 110-129, 1970.
19. Lax, M.: Quantum noise IV. Quantum theory of noise sources. Phys.Rev., vol. 145, pp. 1599-1611, 1965.
20. Belavkin,V.P.: Optimal linear randomized filtration of quantum boson signals. Problems of control and inform. theory, vol.3 (1), pp. 47-62, 1974.

QUANTUM NONDEMOLITION FILTERING

J.W. Clark*, T.J. Tarn**
*McDonnell Center for the Space Sciences
and Department of Physics

and

**Department of Systems Science and Mathematics
Washington University, St. Louis, Missouri, U.S.A.

ABSTRACT

A continuous-time filter is formulated for a physical system modeled as an infinite-dimensional bilinear system. Efforts focus on a quantum system with Hamiltonian operator of the form $H_0 + u(t)H_1$, where H_0 is the Hamiltonian of the undisturbed system, H_1 couples the system to an external classical field, and $u(t)$ represents the time-varying signal carried by this field. An important problem is to determine when and how the signal $u(t)$ can be extracted from the time development of the measured value of a suitable system observable C. There exist certain observables, called quantum nondemolition observables (QNDO), which have the property that their expected and measured values coincide. The invertibility problem has been posed and solved for these quasiclassical observables. Since one is addressing an *infinite-dimensional* bilinear system, the domain issue for the operators H_0, H_1, and C becomes nontrivial. An additional complication is that the input observable C is in general time dependent. Having derived conditions for invertibility, necessary and sufficient conditions are developed for an observable to qualify as a QNDO. If an observable meets both sets of criteria, it is said to constitute a quantum nondemolition filter (QNDF). By construction, the associated filtering algorithm separates cleanly into the choice of output observable (a QNDO) and the choice of procedure for processing measurement outcomes. This approach has the advantage over previous schemes that no optimization is necessary. QNDF's may see practical application in the demodulation of optical signals and the detection and monitoring of gravitational waves.

1. INTRODUCTION

Consider a quantum-mechanical control system described by the state-evolution equation

$$i\hbar \frac{d\psi(t)}{dt} = [H_0 + u(t)H_1]\psi(t) \quad , \quad \psi(0) = \psi_o \quad , \tag{1}$$

and the output function

$$y(t) = \langle\psi(t)\,|\,C(t)\psi(t)\rangle \quad . \tag{2}$$

The state of the system is represented by a vector ψ in a suitable Hilbert space $\boldsymbol{H}$. The time evolution of ψ is governed by a Hamiltonian operator $H = H_0 + u(t)H_1$ acting in $\boldsymbol{H}$. This Hamiltonian is made up of self-adjoint operators H_0 and H_1 in $\boldsymbol{H}$, the coefficient $u(t)$ being a bounded, real, analytic function of the time t. One may interpret H_0 as the unperturbed Hamiltonian of the system and H_1 as the system observable which couples to an external classical field whose strength is represented by $u(t)$. Thus $u(t)$ plays the role of (a) a signal to be extracted or, alternatively, (b) a control which may be adjusted to guide the time development of the system. Here we shall be concerned with role (a).

Due to the self-adjointness of the Hamiltonian operator, the evolution of $\psi(t)$ is confined to a manifold of constant norm, chosen for convenience to be the unit sphere $S_{\boldsymbol{H}}$. The time-development of the system is monitored via the function $y(t)$, taken as the Hilbert-space inner product of the vector $C(t)\psi(t)$ with the state representative $\psi(t)$. The system observable C, generally time-varying, is also a self-adjoint operator in $\boldsymbol{H}$. It is well-known principle of quantum theory [2] that the inner product $y(t)$ gives the expected value of the dynamical quantity $\tilde{C}(t)$ associated with the operator $C(t)$, for the system in state $\psi(t)$. In an earlier paper [1], we have determined necessary and sufficient conditions for invertibility of the control system (1)-(2) under the assumption that the chosen output $y(t)$ yields not merely the expected value of $\tilde{C}(t)$ but in fact the actual measurement result. This assumption is not justified unless the dispersion in measurement results obtained for an ensemble of copies of the system is zero. The latter requirement implies in turn (see e.g., [2]) that the state $\psi(t)$ evolves on an eigenmanifold of C, once an initial precise measurement of $\tilde{C}$ has been performed. Accordingly, the results of [1] were obtained under the assumption that the output observable C is a very special sort of operator, namely a quantum nondemolition observable (QNDO) [3].

For invertibility to hold it is further required of C that it allows the input $u(t)$ to be calculated from a knowledge of the output $y(t)$. A QNDO which so qualifies will be called a quantum nondemolition filter (QNDF) [4,5]. The aim of this paper is to explore the concepts of QNDO and QNDF within the systems-theoretic framework. Appealing to the existence of an analytic domain in the sense of Nelson [6], we shall, in particular, establish necessary and sufficient conditions for an observable to be a QNDF. Such criteria may be valuable in the design of quantum nondemolition techniques for the detection and accurate reconstruction of signals which are so weak that the quantum nature of the detection apparatus becomes important. Space limitations preclude the inclusion of proofs of our results. Detailed proofs are available in [1,5].

2. QUANTUM NONDEMOLITION MEASUREMENT

Quantum nondemolition (QND) measurements--or "back-action-evading measurements"--were first proposed by physicists, as a means for monitoring a weak classical force acting on a one-dimensional quantum-mechanical harmonic oscillator [7-10]. QND measurement involves a time sequence of precise, instantaneous measurements of a suitably chosen observable, known as a quantum nondemolition observable (QNDO). The key feature of a QND observable is that its Heisenberg time development [2] proceeds independently of observables with which it does not commute. Accordingly, if the Hamiltonian governing the evolution is known, repeated precise measurement of the QNDO may be carried out with results which are free from uncertainty and completely predictable.

These notions have been formalized by Unruh [11] and Caves *et al.* [7]. The definition of QNDO adopted in our work is:

Definition 1. Given the system Hamiltonian $H = H_0 + u(t)H_1$, an observable C is a quantum nondemolition observable (QNDO) if, in a sequence of precise measurements of $\tilde{C}$, the result of each measurement of $\tilde{C}$ after the first is uniquely determined by the outcome of the first measurement of $\tilde{C}$ together with the outcomes $b_o^{(1)},...,b_o^{(g)}$ of measurements of an additional set of g quantities, $0 \le g < \infty$, performed at the initial time.

In fact, this is what Caves *et al.* [7] call a generalized QNDO. The additional g measurements are needed to deal with the case that C depends, in its Heisenberg evolution, on g observables $B_1, \dots, B_g$ which commute with one another and with the initial value of C. It is assumed that there has been no preparation of the state of the system prior to the first measurement.

Remark 1. From Definition 1, it follows that a necessary and sufficient condition for $C(t)$ to be a QNDO is that after the initial measurement(s) the system remains in an eigenstate of C, the eigenvalue $c = c(c_o, b_o^{(1)}, \dots, b_o^{(g)}; t, t_o)$ being determined uniquely as a function of time by the outcome $(c_o, b_o^{(1)}, \dots, b_o^{(g)})$ of the first measurement(s). Accordingly, we may state this essential property of a QNDO: there is no dispersion in any subsequent measurement of $\tilde{C}$, the expected value $y(t) = \langle\psi(t)|C(t)\psi(t)\rangle$ being coincident with the certain outcome $c(c_o, b_o^{(1)}, \dots, b_o^{(g)}; t, t_o)$. In this sense a QNDO may be thought of as a "quasiclassical" observable. We note that in the case of QNDO's, measurements may be performed and predictions given for a single system, without reference to an ensemble.

An alternative statement of necessary and sufficient condition has been given by Caves *et al.* [7], in terms of the Heisenberg picture of quantum dynamics: The observable C is a (generalized) QNDO if and only if

$$[C_{\mathrm{H}}(t), C_{\mathrm{H}}(t')] = 0 \quad , \quad \forall t, t' \quad , \tag{3}$$

where C_{H} is the Heisenberg transform [2] of the Schrödinger operator C.

For C to be a QNDO it is evidently sufficient (but not necessary!) that C be a constant of the motion, $dC_{\mathrm{H}}(t)/dt = 0$.

We follow Caves *et al.* [7] in imposing some rather broad requirements on the interaction between the measuring apparatus and the system:

(a) The measuring instrument responds to the observable C and hence the interaction Hamiltonian H_I depends on C as well as one or more dynamical quantities referring to the measuring apparatus alone.

(b) The measuring instrument does not respond to observables of the system other than C.

Most simply, we might suppose $H_I = kC\overline{A}$, where k is a coupling constant and $\overline{A}$ is some observable of the measuring apparatus.

Proposition 1 [7]. Given that conditions (a) and (b) are fulfilled, the evolution of a QND observable C in the Heisenberg picture is completely unaffected by the interaction with the measuring instrument.

This proposition means that the expectation and variance of C evolve during and after measurement just as if the measuring apparatus is disconnected. It will allow us to suppress any details of the interaction H_I in the formal analysis below, although in practice these details are extremely important.

3. QUANTUM NONDEMOLITION FILTERS

Quantum nondemolition measurements were first conceived as a possible means for detecting gravitational waves, traveling to us through space from energetic cosmic sources (supernovae, neutron-star binaries, etc.). The difficulty of detecting gravitational waves lies in the weakness of the coupling to material detectors and the vast distances of proposed sources. The essential ideas of QND or back-action-evading measurement may be exploited if one imagines that the gravitational wave couples (weakly) to a quantum harmonic oscillator (in practice, a resonant bar of macroscopic size and suitable composition). The Hamiltonian for this detection system takes the standard form $H = H_0 + u(t)H_1$, where $u(t)$ represents the unknown classical gravitational signal, H_0 is the Hamiltonian for the free one-dimensional oscillator, and H_1 is the oscillator coordinate x. Another practical context in which QND ideas may be useful is optical communication. In this case the components of the standard form of H assume the following interpretations: $H_0 = \hbar\omega(a^\dagger a + \frac{1}{2})$ is the Hamiltonian of the unperturbed, single-mode radiation field, H_1 is a function of the photon annihilation and creation operators a and $a^\dagger$ which depends on the modulation scheme in effect, and $u(t)$ is the impressed optical signal.

Consider now the problem of signal detection and reconstruction for one or another such system, in terms of the results of measurement of a dynamical quantity $\tilde{C}$. If, in the absence of any signal ($u(t) \equiv 0$) the corresponding observable C is a QNDO, then by definition the results of a sequence of precise measurements of $\tilde{C}$ will be exactly predictable from the results of an initial set of measurements. If C remains a QNDO in the presence of the $u(t)$ term in the system Hamiltonian, then there is the possibility of determining $u(t)$ by the changes it produces in the otherwise precisely predictable results of a series of $\tilde{C}$-measurements. Clearly, not all QNDO's will permit $u(t)$ to be monitored in this way. QNDO's for which this possibility can be realized are encompassed by the following definition.

Definition 2. An observable C is a quantum nondemolition filter (QNDF) iff in the presence of an arbitrary analytic signal a sequence of measurements of $\tilde{C}$ can reveal with arbitrary accuracy the time dependence of the signal.

We may already note the trivial fact that a QNDF must be a QNDO in the presence (or absence) of the external force or signal. A further requirement mentioned by Caves *et al.* [7] is that measurements of $\tilde{C}$ can be carried out at arbitrarily closely-spaced times. This latter requirement may actually be more stringent than necessary, since interpolating functions can be used to reconstruct the output $y(t)$ provided $y(t)$ has a finite spectrum and its Nyquist rate does not exceed the sampling rate. Given that C is a QNDF, the signal $u(t)$ can be recovered by suitably processing the measurement outcomes of $\tilde{C}$ furnished by $y(t)$; usually one expresses $u(t)$ in terms of an appropriate derivative of $y(t)$ [1]. As a practical matter, we note that the implied filtering algorithm separates neatly into the choice of output observable (a QNDO) and the choice of procedure for processing the measurement outcomes.

The QNDO condition (3) given by Caves *et al.* [7], together with the invertibility criteria established in [1], provide necessary and sufficient conditions for an observable C to be a QNDF. We proceed now to sketch certain mathematical underpinnings of this statement, and to offer examples of its realization. For a cleaner notation, the Dirac constant $\hbar$ in (1) is henceforth taken as unity.

4. ASSUMED PROPERTIES OF THE CONTROL SYSTEM

The assumptions on our control system (1)-(2) which were imposed in [1] will again be adopted:

(a) The operators $K_0 = -iH_0$ and $K_1 = -iH_1$ are skew-adjoint.

(b) The observable C is supposed to have the structure $C(t) = \Sigma_{r=1}^{q<\infty} \gamma_r(t) iQ_r$, where the functions $\gamma_r(t)$ are real analytic in t and the Q_r are time-independent skew-adjoint operators.

(c) To ensure a well-defined output, we take dom $C \supset$ dom $H_0 \cap$ dom H_1.

(d) We assume that there exists an analytic domain D_ω (as defined below) for the Lie algebra A' generated by $K_0, K_1, Q_1, \ldots, Q_q$ under the bracket operation $[\bullet,\bullet]$.

(e) It is further assumed that the tangent space $A(\phi) \equiv \{X(\phi), X \in A\}$ of the Lie algebra A generated by K_0, K_1 has constant, finite dimension for all $\phi \in D_\omega$.

Definition 3. System (1)-(2) is said to admit an analytic domain D_ω if there exists a common, dense invariant domain D_ω for the Lie algebra A' such that (i) D_ω is invariant under the corresponding unitary Lie group G as well as A', (ii) on D_ω, an arbitrary element g of G can be written locally in a group parameter t as $g = \exp Xt$, where $X \in A'$, and (iii) this exponential expression can in fact be extended globally to all $t \in \mathbf{R}^+$.

Sufficient conditions for the existence of an analytic domain are given in [6] and [12]. The analytic domain plays a central role in the rigorous study of fundamental examples of quantum-mechanical control systems, especially the simple harmonic oscillator. For our purposes, the crucial point is that on such a domain, standard techniques, developed for the treatment of finite-dimensional bilinear systems, are applicable [13-15].

Consider the maximal integral manifold $M \subset S_{\mathcal{H}}$ of solution curves of (1), characterized by the Lie algebra A through Frobenius' theorem [13]. This manifold inherits the Hilbert space topology and therefore it is paracompact and connected. Referring to the assumptions (a)-(e) set forth above, we observe that, in general, M need not belong to D_ω. However, $D_\omega \cap M$ is dense in M, and since D_ω is invariant under $A'(K_0, K_1, Q_1, \ldots Q_q)$, the submanifold $D_\omega \cap M$ will actually be the main scene of action. Assumption (e) implies that the manifold M and hence $D_\omega \cap M$ is finite-dimensional [13]. In addition, part (ii) of Definition 3 implies that

$D_\omega \cap M$ is an analytic manifold.

5. ANALYTIC INVERTIBILITY

If invertible on D_ω, the system (1)-(2) will be called analytically invertible. Since the system is nonlinear, we take the lead of Hirschorn [16] in formulating suitable definitions.

Definition 4.

(i) System (1)-(2) is analytically invertible at $\psi_o = \psi(t=0) \in M \cap D_\omega$ if distinct inputs $u_1(t)$, $u_2(t)$ give rise to distinct outputs, i.e.,

$$y(t,u_1,\psi_o) \not\equiv y(t,u_2,\psi_o) \quad .$$

(ii) System (1)-(2) is strongly analytically invertible at ψ_o if there exists an open neighborhood N of ψ_o such that analytic invertibility holds at ξ for all $\xi \in N \cap D_\omega$.

(iii) System (1)-(2) is strongly analytically invertible if there exists an open submanifold M_o of M, dense in M, such that strong analytic invertibility holds at ζ, for all $\zeta \in M_o \cap D_\omega$.

Definition 5. Given system (1)-(2), we define a sequence of operators C_p, where p is a positive integral index, by the recursive relation

$$C_p(t) = [C_{p-1}(t), K_0] + \frac{d}{dt} C_{p-1}(t) \quad ,$$

with $C_{p=0}(t) = C_0(t) = C(t)$. If the output observable C is independent of time, then

$$C_p = (-1)^p \, \mathrm{ad}_{K_0}^p C \quad .$$

(We adopt the conventional notation $\mathrm{ad}_X Y = [X,Y]$ and more generally, for p a positive integer, $\mathrm{ad}_X^p Y = [X, \mathrm{ad}_X^{p-1} Y]$, $\mathrm{ad}_X^0 Y = Y$.)

Noting that the output map is time-varying, a definition for the relative order of the system may be formulated as follows:

Definition 6. The relative order μ of system (1)-(2) is the smallest positive integer p such that $[C_{p-1}(t), K_1] \neq 0$ for almost all t, where "for almost all t" means the Lebesgue measure of $\{t : [C_{p-1}(t), K_1] = 0\}$ is zero.

If a system is invertible, one can in principle recover the control $u(t)$ by constructing an inverse system such that when driven by an appropriate derivative of

y, the inverse system produces u as its output. As in finite-dimensional systems, the relative order gives (essentially) the lowest order of differentiation necessary to recover u.

Since $C(t)$ is dependent on time, we could think of making observations on the time derivatives of C. However, as far as invertibility goes there is nothing to be gained from this; for if $d^\rho C/dt^\rho$, $\rho \geq 1$, is substituted for the observable C in the above, the relative order is greater than or equal to that found for C.

If $\mu < \infty$, it may be shown that system (1)-(2) is invertible and an inverse system can be constructed. In general the state space for the inverse system is a submanifold of M, denoted by M_μ.

Definition 7. The <u>inverse submanifold</u> for the system (1)-(2) having relative order μ is defined as

$$M_\mu = \{\zeta \in M \cap D_\omega : <\zeta | [C_{\mu-1}(t), H_1]\zeta> \neq 0 \text{ for almost all } t\} \quad .$$

Verification of the statement that M_μ is a submanifold of M runs thus: On the analytic domain D_ω, $[C_{\mu-1}(t), K_1]\zeta$ is an analytic vector; hence $g(\zeta) = <\zeta | [C_{\mu-1}(t), K_1]\zeta>$ is a nonzero real analytic function of ζ for almost all t. The analyticity of $g(\zeta)$ means that it cannot vanish on any open subset of M. This fact together with the continuity of $g(\zeta)$ implies that M_μ is an open dense subset of M and therefore a submanifold of M.

We are now prepared to state the main results currently available on the invertibility problem, derived in [1].

Theorem 1. Given that system (1)-(2) admits an analytic domain D_ω, the system is strongly analytically invertible if its relative order μ is finite. If indeed μ is finite and if the initial state ψ_o belongs to the inverse submanifold M_μ, the system specified by

$$\frac{d}{dt}\hat{\psi}(t) = a(\hat{\psi}(t)) + \hat{u}(t) b(\hat{\psi}(t)) \quad , \quad \hat{\psi}(0) = \psi_o \quad , \tag{1'}$$

$$\hat{y}(t) = d(\hat{\psi}(t)) + \hat{u}(t) e(\hat{\psi}(t)) \quad , \tag{2'}$$

provides an acceptable inverse for the quantum control system (1)-(2) with

$$\begin{aligned} a(\hat{\psi}(t)) &= K_0\hat{\psi} - <\hat{\psi} | [C_{\mu-1}, K_1]\hat{\psi}>^{-1} <\hat{\psi} | C_\mu \hat{\psi}> K_1\hat{\psi} \quad , \\ b(\hat{\psi}(t)) &= <\hat{\psi} | [C_{\mu-1}, K_1]\hat{\psi}>^{-1} K_1\hat{\psi} \quad , \\ d(\hat{\psi}(t)) &= -<\hat{\psi} | [C_{\mu-1}, K_1]\hat{\psi}>^{-1} <\hat{\psi} | C_\mu \hat{\psi}> \quad , \\ e(\hat{\psi}(t)) &= <\hat{\psi} | [C_{\mu-1}, K_1]\hat{\psi}>^{-1} \quad . \end{aligned} \tag{4}$$

Corollary 1.1. Suppose that (1)-(2) admits an analytic domain and also that $d^{\rho}C/dt^{\rho}$ vanishes for some positive integer ρ. Then $\mu < \infty$ provides a necessary as well as sufficient condition for the system to be strongly analytically invertible.

Corollary 1.2. Given that the system (1)-(2) admits an analytic domain, the condition $[C, \mathrm{ad}_{K_0}^{p-1} K_1] \neq 0$ for some positive $p < \infty$ is necessary and sufficient for analytic invertibility. If this condition is met and ν is the minimum such p, the inverse system is specified by (1'), (2'), (4) with the replacements $\mu \to \nu$, $[C_{\mu-1}, K_1] \to [C, \mathrm{ad}_{K_0}^{\nu-1} K_1]$ and $C_\mu \to (-1)^{\nu-1}\, \mathrm{ad}_{K_0}^{\nu-1} C$ in (1'), (2'), (4) and the definition of the inverse submanifold.

6. CONDITIONS FOR A QUANTUM NONDEMOLITION FILTER

Let E_C be the subspace of $\boldsymbol{H}$ spanned by the union over all $t \in \boldsymbol{R}$ of the set of all eigenstates of $C(t)$. According to Remark 1 of Section 1, a crucial requirement for $C(t)$ to provide a QNDO is that (after the prescribed initializing measurement of $\tilde{C}$ and other dynamical quantities), the state $\psi(t)$ of the system remain in the subspace E_C. Further, since the analytic domain D_ω is dense in the unit sphere $S_{\boldsymbol{H}}$, it is sufficient to carry out the analysis on $D_\omega \cap E_C$. For $\psi(t) \in D_\omega$, the map $t \to C(t)\psi(t)$ is analytic in $\boldsymbol{R}$ and therefore the function $y(t) = \langle\psi(t)|C(t)\psi(t)\rangle$ is also analytic for all real t. If $\psi(t)$ is outside D_ω, we approximate it by some $\psi_\omega(t) \in D_\omega$ (see [1] for details, especially Remark 3 of that paper), in such a way that we may always consider $y(t)$ as analytic in t, $\forall\, t \in \boldsymbol{R}$. This ensures that complete information for constructing the output at some future time t is contained in the time derivatives of the output function, evaluated at the referential time t_o.

Also calling on Remark 1, it is clear that for C to be a QNDO, the variance associated with $y(t)$ must be zero for all t. This in turn implies certain commutation relations among $H_0 = iK_0$, $H_1 = iK_1$, $C(t)$, and the partial derivatives of $C(t)$. The following definition facilitates the explication of these relations.

Definition 8. For any positive integer p, a family $\Lambda^{[p]}$ of operators is specified recursively by

$$\Lambda^{[p]} = \begin{cases} \text{the single element } C_p \quad , & 0 \le p < \mu \quad , \\ \{\dot{L} + [L, K_0],\ [L, K_1] \mid L \in \Lambda^{[p-1]}\} \quad , & p \ge \mu \quad , \end{cases}$$

where μ is the relative order of the system (1)-(2) and

$$C_p = [C_{p-1}, K_0] + \frac{\partial C_{p-1}(t)}{\partial t} \quad , \quad C_0(t) = C(t) \quad .$$

We also construct the set Λ as the union of all the $\Lambda^{[p]}$. The elements of Λ, being self-adjoint operators in $\mathbb{H}$, are presumed to define physical observables [2].

Having introduced the operators $\Lambda^{[p]}$, it becomes convenient (as well as natural) to replace the condition (3) of Caves *et al.* by an essentially equivalent condition in terms of the operators C_p which played a key role in our approach to the invertibility problem. The reader should consult [5] for the details of this process. We arrive at an economical formulation of necessary and sufficient conditions for a time-dependent observable to provide a quantum nondemolition filter, and a consequent corollary for the time-independent case:

Theorem 2. The observable $C(t)$ qualifies as a QNDF for the quantum control system (1)-(2) iff

(a) the system is invertible, and

(b) the commutation relations

$$[C(t), L(t)] = 0 \tag{5}$$

hold for all $L \in \Lambda$, $\forall t$.

Corollary 2.1. If C is a time-independent observable, it is a QNDF iff

(a) $[C, \mathrm{ad}_{K_0}^{p-1} K_1] \neq 0$ for some positive integer $p < \infty$, and

(b) $[C, \mathrm{ad}_{K_0}^{p-1} C] = 0$ and $[C, [\mathrm{ad}_{K_0}^{p-1} C, K_1]] = 0 \ \forall\, p > 0$. (6)

Remark 2. The requirement that the relative order of the system be finite ensures that C does respond to $u(t)$. When $u \equiv 0$, conditions (5) and (6) reduce, respectively, to corresponding conditions given by Caves *et al.* [7] and Unruh [11].

Remark 3. As we mentioned earlier in this section, if $\psi(t)$ is not in the analytic domain D_ω, it can be approximated by $\psi_\omega(t) \in D_\omega$. This option rests on the fact [1] that $\psi(t)$ belongs to the closure of $R^t_{pc}(\phi)$, the reachable set of $\psi(0) = \phi$ at time t for piecewise-constant controls. The corresponding approximation $y_\omega(t) = \langle\psi_\omega(t)|C(t)\psi_\omega(t)\rangle$ to the measurement result $y(t) = \langle\psi(t)|C(t)\psi(t)\rangle$ converges uniformly to $y(t)$. This can be seen from the following sequence of manipulations, the Cauchy-Schwarz inequality being applied in the last step:

$$\begin{aligned}
|y - y_\omega| &= |\langle\psi|C\psi\rangle - \langle\psi_\omega|C\psi_\omega\rangle| \\
&= |\langle\psi|C\psi\rangle - \langle\psi_\omega|C\psi\rangle + \langle\psi_\omega|C\psi\rangle - \langle\psi_\omega|C\psi_\omega\rangle| \\
&= |\langle\psi - \psi_\omega|C\psi\rangle + \langle\psi_\omega|C(\psi - \psi_\omega)\rangle| \\
&\leq |\langle\psi - \psi_\omega|C\psi\rangle| + |\langle\psi_\omega|C(\psi - \psi_\omega)\rangle| \\
&\leq \|\psi - \psi_\omega\|^2\|C\psi\|^2 + \|C\psi_\omega\|^2\|\psi - \psi_\omega\|^2 \quad .
\end{aligned}$$

Since $\|C\psi\|$ and $\|C\psi_\omega\|$ are bounded, and since $\psi_\omega \to \psi$ uniformly, $y_\omega \to y$ uniformly.

7. EXAMPLES

Finally, we illuminate certain important aspects of quantum nondemolition filters through a few simple explicit realizations.

Example 1. (Electrooptic Amplitude Modulation) Consider the Hamiltonian

$$H = \omega a^\dagger a + iu(t)(a^\dagger - a)$$

defined on the Schwarz space $S(\boldsymbol{R})$ of infinitely differentiable functions, where a and $a^\dagger$ are the annihilation and creation operators introduced previously. The set of finite linear combinations of the Hermite functions

$$\psi_n(x') = \pi^{-1/4}(n!)^{1/2}(-1)^n\, 2^{-n/2} \exp(-x'^2/2)h_n(x') \;, \quad x' \in \boldsymbol{R} \quad, \; n = 0,1,2,\ldots \;,$$

where $h_n(x')$ is the nth-order Hermite polynomial, provides a dense set of analytic vectors invariant under a, $a^\dagger$ and $a^\dagger a$ (see [12]). We note that $H_0 = iK_0 = \omega a^\dagger a$ is just the unperturbed oscillator Hamiltonian, apart from an additive constant $-\omega/2$, while $H_1 = iK_1 = (2/m\omega)^{1/2}p$.

For the Hamiltonian H, the observable

$$C = a\, e^{i\omega t} + a^\dagger\, e^{-i\omega t} \tag{7}$$

is a QNDF according to the criteria laid down in Theorem 2 of Section 6. In particular:

(i) We have $[C,K_1] = e^{i\omega t} + e^{-i\omega t} = 2\cos\omega t \neq 0$ except at $t = l\pi/2$, $l = 1,3,5,\ldots$. Consequently the invertibility condition (a) of Theorem 2 is met.

(ii) We have $C_1 = [C,K_0] + \partial C/\partial t = 0$, and therefore $C_j = 0 \;\forall\; j \geq 1$. From this it is seen that condition (b), i.e., Eq. (5), is satisfied.

As a check on the claim that C is a QNDO, we may integrate the Heisenberg equation of motion for $C_{\rm H}(t)$, beginning at time t_o. The result is

$$C_{\rm H}(t) = 2C_{\rm H}(t_o)\cos\omega t_o + 2I \int_{t_o}^{t} u(s)\cos\omega s\; ds \quad .$$

Hence an eigenstate of $C_{\rm H}(t_o)$ remains an eigenstate of $C_{\rm H}(t)$, $\forall t > t_o$, and C is indeed a QNDO. The input $u(t)$ is obtained trivially as

$$u(t) = \frac{dy(t)}{dt}\,(2\cos\omega t)^{-1} \quad ;$$

avoiding the zero of $\cos \omega t$, we can reconstruct $u(t)$ from $y(t)$. Hence C is a QNDF as well as a QNDO.

For the given Hamiltonian, Baras [17] arrived at the observable $C_B = \frac{1}{2}(a + a^\dagger)$ as a representation of the receiver structure. Working in the discrete-time case, this form was obtained after making assumptions which permit the choice of optimal quantum observables for measurement to be separated from the choice of optimal classical post-processing of the measurement outcomes (in this case Kalman filtering). Baras *et al.* [18] call this filter separation. Within the QND approach, filter separation is quite natural. Choosing an optimal quantum observable corresponds to choosing a QNDF, and classical postprocessing is the invertibility procedure. Note that C_B, aside from a multiplicative constant, is a special case of (7) at times $t = 0, \pi, 2\pi, \ldots$.

Example 2. Next consider the Hamiltonian

$$H = J^o + u(t)(J^+ + J^-) \quad ,$$

where (suppressing dimensional constants)

$$J^o = (-d^2/dx^2 + x^2)/2 \quad , \quad J^\pm = \pm(2)^{-1/2}(d/dx \mp x) \quad .$$

The operators J^o, $J^\pm$ have commutation relations

$$[J^o, J^\pm] = \pm J^\pm \quad , \quad [J^+, J^-] = -I \quad .$$

Moreover, they share a common dense invariant domain of analytic vectors, in fact the same as that involved in the first example. As in Example 1, the given Hamiltonian may be interpreted as that of a simple harmonic oscillator coupled to an external classical force; however, in the present case the coupling is through the position operator x rather than the momentum operator p.

From the commutation relations, it can be seen that either J^+ or J^- qualifies as a QNDF if we widen the definition of QNDF to allow C to be non-selfadjoint. (Of course, such operators in general do not correspond to physically measurable quantities, and accordingly are not termed observables.) With the obvious identifications,

$$K_0 = -iJ^o \quad , \quad K_1 = -i(J^+ + J^-) \quad , \quad C = J^\pm$$

we readily find that

$$\text{(i)} \quad [C, K_1] = -i[J^\pm, J^+ + J^-] = \pm iI$$

and

$$\text{(ii)} \quad [C, K_0] = -i[J^\pm, J^o] = \pm iJ^\pm \quad .$$

From (i) it follows that the relative order is finite and the system is invertible; from (i) and (ii) it follows that $[C, \mathrm{ad}_{K_0}^{p-1} C] = 0 \ \forall \ p$ and $[C, [\mathrm{ad}_{K_0}^{p-1} C, K_1]] = 0 \ \forall \ p$. Therefore according to Corollary 2.1 of the last section, C is a QNDF (in the wider sense).

In the Heisenberg picture, $C = J^{\pm}$, denoted $C_{\mathrm{H}}(t)$, evolves according to

$$\frac{dC_{\mathrm{H}}(t)}{dt} = i[J_{\mathrm{H}}^{o}, C_{\mathrm{H}}] + iu(t)[J_{\mathrm{H}}^{+} + J_{\mathrm{H}}^{-}, C_{\mathrm{H}}]$$

$$= iC_{\mathrm{H}}(t) + iu(t)I \quad .$$

Thus the input may be formally extracted as

$$u(t) = y(t) - i\,\frac{dy(t)}{dt} \quad .$$

Example 3. As another example, consider the Hamiltonian (again with dimensional constants suppressed)

$$H = p^2/2 + u(t)(px + xp) \quad , \tag{8}$$

where p and x are respectively the momentum and position operators for a one-dimensional quantum system. The latter operators possess a common, dense, invariant domain of analytic vectors, which may be constructed according to the same prescription as in Examples 1 and 2 [12]. Here $K_0 = -ip^2/2$, $K_1 = -i(px + xp)$, and we claim that $C = p^2$ is a QNDF. The invertibility condition is satisfied due to

$$[C, K_1] = -i[p^2, px + xp] = -4p^2 \neq 0 \quad .$$

Furthermore, $[C, K_0] = 0$ and hence $[C, \mathrm{ad}_{K_0}^{m-1} C] = 0$, $[C, [\mathrm{ad}_{K_0}^{m-1} C, K_1]] = 0, \ \forall \ m$. In the Heisenberg picture p^2 evolves according to

$$\frac{dp_{\mathrm{H}}^2(t)}{dt} = -\,4u(t)\,p_{\mathrm{H}}^2 \quad .$$

Hence the input can be retrieved via

$$u(t) = -\,\frac{1}{4}\,\frac{1}{y(t)}\,\frac{dy(t)}{dt} \quad . \tag{9}$$

Note that the choice of $C = p$ would have provided an equally good QNDF, leading to (9) with a factor 1/2 in place of 1/4.

Example 4. We may elaborate on Example 3 by adding a term $+\lambda x$ to H of (8), where λ is a real constant. The modified Hamiltonian corresponds (for example) to a particle in a uniform gravitational field. Again both p and p^2 qualify as QNDF's. The operators remain QNDF's if the interaction $u(t)(px + xp)$ with the external field is replaced by $u(t)x$.

Example 5. (Dual inputs) The Hamiltonian

$$H = \omega a^\dagger a + \omega b^\dagger b + \alpha(ab^\dagger + ba^\dagger) + iu_1(t)(a^\dagger - a) + u_2(t)(a + a^\dagger) \tag{10}$$

may be used to describe a physical system consisting of two interacting one-dimensional oscillators, with the same angular frequency ω and with respective pairs of annihilation and creation operators $(a,a^\dagger)$ and $(b,b^\dagger)$. As usual $[a,a^\dagger] = [b,b^\dagger] = I$, and each of $(a,a^\dagger)$ is supposed to commute with each of $(b,b^\dagger)$. There is an interaction between the two oscillators, represented by the term $\alpha(ab^\dagger + ba^\dagger)$, where α is a real constant. The "a" oscillator is coupled to two external fields (or two signals) $u_1(t)$ and $u_2(t)$, via its coordinate and momentum operators, respectively.

It is asserted that the observables

$$C = ae^{i\omega t} + a^\dagger e^{-i\omega t} - i(be^{i\omega t} - b^\dagger e^{-i\omega t}) \quad ,$$
$$D = \omega^{-1}\, \partial C/\partial t$$

are commuting QNDF's for the Hamiltonian (10). That C and D commute is easily verified by direct computation. To see that C is a QNDF, let us construct its Heisenberg equation of motion. We have

$$[H,C] = iu_1(t)(e^{-i\omega t} + e^{i\omega t}) + u_2(t)(e^{i\omega t} - e^{-i\omega t}) + \omega[a^\dagger a,C]$$
$$+ \omega[b^\dagger b,C] + \alpha(be^{i\omega t} - b^\dagger e^{-i\omega t} - iae^{i\omega t} - ia^\dagger e^{-i\omega t}) \quad .$$

Since

$$i\omega[a^\dagger a,C] + i\omega[b^\dagger b,C] + \partial C/\partial t = 0 \quad ,$$

we arrive at

$$dC_H/dt = 2Iu_1(t)\cos\omega t - 2Iu_2(t)\sin\omega t + \alpha C_H$$

and conclude that C is a QNDO. For $D = \omega^{-1}\,\partial C/\partial t$ we obtain straightaway the Heisenberg equation

$$dD_H/dt = -2Iu_1(t)\sin\omega t - 2Iu_2(t)\cos\omega t + \alpha D_H \quad ,$$

which implies that D is likewise a QNDO. The measurement results for $C(t)$ and $D(t)$, denoted respectively by $c(t)$ and $d(t)$, evidently satisfy

$$\dot{c} = 2u_1(t)\cos\omega t - 2u_2(t)\sin\omega t + \alpha c \quad ,$$

$$\dot{d} = -2u_1(t)\sin\omega t - 2u_2(t)\cos\omega t + \alpha d \quad .$$

Knowing $c(t)$ and $d(t)$, these two equations can be solved simultaneously for $u_1(t)$ and $u_2(t)$. The explicit result is

$$u_1(t) = \frac{1}{2}\left[(\dot{c} - \alpha c)\cos\omega t - (\dot{d} - \alpha d)\sin\omega t\right] \quad ,$$

$$u_2(t) = -\frac{1}{2}\left[(\dot{c} - \alpha c)\sin\omega t + (\dot{d} - \alpha d)\cos\omega t\right] \quad .$$

Although it does not provide a direct illustration of the formal results of Sections 5 and 6, this last example, involving dual inputs, is included to show in concrete terms that the QNDF scheme is not restricted to the single-input case. The general multi-signal case remains an open problem.

Acknowledgments

This work was supported in part by the National Science Foundation under Grant Nos. DMR-8519077 and ECS-8515899. We express our sincere thanks to C. K. Ong and G. M. Huang for fruitful collaboration on the problems discussed herein.

References

[1] Ong, C. K., G. M. Huang, T. J. Tarn and J. W. Clark: Invertibility of quantum-mechanical control systems, Mathematical Systems Theory, 17 (1984), 335-350.

[2] Messiah, A.: Quantum Mechanics, Vol. 1, John Wiley and Sons, New York 1962.

[3] Braginsky, V. B., Y. I. Vorontsov and K. S. Thorne: Quantum nondemolition measurements: Science, 209 (1980), 547-557.

[4] Tarn, T. J., J. W. Clark, C. K. Ong and G. M. Huang: Continuous-time quantum mechanical filter, in: Proceedings of the Joint Workshop on Feedback and Synthesis of Linear and Nonlinear Systems, Bielefeld and Rome (Ed. D. Hinrichsen and A. Isidori), Springer-Verlag, Berlin 1982.

[5] Clark, J. W., C. K. Ong, T. J. Tarn and G. M. Huang: Quantum nondemolition filters, Mathematical Systems Theory, 18 (1985), 33-55.

[6] Nelson, E.: Analytic vectors, Annals of Mathematics 70 (1959), 572-615.

[7] Caves, C. M., K. S. Thorne, R. W. P. Drever, V. D. Sandberg and M. Zimmermann: On the measurement of a weak classical force coupled to a

quantum-mechanical oscillator. I. Issues of principle, Reviews of Modern Physics, 52 (1980), 341-392.

[8] Braginsky, V. B.: The prospects for high sensitivity gravitational antennae, in: Gravitational Radiation and Gravitational Collapse (Ed. C. DeWitt-Morette), Reidel, Dordrecht 1974, 28-34.

[9] Thorne, K. S., R. W. P. Drever, C. M. Caves, M. Zimmermann and V. D. Sandberg: Quantum nondemolition measurements of harmonic oscillators, Physical Review Letters, 40 (1978), 667-671.

[10] Caves, C. M.: Quantum nondemolition measurements, in: Quantum Optics, Experimental Gravitation and Measurement Theory (Ed. P. Meystre and M. O. Scully), Plenum Press, New York 1982.

[11] Unruh, W. G.: Quantum nondemolition and gravity-wave detection, Physical Review D 19 (1979), 2888-2896.

[12] Barut, A. O. and R. Raczka: Theory of Group Representations and Applications, Polish Scientific Publishers, Warsaw 1977.

[13] Huang, G. M., T. J. Tarn and J. W. Clark: On the controllability of quantum-mechanical systems, Journal of Mathematical Physics, 24 (1983), 2608-2618.

[14] Wei, J. and E. Norman: Lie algebraic solution of linear differential equations, Journal of Mathematical Physics, 4 (1963), 575-581.

[15] Wei, J. and E. Norman: On global representations of the solutions of linear differential equations as a product of exponentials, Proc. Am. Math. Soc., 15 (1964), 327-334.

[16] Hirschorn, R. M.: Invertibility of nonlinear control systems, SIAM Journal of Control and Optimization, 17 (1979), 289-297.

[17] Baras, J.: Continuous quantum filtering, in: Proceedings of the 15th Allerton Conference (1977), 68-77.

[18] Baras, J. S., R. O. Hargar, Y. H. Park: Quantum-mechanical linear filtering of random signal sequences, IEEE Trans. on Information Theory, IT-22 (1976), 59-64.

THE METHOD OF SEEKING FINITE CONTROL FOR QUANTUM MECHANICAL PROCESSES

Anatoliy G. Butkovskiy, Ye.I. Pustil'nykova
Institute of Control Sciences, Moscow, U.S.S.R.

ABSTRACT

A new method for solving of controllability and finite control problems in quantum processes is presented. This method uses finite integral transformations and theory of finite control.

1. INTRODUCTION

This paper is devoted to the investigation of the purposive action on the quantum state of substance. Such study is ultimately motivated by the long-term exerting in such diverse contexts as plasma physics, nucleus-magnetic resonance, electron microscopy, quantum golographics, solid-state technology and other [1-3]. Up to the present the problems in governing the physical processes have been primarily considered in average physical quantity space, but, nowadays the problems of guiding the quantum state itself in pure and confounded ensembles are of more significance [1], [4-6]. It is one of the types of the controllability that concerns us here.

2. SERVO-PROBLEM

Let us consider a servo-problem where the dynamics of the control system is determined by an equation

$$\frac{1}{a}\frac{\partial Q}{\partial t} = \frac{\partial^2 Q}{\partial x^2} + w(x,t), \quad x \in [x_1, x_2], \tag{1}$$

with boundary conditions

$$\ell_1 Q(x_1,t) = m_1(t), \quad \ell_2 Q(x_2,t) = m_2(t), \tag{2}$$

where $Q(x,t)$ is the state of the system, ℓ_1, ℓ_2 are given operators, $1/a$ factor may be real or imaginary, $w(x,t)$ is the control, i.e. the external input agents with the help of which the system state (1), (2) can be purposely changed. We must obtain such value $w(x,t) = w^0(x,t)$ when the system (1), (2) can be evolved under influence of $w^0(x,t)$ from its initial state

$$Q(x,0) = Q_0(x) \tag{3}$$

to its finishing state

$$Q(x,T) = Q_0^*(x) \tag{4}$$

during the period of time T.

Let us notice that boundary conditions and initial condition may be considered as homogeneous without loss of generality transfering if necessary to a standard form of the boundary value problem [7].

Such a finite integral transformation may be applied to the equations (1) - (4),

$$\bar{f}(\lambda_n) = \int_{x_1}^{x_2} f(x)\,\varphi(\lambda_n, x)\,r(x)\,dx\,, \quad n = 1, 2, \dots, \tag{5}$$

in order to the system has no differentiation operators on the variable x in the transformation space. Here all values λ_n, the kernel of transformation $\varphi(\lambda_n, x)$, $n = 1, 2, \dots$ and the function r(x) are correspondingly proper values, orthonormal proper function and weighting function of a limiting problem (Shturm-Liuvill problem) for the system (1), (2) [8]. Therefore an inverse transformation of (5) will be

$$f(x) = \sum_{\lambda_n} \varphi(\lambda_n, x)\,\bar{f}(\lambda_n) \tag{6}$$

The transformation (5) correlates the control problem (1) - (4) to the modal representation of this problem in transformation space

$$\frac{1}{a}\,\frac{d\bar{Q}(\lambda_n, t)}{dt} = -\lambda_n^2\,\bar{Q}(\lambda_n, t) + \bar{w}(\lambda_n, t), \tag{7}$$

$$\bar{Q}(\lambda_n, 0) = 0\,, \quad \bar{Q}(\lambda_n, T) = \bar{Q}_0^*(\lambda_n). \tag{8}$$

Solving (7), (8) one can obtain the N integral equations as follows

$$\frac{1}{a}\,\bar{Q}_0^*(\lambda_n)\,e^{a\lambda_n^2 T} = \int_0^T \bar{w}(\lambda_n, \tau)\,e^{a\lambda_n^2 \tau}\,d\tau\,, \tag{9}$$

$$n = 1, 2, \dots, N\,; \quad N \to \infty\,.$$

Each of (9) equations can be interpreted as a moment problem consisting of one moment equation only. I.e. for each fixed λ_n one can obtain only one moment equation from which its own function $\bar{w}(\lambda_n, t)$ must be received. Let have it.

As it is shown in [9], it is easy to perform the reduction of the moment equation (9) to some interpolation problem. Indeed, the equation (9) can be written in the other way

$$\int_0^T e^{-j(ja\lambda_n^2)\tau}\,\bar{w}(\lambda_n, \tau)\,d\tau = \frac{1}{a}\,\bar{Q}_0^*(\lambda_n)\,e^{a\lambda_n^2 T}, \tag{10}$$

where λ_n is fixed.

But the integral (10) is a sense of the Fourier transformation $\tilde{\bar{w}}(\lambda_n, \omega)$ in the point

$$\omega = j\,a\,\lambda_n^2 \tag{11}$$

of the function $\overline{w}(\lambda_n, t)$. Thus one has following interpolation problem

$$\tilde{\overline{w}}(\lambda_n, j a \lambda_n^2) = \frac{1}{a} \overline{Q}_o^*(\lambda_n) e^{a\lambda_n^2 T} \qquad (12)$$

with a single node (11). We shall obtain the description of general class of finite function $\overline{w}(\lambda_n, t)$ with support [0,T], satisfying (10), if we shall seek solution of problem (12), in the entire function class of T degree [9].

Applying the known finite control methods [7,9] to (12) it becomes of no difficulty to find the solution

$$\overline{w}(\lambda_n, t) = \frac{\overline{Q}_o^*(\lambda_n)}{aT} e^{a\lambda_n^2 (T-t)} \qquad (13)$$

which is correct for every λ_n, $n = 1, 2, \ldots$. Then in conformity with (6) we obtain the control of interest

$$w^c(x,t) = \sum_{\lambda_n} \varphi(\lambda_n, x) \frac{\overline{Q}_o^*(\lambda_n)}{aT} e^{a\lambda_n^2 (T-t)} \qquad (14)$$

Series in (14) is convergent only for some set $\{\overline{Q}_o^*(\lambda_n)\}$ that makes it possible to receive, for example, sufficient controllability conditions for the system (1) - (4).

3. BILINEAR SYSTEMS

In majority of quantum-mechanical problems the external action is included into a system as bilinear, so, we deal with the case when

$$w(x,t) = u(x,t)\, Q(x,t) \qquad (15)$$

and the equation (1) can be specified through

$$\frac{1}{a} \dot{Q} = Q'' + u(x,t) Q \quad , \quad x \in [x_1, x_2] \qquad (16)$$

The boundary and initial conditions are still considered to be homogeneous.

As it was above the problem is to find such a control $u(x,t) = u^o(x,t)$, under which the system can be evolved from its initial state (3) to the finishing one (4) during T-time. We shall apply one of the possible methods of solving similar control problems reviewed in [10] for the exceptional quantum harmonic oscillator case.

Just, taking (15) into consideration we are to change

$u(x,t)\,Q$ in control of (16) to the obtained above control $w^{o}(x,t)$. Then $Q = Q^{c}(x,t)$ corresponds to this controlling action and is in its turn the solution of the Cauchy problem as follows

$$\frac{1}{a}\dot{Q} = Q'' + \sum_{\lambda_n} \varphi(\lambda_n, x)\,\frac{\overline{Q}_o^{*}(\lambda_n)}{aT}\,e^{a\lambda_n^2(T-t)} \tag{17}$$

$$\ell_1 Q(x_1,t) = \ell_2 Q(x_2,t) = 0\,, \quad Q(x,0) = 0 \tag{18}$$

Moreover, it is essentially that evidently the same solution satisfies the requirements of (4).

For example, if the system is defined by the equation (16) with gomogeneous boundary conditions of I type $x \in [0, \ell]$, then the solution $Q^{c}(x,t)$ which meets the requirements of the given Cauchy problem with the action $w^{c}(x,t)$ in (14), will be written as follows

$$Q^{o}(x,t) = \frac{2}{\pi\ell^2}\sum_n \frac{\overline{Q}_o^{*}(n)}{n^2}\,\sin\frac{n\pi x}{\ell}\left\{\exp\left[\left(\frac{n\pi}{\ell}\right)^2(T-2t)\right] - \exp\left[\left(\frac{n\pi}{\ell}\right)^2 T\right]\right\} \tag{19}$$

Now knowing $w^{o}(x,t)$ and $Q^{c}(x,t)$ corresponding to it, it is easy to define the $u = u^{o}(x,t)$ control of interest from (15). The latter exists when $Q^{c}(x,t) \neq 0$, that can give some additional information about controllability of the system.

4. CONTROL IN SCHRÖDINGER EQUATION

Now, let us turn to the system describing by the Schrödinger equation

$$j\hbar\dot{\psi} = -\frac{\hbar^2}{2m}\psi'' + [H_o(x,t) + H_1(x,t)]\psi\,, \quad x \in (-\infty, \infty) \tag{20}$$

Without specifying a Hamiltonian for some time let us consider the following control problem. Let $H_o(x,t)$ be the fixed given and structurally unchangable part of a Hamiltonian. In the majority of problems it may be a coordinate function only. Another part of the Hamiltonian denote $H_1(x,t)$ symbol is associated with external action, or, in other words, it is a mutual effect Hamiltonian of an external field with the system. For example, we may speak about field which we can change, performing the role of control.

Further we shall define all this part of the Hamiltonian, i.e. $H_1(x,t)$ as the control.

Assume that in $t=0$, when the state of the system is

$$\psi(x,0) = \psi_0(x)\,, \tag{21}$$

the external field is switched, which has effect on the system during time T, i.e. the $H_1(x,t)$ control is finite with support $[0, T]$. It is necessary to obtain such $H_1(x,t) = H_1^0(x,t)$, under influence of which the system will evolve from initial state (21) to the given finishing state

$$\psi(x,t) = \psi_1(x) \tag{22}$$

during T time.

Let denote

$$[H_0(x,t) + H_1(x,t)]\psi \equiv w(x,t)\,, \tag{23}$$

as the result of this, we obtain an equation, similar to (1), however $x \in (-\infty, \infty)$ in it. So, instead of the finit integral transformation we will use to the system (16)-(19) Fourier transformation with respect to space coordinate

$$\tilde{\psi}(q,t) = \int_{-\infty}^{\infty} \psi(\xi,t)\, e^{-jq\xi}\, d\xi \tag{24}$$

Note, that in quantum mechanics it corresponds to an transition of impulse representation Solving the transformed system

$$j\hbar \frac{d\tilde{\psi}(q,t)}{dt} = -\frac{\hbar^2}{2m} q^2 \tilde{\psi}(q,t) + w(q,t)\,, \tag{25}$$

$$\tilde{\psi}(q,0) = \tilde{\psi}_0(q) \tag{26}$$

$$\tilde{\psi}(q,T) = \tilde{\psi}_1(q) \tag{27}$$

we come to the integral equation

$$j\hbar\left\{\tilde{\psi}_1(q)\exp\left[-\frac{j\hbar q^2}{2m}T\right] - \tilde{\psi}(q)\right\} = \int_0^T \exp\left[-\frac{j\hbar q^2}{2m}\tau\right]\tilde{w}(q,\tau)\,d\tau \tag{28}$$

This equation may be interpretated as a moment problem, consisting only of one moment equation for every fixed q. As earlier, we will come from the moment problem to an interpolation problem consisting of a single equality only

$$\tilde{\tilde{w}}\left(q, \frac{\hbar}{2m} q^2\right) = j\hbar\left\{\tilde{\varphi}(q)\exp\left[-\frac{j\hbar q^2}{2m}T\right] - \tilde{\varphi}_0(q)\right\}, \tag{29}$$

with one node $-\frac{\hbar}{2m}q^2$. We should also point out that $\tilde{f}(q,t)$ is qualified everywhere in this article, as the Fourier transformation $f(x,t)$ with respect to space variable, but $\tilde{\tilde{f}}(q,q')$ denotes the Fourier transformation of $\tilde{f}(q,t)$ with respect to time variable.

Solution of this interpolation problem is

$$\tilde{w}^c(q,t) = \frac{j\hbar}{T}\left\{\tilde{\varphi}_1(q)\exp\left[\frac{j\hbar q^2}{2m}(t-T)\right] - \tilde{\varphi}_0(q)\exp\left[\frac{j\hbar q^2}{2m}t\right]\right\} \tag{30}$$

In order to define $w^c(x,t)$, we must apply the inversed Fourier transformation to (30), i.e. come back to a position representation. Thus we shall have

$$w^c(x,t) = \frac{j\hbar}{2\pi T}\int_{-\infty}^{\infty}\left\{\tilde{\varphi}_1(q)\exp\left[\frac{j\hbar q^2}{2m}(t-T)\right] - \right.$$
$$\left. - \tilde{\varphi}_0(q)\exp\left[\frac{j\hbar q^2}{2m}t\right]\right\}\exp[jqx]\,dq. \tag{31}$$

Using the Cauchy problem (20), (21), (23) and taking (31) into consideration, function $\psi^c(x,t)$ can be easily defined. This function corresponds to the action $w^o(x,t)$ and can be written as

$$\psi^o(x,t) = \frac{t}{2\pi T}\int_{-\infty}^{\infty}\tilde{\varphi}_0(q)\exp\left[\frac{j\hbar q^2}{2m}(t-T)\right]\exp[jqx]\,dq +$$
$$+ \frac{T-t}{2\pi T}\int_{-\infty}^{\infty}\tilde{\varphi}_0(q)\exp\left[\frac{j\hbar q^2}{2m}t\right]\exp[jqx]\,dq \tag{32}$$

It is evident, that $\psi^o(x,t)$ satisfies the requirements (22).

The control $H_1^o(x,t)$ may be obtained from (23) in which now $w(x,t) = w^o(x,t)$ and $\psi(x,t) = \psi^o(x,t)$. So, in common case (23) is an operator equation in respect to $H_1(x,t)$:

$$H_1(x,t)\dot{\psi}^o(x,t) = w^o(x,t) - H_0(x,t)\psi^o(x,t) \tag{33}$$

This equation can be solved when one has the sufficient additional supposition about a class, which the operator $H_1(x,t)$ looking for must belong to.

For example, for the majority of problems, from the point of view of physics, it follows that $H_1(x,t)$ must be a multiplying operator.

It is evident that in this case we obtain

$$H_1^0(x,t) = \frac{w^0(x,t)}{\psi^0(x,t)} - H_0(x,t). \tag{34}$$

In the common case, denoting

$$H_0(x,t) + H_1(x,t) \equiv H(x,t), \tag{35}$$

$$L(x,t)\tilde{\varphi} \equiv \int_{-\infty}^{\infty} \tilde{\varphi}(q) R(x,t,q)\, dq, \tag{36}$$

where $R(x,t,q)$ is an integral operator kernel of $L(x,t)$, i.e.

$$R(x,t,q) = \frac{1}{2\pi T} \exp\left[\frac{j\hbar q^2}{2m}\right] \exp[jqx], \tag{37}$$

in order to investigate $H(x,t) = H^0(x,t)$, we obtain the following equation

$$\begin{aligned} H(x,t)[tL(x,t-T)\tilde{\varphi}_1(q) - (t-T)L(x,t)\tilde{\varphi}_0(q)] = \\ = j\hbar[L(x,t-T)\tilde{\varphi}_1(q) - L(x,t)\tilde{\varphi}_0(q)] \end{aligned} \tag{38}$$

This equation is correct for each given $\psi_0(x)$ and $\psi_1(x)$ states.

Example 1

Let us consider the quantum system with two stationary states $\psi_1(x)$ and $\psi_2(x)$ only with the energies $\hbar\omega_1$ and $\hbar\omega_2$ respectively $(\hbar\omega_1 < \hbar\omega_2)$.

Let $H_0(x)$ be the Hamiltonian of the unperturbed system and two of its possible stationary states satisfy the equations

$$H_0\psi_1(x) = \hbar\omega_1\psi_1(x),\ H_0\psi_2(x) = \hbar\omega_2\psi_2(x) \tag{39}$$

and results in a complete orthonormal state set [11,12].

At the moment time $t=0$, when the system is in its primary state

$$\psi(x,0) = \psi_1(x), \tag{40}$$

the external perturbation $H_1(x,t)$ is switched and is able to change the system state. At the moment time $t = T$ the perturbation is switched off. Taking the perturbation into consideration the Schrödinger equation will be

$$j\hbar \frac{d}{dt}\psi(x,t) = [H_0(x) + H_1(x,t)]\psi(x,t), \quad x \in (-\infty,\infty) \quad (41)$$

Such a perturbation $H_1(x,t) = H_1^0(x,t)$ must be sought when the system to the moment $t = T$ has changed its initial state (40) to its finishing state

$$\psi(x,T) = \psi_2(x) \quad (42)$$

It is evident, that $H_1(x,t)$ is the control.

Let us apply follows integral transformation to the system (40) - (42)

$$\tilde{f}(n,t) = \int_{-\infty}^{\infty} f(\xi,t)\,\varphi(\lambda_n,\xi)\,d\xi \quad (43)$$

where λ_n and $\varphi(\lambda_n,x)$ are the proper values and proper functions of the unperturbed Hamiltonian $H_0(x)$ satisfying the equations (39), i.e.

$$\lambda_n^2 = -\hbar\omega_n \quad , \quad \varphi(\lambda_n x) = \psi_n(x) \quad (44)$$

and n has two values only: $n=1$ and $n=2$. An inverse transformation to it is

$$f(x,t) = \sum_{n=1}^{2} \psi_n(x)\bar{f}(n,t) \quad (45)$$

Using the notation

$$H_1(x,t)\psi(x,t) \equiv w(x,t) \quad (46)$$

and taking into consideration that is a Hermitian operator, we obtain in the transformation space

$$j\hbar\frac{d}{dt}\bar{\psi}(n,t) = -\lambda_n^2\bar{\psi}(n,t) + \bar{w}(n,t), \quad (47)$$

$$\bar{\psi}(n,0) = \bar{\psi}_1(n) = \delta_{1n} \quad , \quad (48)$$

$$\bar{\psi}(n,T) = \bar{\psi}_2(n) = \delta_{2n} \quad (49)$$

where δ_{mn} is a Cronicker symbol,

Further solving the system (47) - (49), we come to two moment problems, each of them consists of one moment equation

$$j\hbar\left[e^{-j\frac{\lambda_n^2}{\hbar}T}\delta_{2n} - \delta_{1n}\right] = \int_0^T \bar{w}(n,\tau)e^{-j\frac{\lambda_n^2}{\hbar}\tau}d\tau \quad (50)$$
$$n = 1, 2$$

As it was in our priories considerations we pass from (50)

to a corresponding interpolation problem

$$\tilde{\tilde{w}}\left(n, \frac{\lambda_n^2}{\hbar}\right) = j\hbar\left[e^{-j\frac{\lambda_n^2}{\hbar}T}\delta_{2n} - \delta_{1n}\right], \quad n=1,2, \tag{51}$$

and obtain its solution

$$\bar{w}(n,t) = \frac{j\hbar}{T}\left[e^{-j\frac{\lambda_n^2}{\hbar}(T-t)}\delta_{2n} - e^{j\frac{\lambda_n^2}{\hbar}t}\delta_{1n}\right]. \tag{52}$$

Taking into consideration (44) and using inversion formula (45) the solution $w = w^c(x,t)$ can be written as follows

$$w^0(x,t) = -\psi_1(x)\frac{j\hbar}{T}e^{-j\omega_1 t} + \psi_2(x)\frac{j\hbar}{T}e^{j\omega_2(T-t)} \tag{53}$$

Then, by analogy with the observed above it is easy to find that the function, solving the Cauchy problem (40) - (41), where $H_1(x,t)\psi(x,t) \equiv w^0$ is specified through

$$\psi^0(x,t) = \psi_1(x)\frac{T-t}{T}e^{-j\omega_1 t} + \psi_2(x)\frac{t}{T}e^{-j\omega_2(t-T)} \tag{54}$$

Let us introduce the notation

$$\omega_0 = \omega_2 - \omega_1 \tag{55}$$

where $\hbar\omega_0$ is difference energy between the levels. Thus, the control of interest $H_1 = H_1^D(x,t)$, as it can be obtain from (46) can be written as

$$H_1(x,t) = j\hbar\,\frac{-\psi_1(x) + \psi_2(x)\exp(j\omega_2 T)\exp(j\omega_0 t)}{(T-t)\psi_1(x) + t\psi_2(x)\exp(j\omega_2 T)\exp(j\omega_0 t)} \tag{56}$$

One of the private interpritations of the discussed above example is as follows. If the Hamiltonian H_c describes mutual effect of S-electron and $\mathcal{H}_0$ magnetic field, directed along the axle z, then the result is the two-leveled system [11,12]. In this case, the $\hbar\omega_0$ value is the distance between the levels with the opposite oriented spin.

Magnetic field, having an external distribution in space and transforming with respect of time, described by Hamiltonian H_1 can perform a role of the control. The corresponding control problem can be investigated with the help of this method.

Example 2

Let the Schrödinger equation of the considering system be

$$j\hbar\dot{\psi} = -\frac{\hbar^2}{2m}\psi'' + H(x,t)\psi, \quad x \in (-\infty,\infty) \tag{57}$$

where $H(x,t)$ is a finite function with the support $[0,T]$. Till the moment $t=0$, $\psi(x,t)$ has described a free moving particle having a impulse $q=\hbar k$. I.e. this is a usual plane wave spreading in k direction. Consequently, initial condition can be written as follows

$$\psi(x,0) = \psi_0(x) = \exp(jkx) \tag{58}$$

In the moment time $t=0$ the external action $H(x,t)$ is switched and it changes the system state.

It is necessary to determine $H(x,t)$, such as to transform the system till time $t=T$ to the state

$$\psi(x,T) = \psi_1(x) = \left(\frac{\lambda}{\pi}\right)^{1/4} \exp\left(-\frac{\lambda}{2}x^2\right), \quad \lambda = \frac{m\omega}{\hbar} \tag{59}$$

that corresponds to a main level of a quantum harmonic oscillator.

In order to use the got above results, let us obtain the Fourier transform of functions $\psi_0(x)$ and $\psi_1(x)$. It is evident that in a impulse representation the plane wave is specified through the Dirack δ-function, i.e.

$$\tilde{\psi}_0(q) = \delta(k-q). \tag{60}$$

Using an error integral

$$\int_{-\infty}^{\infty} \exp(-x^2)\,dx = \sqrt{\pi}, \tag{61}$$

it is easy to show that the Fourier transform from (50) functionally is the same as in the origin, i.e. the function's specification of $\psi_1(x)$ coincides in a impulse and coordinate representations.

Indeed

$$\tilde{\psi}_1(q) = \int_{-\infty}^{\infty} \left(\frac{\lambda}{\pi}\right)^{1/4} \exp\left(-\frac{\lambda}{2}x^2\right)\exp(-jqx) = \left(\frac{4\pi}{\lambda}\right)^{1/4} e^{-\frac{q^2}{2\lambda}} \tag{62}$$

Now, let us obtain $w^c(x,t)$. Using the equality (31) we shall have

$$w^{0}(x,t)=\frac{j\hbar}{2\pi T}\int_{-\infty}^{\infty}\left(\frac{4\pi}{\lambda}\right)^{1/4}\exp\left(-\frac{q^{2}}{2\lambda}\right)\exp\left(\frac{j\hbar q^{2}}{2m}(t-T)\right)\times$$

$$\times\exp(jqx)\,dq-\frac{jh}{2\pi T}\int_{-\infty}^{\infty}\delta(k-q)\exp\left(\frac{j\hbar q^{2}}{2m}t\right)\exp(jqx)\,dq \quad (63)$$

Making some simple transformations this equation will be

$$w^{0}(x,t)=\frac{j\hbar}{T}\left(\frac{m\omega}{\pi\hbar}\right)^{1/4}\frac{1}{\sqrt{2+j\omega(T-t)}}\exp\left[-\frac{x^{2}m\omega}{2\hbar(2+j\omega(T-t))}\right]-$$
$$-\frac{j\hbar}{2\pi T}\exp\left(\frac{j\hbar k^{2}}{2m}t\right)\exp(jkx) \quad (64)$$

As it can be seen from (32), function $\psi^{0}(x,t)$ corresponding to $w^{0}(x,t)$, will be specified through

$$\psi^{0}(x,t)=\frac{t}{T}\left(\frac{m\omega}{\pi\hbar}\right)^{1/4}\frac{1}{\sqrt{2+j\omega(T-t)}}\exp\left[-\frac{x^{2}m\omega}{2\hbar(2+j\omega(T-t))}\right]-$$
$$-\frac{t-T}{2\pi T}\exp\left[\frac{j\hbar k^{2}}{2m}t\right]\exp\left[jkx\right]. \quad (65)$$

The control of interest $H^{0}(x,t)$, evidently, is determined by the equality

$$H^{0}(x,t)=\frac{w^{0}(x,t)}{\psi^{0}(x,t)} \quad (66)$$

REFERENCES

1. Butkovskiy A.G., Samoilenko Yu.I. Control of the Quantum-Mechanical Processes. M.: Nauka, 1984.
2. Light Scattering in Solids (Ed. M. Cardona). Springer-Verlag Berlin-Heidelberg-New York, 1975.
3. Slichter C.P. Principles of Magnetic Resonance (Ed. P. Fulde) Springer-Verlag Berlin-Heidelberg-New York, 1980.
4. Ong C.K., Huang G.M., Tarn T.J., Clark J.W. Invertibility of Quantum-Mechanical Control Systems. Mathematical Systems Theory, 17, 355-350, 1984.
5. Huang G.M., Tarn T.J., Clark J.W. On the Controllability of Quantum-Mechanical Systems. Journal of Mathematical Physics, 24, 1983.

6. Malkin I.A., Man'ko V.I. Dynamic Symmetries and Coherent Stats of Quantum Systems. M.: Nauka, 1979.
7. Burkovskiy A.G. Structural Theory of Distributed Systems. M.: Science, 1977. (Ellis Horwood, England, 1983).
8. Butkovskiy A.G., Pustyl'nikov L.M. Theory of Mobile Control of Systems with Distributed Parameters. M.: Nauka, 1980.
9. Butkovskiy A.G. Methods of Control of Systems with Distributed Parameters, M.: Science, 1975.
10. Butkovskiy A.G., Pustyl'nikova Ye.I. Control of Coherent States of Quantum Systems with a Square Hamiltonian. Automation and Telemechanics, No.8, 1984.
11. Blohintsev D.I. Principles of Quantum Mechanics, M.: Science, 1976.
12. Flugge Z. Problems of Quantum Mechanics, v.2, M.: World, 1974.

Printed in the United States
By Bookmasters